Feasible, Robust and Reliable Automation and Control for Autonomous Systems

Feasible, Robust and Reliable Automation and Control for Autonomous Systems

Editors

Umar Zakir Abdul Hamid
Argyrios Zolotas
Chuan Hu

MDPI • Basel • Beijing • Wuhan • Barcelona • Belgrade • Manchester • Tokyo • Cluj • Tianjin

Editors
Umar Zakir Abdul Hamid
CEVT (China Euro Vehicle Technology AB)
Sweden

Argyrios Zolotas
Cranfield University
UK

Chuan Hu
University of Alaska Fairbanks
USA

Editorial Office
MDPI
St. Alban-Anlage 66
4052 Basel, Switzerland

This is a reprint of articles from the Special Issue published online in the open access journal *Electronics* (ISSN 2079-9292) (available at: https://www.mdpi.com/journal/electronics/special_issues/Feasible_Robust_and_Reliable_Automation_and_Control_for_Autonomous_Systems).

For citation purposes, cite each article independently as indicated on the article page online and as indicated below:

LastName, A.A.; LastName, B.B.; LastName, C.C. Article Title. *Journal Name* **Year**, *Volume Number*, Page Range.

ISBN 978-3-0365-5075-6 (Hbk)
ISBN 978-3-0365-5076-3 (PDF)

Contents

About the Editors

Umar Zakir Abdul Hamid

Umar Zakir Abdul Hamid holds a Ph.D. and has been working in the future mobility (connected and autonomous vehicle) field since 2014, with various teams in different countries and continents. Previously, he led a team of 12 engineers (of 10 different nationalities) working on autonomous vehicle software product development with Sensible 4, Finland. Umar is one of the recipients of the Finnish Engineering Award 2020, which he received for his contributions to the development of all-weather autonomous driving solutions with this firm. With more than 30 scientific publications as author and editor under his belt, Umar actively participates in global automotive standardization efforts, where he is Secretary for the Society Automotive Engineers (SAE) Cooperative Driving Automation (CDA) Committee. Since the end of summer 2021, Umar has been working as Senior Lead Product Strategist for CEVT AB (Geely Group) in Sweden.

Argyrios Zolotas

Prof. Dr. Argyrios Zolotas (Senior Member, IEEE) received his B.Eng. degree in electronic and communications engineering from the University of Leeds, Leeds, UK, the M.Sc. degree in electronic and electrical engineering from the University of Leicester, Leicester, UK, and Ph.D. degree in advanced electronic and electrical engineering from Loughborough University, Loughborough, UK. He is a Reader in Systems and Control with Cranfield University, where he leads the Autonomous Systems Dynamics and Control research group (within the Centre for Autonomous and Cyber-Physical Systems). He has held academic posts at University of Lincoln, the University of Sussex, Loughborough University, and a Post-Doctoral Fellowship at Imperial College London. He was previously a Visiting Professor with Grenoble INP (GIPSA Lab, 2018). His research interests include advanced control, autonomous systems and control applications of AI. He is a Fellow of HEA (UK).

Chuan Hu

Chuan Hu earned his Ph.D. degree in mechanical engineering from McMaster University, Hamilton, Ontario, Canada, in 2017. He is a tenure-track assistant professor in the Department of Mechanical Engineering, University of Alaska Fairbanks, Fairbanks, Alaska, USA. His research interests include decision making, path planning, motion control and estimation for autonomous vehicles, vehicle system dynamics and control, human–vehicle interaction, trust dynamics and robust and adaptive control. He has been a guest editor of *Advances in Mechanical Engineering and IET Intelligent Transport Systems.*

Preface to "Feasible, Robust and Reliable Automation and Control for Autonomous Systems"

The past few decades have seen a rapid development toward autonomous systems. Increasing computational power ability and advances in new computing devices currently allow the feasible real-time implementation of autonomous systems. This is further supported by large-scale research in autonomous systems applications, including but not limited to ground, aerial, maritime vehicles, mobile robotics. Differently to automated systems, an autonomous system employs situational awareness information, via perception modules, normally using a multi-layer control strategy to command the effectors driving the system. Given that environments in the real world consist of dynamic and varied conditions, a reliable control strategy for autonomous systems should offer a safe, reliable and robust solution.

Thus, this Special Issue book 'Feasible, Robust and Reliable Automation and Control for Autonomous Systems' aims for wider dissemination of research on control strategy topics for multiple types of autonomous systems that are not constrained to a single platform. It highlights current research in the control field for autonomous systems as well as showcases the state-of-the-art control strategy approaches used for the autonomous platforms. We believe this Special Issue will be of great appeal to researchers in fields related to control systems and their applications typified in the fields of ground, aerial, maritime vehicles and robotics as well as industrial audiences. Thus, in reflecting the most recent progress in control strategies for autonomous platforms, the covered topics include control system design in autonomous systems for different platforms, robustness analysis of control strategy performance of autonomous systems and discussions on the kinematics, dynamics and model nonlinearity effects on the controller performance of the autonomous systems, among many others.

The Special Issue is co-edited by distinguished international control system experts currently based in Sweden, the United States of America, and the United Kingdom. The ten articles published within represent contributions from reputable researchers in China, Austria, France, the United States of America, Poland, and Hungary, among many others.

The editors would like to thank all the contributors to this Special Issue, which include the authors and reviewers as well as the Electronics publishing team.

Umar Zakir Abdul Hamid, Argyrios Zolotas, and Chuan Hu
Editors

 MDPI

Editorial

Feasible, Robust and Reliable Automation and Control for Autonomous Systems

Umar Zakir Abdul Hamid [1,*], Chuan Hu [2] and Argyrios Zolotas [3]

1 CEVT (China Euro Vehicle Technology AB), Pumpgatan 1, 417 55 Gothenburg, Sweden
2 Department of Mechanical Engineering, University of Alaska Fairbanks, Fairbanks, AK 99775, USA; chuan.hu.2013@gmail.com
3 Centre for Autonomous and Cyber-Physical Systems, SATM, Cranfield University, Cranfield MK43 0AL, UK; a.zolotas@cranfield.ac.uk
* Correspondence: umartozakir@gmail.com

The global market for autonomous robotics platforms has grown rapidly due to the advent of drones, mobile robots, and driverless cars, while the mass media coverage examining the progress of robotics and autonomous systems field is widespread. There are instances in which such news may be exaggerated, and to a certain extent, surrounded by potentially misleading hype. This is understood in terms of the valuation standing of automation and robotics industries [1]. With lot of startups raising huge number of investments worldwide, and big corporations launching spin-off companies to expedite the deployment of the emerging technologies, the trajectory of this domain continues to escalate [2].

'Deep Learning', 'Machine Learning', 'Artificial Intelligence', 'Data-Driven', and 'Neural Network' are among the terms frequently used to discuss the rapid developments in the future robotics and autonomous systems technologies sectors. However, robotics and automation are complex fields which are not restricted to the aforementioned phrases. Autonomous systems consist of different domains such as 'mapping', 'localization', 'object detection', and 'embedded systems', as well as 'guidance, navigation and control', among many others. Each of these terms requires cross-disciplinary expertise and talents to facilitate innovation and further developments.

In this Special Issue, the editors aim to provide an in-depth assessment of one of the most important elements of autonomous systems, i.e., automation and control. The Special Issue is co-edited by distinguished international control system experts currently based in Sweden, the United States of America, and the United Kingdom, with contributions from reputable researchers from China, Austria, France, the United States of America, Poland, and Hungary, among many others. The main objective of this Special Issue is to highlight current research and development in the automation and control field for autonomous systems, as well as to showcase state-of-the-art control strategy approaches for the autonomous platforms. The editors believe the ten articles published within this Special Issue will be highly appealing to control-systems-related researchers in applications typified in the fields of ground, aerial, maritime vehicles, and robotics, as well as industrial audiences.

In the first article, Hernandez-Barragen et al. [3] investigated an adaptive single neuron anti-windup PID controller based on the extended Kalman filter algorithm, where the experimental tests are performed on a KUKA® Youbot® omnidirectional platform. This work demonstrated that the proposed adaptive PID controller performed better than the conventional PID and other benchmarked PID approaches.

In the article 'Facilitating Autonomous Systems with AI-Based Fault Tolerance and Computational Resource Economy' [4], Deliparaschos et al. proposed the facilitation of fault-tolerant capability in autonomous systems, with particular consideration of low computational complexity and system interface devices (sensor/actuator) performance.

Citation: Hamid, U.Z.A.; Hu, C.; Zolotas, A. Feasible, Robust and Reliable Automation and Control for Autonomous Systems. *Electronics* **2022**, *11*, 2126. https://doi.org/10.3390/electronics11142126

Received: 29 June 2022
Accepted: 4 July 2022
Published: 7 July 2022

An AI-based control framework enabling low computational power fault tolerance was presented, where the efficacy of the proposed scheme was shown via rigorous analysis of several sensor fault scenarios for an electro-magnetic suspension testbed.

'On-Line Learning and Updating Unmanned Tracked Vehicle Dynamics' by Strawa et al. [5] proposed a method by which to estimate vehicle model parameters using a compound identification scheme utilizing an exponential forgetting recursive least square, generalized Newton–Raphson (NR), and Unscented Kalman Filter methods. The proposed identification scheme facilitates adaptive capability for the control system, improves tracking performance, and contributes to an adaptive path and trajectory planning framework, which is essential for future autonomous ground vehicle missions and traversability.

Meng et al. presented 'High Velocity Lane Keeping Control Method Based on the Non-Smooth Finite-Time Control for Electric Vehicle Driven by Four Wheels Independently' [6], in which two kinds of tracking error computing methods of lane keeping control for electric vehicles were proposed to tackle different conditions, and a Non-FT lane keeping controller was designed to keep the EV-DFWI running in the desired lane suffering external disturbances.

The topics of path planning and tracking algorithms were studied by Rumetshofer et al. in 'A Generic Interface Enabling Combinations of State-of-the-Art Path Planning and Tracking Algorithms' [7]. A generic interface design between the local path planning and path tracking systems was examined. This topic holds significant importance for autonomous driving applications.

Tran et al. [8] authored 'Integrated Comfort-Adaptive Cruise and Semi-Active Suspension Control for an Autonomous Vehicle: An LPV Approach', which presents an integrated linear parameter-varying (LPV) control approach of an autonomous vehicle intending to guarantee driving comfort, consisting of cruise and semi-active suspension control. Simulation-based experiments were conducted using a realistic nonlinear vehicle model validated from experimental data. The simulation results demonstrate the proposed approach's capability to improve driving comfort (a topic that will remain important for future people-carrier autonomous vehicles).

In 'Development and Verification of Infrastructure-Assisted Automated Driving Functions' [9], Rudigier et al. extensively discussed the autonomous driving and ADAS applications of control systems. Their paper presents specific use cases in the said context, and the verification results for a proposed system utilizing a simulation framework are reported.

In 'Automatically Learning Formal Models from Autonomous Driving Software' [10], Selvaraj et al. shared their Autonomous Driving expertise and applied active learning techniques to obtain formal models of an existing (though still in development) autonomous driving software module implemented in MATLAB. This demonstrates the feasibility of automated learning for automotive industrial use. Practical challenges in applying automata learning, and possible directions for integrating automata learning into the automotive software development workflow, are also discussed in this work.

In 'Practical Nonlinear Model Predictive Controller Design for Trajectory Tracking of Unmanned Vehicles', Pang et al. explored the trajectory tracking issue of unmanned vehicles [11]. The authors proposed an improvement of the nonlinear model predictive controller (NMPC) for the trajectory tracking application of an unmanned vehicle (UV). The simulation results confirm that the proposed NMPC scheme reveals better control accuracy and computational efficiency than the standard MPC controller under two different prescribed roads.

Finally, in 'Leader-Based Trajectory Following in Unstructured Environments—From Concept to Real-World Implementation', Nestlinger et al. described the issue of guiding a vehicle by means of an external leader [12]. A system was proposed and tested in a simulation framework and then deployed in a demonstrator vehicle for validation under real operating conditions.

The editors would like to thank all the contributors to this Special Issue, which includes the authors and reviewers, as well as the *Electronics* publishing team. The editors believe the findings presented in this Special Issue will be beneficial for the reading of interested researchers and general audiences.

Author Contributions: U.Z.A.H., C.H. and A.Z. have contributed equally to this editorial for the Special Issue, 'Feasible, Robust and Reliable Automation and Control for Autonomous Systems', which is published by the journal *Electronics*. All authors have read and agreed to the published version of the manuscript.

Funding: This research received no external funding.

Institutional Review Board Statement: Not applicable.

Informed Consent Statement: Not applicable.

Data Availability Statement: Not applicable.

Acknowledgments: The editors would like to thank all the contributors to this Special Issue, which include the authors and reviewers as well as the *Electronics* publishing team.

Conflicts of Interest: The authors declare no conflict of interest.

References

1. Pagliarini, L.; Lund, H.H. The future of Robotics Technology. *J. Robot. Netw. Artif. Life* **2017**, *3*, 270–273. [CrossRef]
2. Herrmann, A.; Walter, B.; Stadler, R. *Autonomous Driving: How the Driverless Revolution Will Change the World*; Emerald Group Publishing: Bingley, UK, 2018.
3. Hernandez-Barragan, J.; Rios, J.D.; Alanis, A.Y.; Lopez-Franco, C.; Gomez-Avila, J.; Arana-Daniel, N. Adaptive Single Neuron Anti-Windup PID Controller Based on the Extended Kalman Filter Algorithm. *Electronics* **2020**, *9*, 636. [CrossRef]
4. Deliparaschos, K.M.; Michail, K.; Zolotas, A.C. Facilitating Autonomous Systems with AI-Based Fault Tolerance and Computational Resource Economy. *Electronics* **2020**, *9*, 788. [CrossRef]
5. Strawa, N.; Ignatyev, D.I.; Zolotas, A.C.; Tsourdos, A. On-Line Learning and Updating Unmanned Tracked Vehicle Dynamics. *Electronics* **2021**, *10*, 187. [CrossRef]
6. Meng, Q.; Zhao, X.; Hu, C.; Sun, Z.-Y. High Velocity Lane Keeping Control Method Based on the Non-Smooth Finite-Time Control for Electric Vehicle Driven by Four Wheels Independently. *Electronics* **2021**, *10*, 760. [CrossRef]
7. Rumetshofer, J.; Stolz, M.; Watzenig, D. A Generic Interface Enabling Combinations of State-of-the-Art Path Planning and Tracking Algorithms. *Electronics* **2021**, *10*, 788. [CrossRef]
8. Tran, G.; Pham, T.-P.; Sename, O.; Costa, E.; Gaspar, P. Integrated Comfort-Adaptive Cruise and Semi-Active Suspension Control for an Autonomous Vehicle: An LPV Approach. *Electronics* **2021**, *10*, 813. [CrossRef]
9. Rudigier, M.; Nestlinger, G.; Tong, K.; Solmaz, S. Development and Verification of Infrastructure-Assisted Automated Driving Functions. *Electronics* **2021**, *10*, 2161. [CrossRef]
10. Selvaraj, Y.; Farooqui, A.; Panahandeh, G.; Ahrendt, W.; Fabian, M. Automatically Learning Formal Models from Autonomous Driving Software. *Electronics* **2022**, *11*, 643. [CrossRef]
11. Pang, H.; Liu, M.; Hu, C.; Liu, N. Practical Nonlinear Model Predictive Controller Design for Trajectory Tracking of Unmanned Vehicles. *Electronics* **2022**, *11*, 1110. [CrossRef]
12. Nestlinger, G.; Rumetshofer, J.; Solmaz, S. Leader-Based Trajectory Following in Unstructured Environments—From Concept to Real-World Implementation. *Electronics* **2022**, *11*, 1866. [CrossRef]

Article

Adaptive Single Neuron Anti-Windup PID Controller Based on the Extended Kalman Filter Algorithm

Jesus Hernandez-Barragan, Jorge D. Rios, Alma Y. Alanis, Carlos Lopez-Franco, Javier Gomez-Avila and Nancy Arana-Daniel *

Centro Universitario de Ciencias Exactas e Ingenierías, Universidad de Guadalajara, Blvd. Marcelino García Barragán 1421, Guadalajara C.P. 44430, Jalisco, Mexico; josed.hernandezb@academicos.udg.mx (J.H.-B.); jorge.rarranaga@academicos.udg.mx (J.D.R.); alma.alanis@academicos.udg.mx (A.Y.A.); carlos.lfranco@academicos.udg.mx (C.L.-F.); jenrique.gomez@academicos.udg.mx (J.G.-A.)

* Correspondence: nancyaranad@gmail.com or nancy.arana@academicos.udg.mx

Received: 7 March 2020; Accepted: 7 April 2020; Published: 11 April 2020

Abstract: In this paper, an adaptive single neuron Proportional–Integral–Derivative (PID) controller based on the extended Kalman filter (EKF) training algorithm is proposed. The use of EKF training allows online training with faster learning and convergence speeds than backpropagation training method. Moreover, the propose adaptive PID approach includes a back-calculation anti-windup scheme to deal with windup effects, which is a common problem in PID controllers. The performance of the proposed approach is shown by presenting both simulation and experimental tests, giving results that are comparable to similar and more complex implementations. Tests are performed for a four wheeled omnidirectional mobile robot. Tests show the superiority of the proposed adaptive PID controller over the conventional PID and other adaptive neural PID approaches. Experimental tests are performed on a KUKA® Youbot® omnidirectional platform.

Keywords: neuron PID; Kalman filtering; omnidirectional mobile robot; implementations; anti-windup

1. Introduction

Presently, Proportional–Integral–Derivative (PID) controllers are still among the most popular controllers used in the industry [1–3]. However, a PID is just adequate for a nominal process; it performs poorly for a system with uncertainties in operating conditions or environmental parameters [2,4,5]. It is well known that if the mathematical model of a plant is available, various techniques to determine PID controller parameters exist. Based on that, improvements and tuning mechanisms were proposed in the literature for conventional PID controllers. However, those techniques are mainly offline methodologies, and in most cases, they require a model of the system, which is commonly not available [1–3]. Among these techniques, adaptive neural PID controllers are presented as an option due to neural networks characteristic that allows them to adapt themselves to changes in operating conditions and environmental parameters, giving the controller the capability of adapting its parameters online [5–7]. Adaptive control techniques are important to solve problems in robotics research, such as control of robot manipulators [8,9], control of mobile robots [10,11] and formation control [12], control of underwater vehicles [13,14], control for teleoperation systems [15] and industrial applications [16,17].

Adaptive neural PID controllers have been presented mainly in three different forms:

- **Single neuron PID controllers**
  Examples of this group are works [6,18–21].
  These adaptive controllers are base on a single neuron whose inputs are the proportional error (P), integral of the error (I), and derivative of the error (D) (see Figure 1).

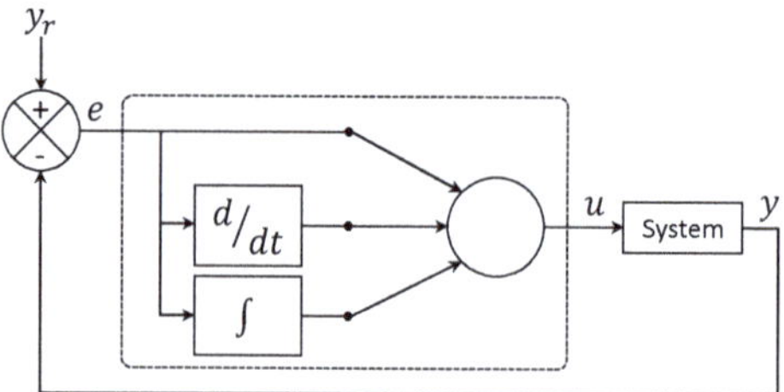

Figure 1. Single neuron PID controller scheme.

- **Multi-layer neural PID controllers**
  Examples of this group are works [4,22–26].
  These works are mainly based on an architecture as shown in Figure 2, where the output layer only has one neuron, and the immediately before layer to this output layer has three neurons, each one dedicated to the proportional gain (K_P), the integral gain (K_I), and the derivative gain (K_D), respectively. The inputs are the proportional error (P), integral of the error (I), and derivative of the error (D). In some works, the inputs include the system output and its reference.

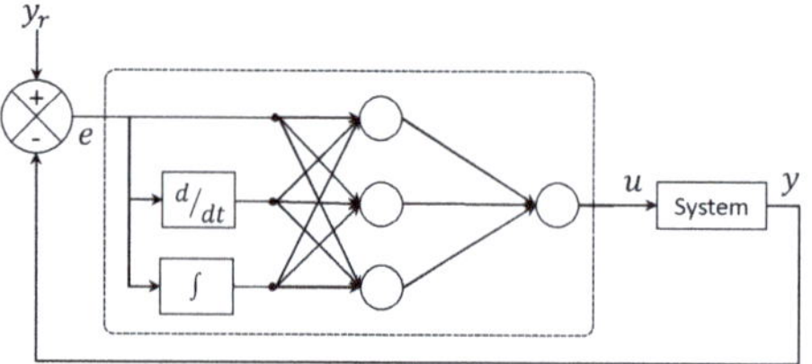

Figure 2. Multilayer PID controller scheme.

- **Hybrid neural PID controllers**
  Examples of this group are [5,27,28].
  These adaptive neural PID controllers are based on improving the performance of a conventional PID controller (Figure 3). Typically, the neural network chooses online the proportional gain (K_P), the integral gain (K_I), and the derivative gain (K_D) parameters of the controller. However, they tend to have slow learning rates, complex architectures, and high computational cost [4,23].

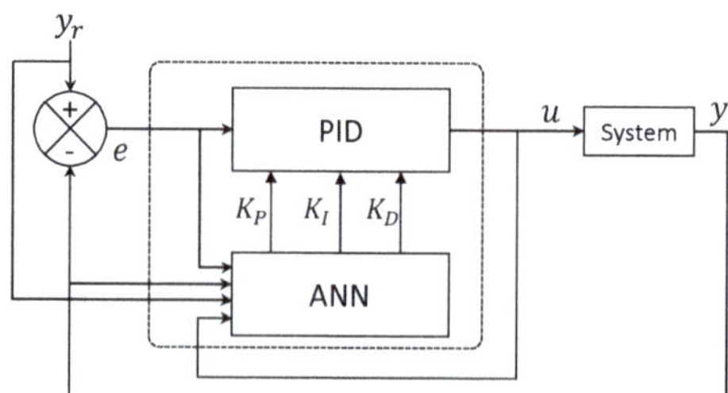

Figure 3. Hybrid PID controller scheme. ANN: Artificial Neural Networks.

Training methods for the previous mentioned adaptive neural PID controllers are mostly based on backpropagation. Also, most of them only report simulation results using benchmark systems that do not necessarily represent real-world problems. Another common problem with PID controllers is the windup effect [29]. The windup effect occurs when the cumulative error in the integral control action produces a saturation on the actuators, which while on the saturation zone, the system loses controllability [30,31]. The windup effect contributes to poor performance, overshoot, high settling time, and instability [32,33]. Anti-windup methods were proposed to deal with these issues, among them, the limiter integrator, conditional integration, back-calculation, the observer approach, modified tracking anti-windup, and others [31,33]. The inclusion of anti-windup strategies in control designs is essential to be considered. The previously mentioned adaptive neural PID controllers do not include it.

This work proposes an adaptive single neuron PID controller trained with the Extended Kalman Filter (EKF) training algorithm. The EKF for neural networks were proven to have faster learning speeds and convergence times than training based on backpropagation, which make them ideal for experimental and real-time tests [34,35]. Moreover, the neuron PID controller includes a back-calculation scheme to deal with windup effects. The proposed anti-windup scheme uses the information of the cumulative error and the input saturation to compute the integral action dynamically. No special tuning is required to adjust the contribution of input saturation, which is needed by conventional back-calculation methods. The performance of the proposed controller is shown through simulation and experimental results using an omnidirectional robot. For simulation tests, the robot is simulated on Matlab® (Matlab is a registered trademark of the MathWorks, Inc.), on the other hand, for the experimental case a KUKA® (KUKA is a registered trademark of KUKA Aktiengesellschaft Germany) Youbot® (Youbot is a registered trademark of KUKA Aktiengesellschaft Germany) is used. Moreover, comparative results are presented with respect to the conventional PID controller, an adaptive single neuron PID controller, and an adaptive multilayer PID controller. A case of study is also presented to compare the performance of a PID controller with an anti-windup method against the proposed adaptive controller. The control objective is set as trajectory tracking, where through the tests and a smaller tracking error is obtained with the proposed controller compared to the other implemented techniques. Since the proposal is based on a neuron and an the EKF training algorithm, then, it is essential to mention the works (Sento, 2016, [24] and Gomez, 2016, [26]) where multilayer PID controllers are presented. However, those works only present simulation results; the first one uses an inverted pendulum system and a DC motor system. The second one presents results using a quadrotor system and a control scheme that presents a neural PD controller. Moreover, both approaches do not include an anti-windup scheme.

The main contribution of this work is an adaptive single neuron PID controller trained with an extended Kalman filter training algorithm, which includes an anti-windup scheme to overcome the overshoot and settling time inconveniences. The proposed controller scheme offers excellent results compared with other similar proposals and also compared to more complex ones.

A mobile system is considered to be autonomous if its motion is based on its own knowledge of the environment. Mobile robots can make decisions and perform appropriate actions without the intervention of a human user. To perform an autonomous motion, control techniques are required. The use of controllers that require the tuning of gains is not appropriated because the intervention of a user is needed every time the model changes. However, the proposed adaptive neuron controller provides an auto-tuning mechanism that contributes to robot autonomy.

Path planning is an important task to be solved for autonomous robot systems. Path planning consists of the determination of a path from an initial point to an endpoint, without collision with obstacles in the environment. Moreover, the autonomous guided vehicle (AGV) robots are widely used to transport materials between assembly stations. The AGV robots need to use a controller technique to follow special paths, which are usually electrical guide wires based on sensors. A trajectory tracking control algorithm is crucial to follow the given paths successfully. The proposed approach proved to be robust and reliable for control tracking tasks that are beneficial for autonomous mobile systems.

The work outline is as follows:

- in Section 2 the proposed adaptive neuron anti-windup PID controller is presented, where EKF neural PID controller is presented, where Section 2.1 explains the EKF training algorithm.

- Section 3 presents the implementation of the propose adaptive controller for position tracking of an omnidirectional mobile robot. In Section 3.1 the robot kinematics are presented for illustrative purposes, Section 3.2 presents the conventional PID controller implementation scheme, and Section 3.3 presents the adaptive control scheme.
- in Section 4, both simulation (Section 4.1) and experimental results (Section 4.3) are presented, showing the performance of the proposed single neuron PID controller against the conventional PID and others neural PID controllers.
- in Section 5, important conclusions are presented and discussed.

2. Adaptive Single Neuron PID Controller Based on the Extended Kalman Filter Algorithm

An artificial neuron is a mathematical simplification of a biological neuron; however, simple as it is, an artificial neuron is capable of accomplishing multiple tasks. Many artificial neurons can then be put together to achieve even more complex tasks [7]. Similar to a neuron and a neural network, they learn a task of function obtaining information of the environment, such knowledge or information is stored in the synaptic weights ω, the output of a neuron is computed as represented in Equation (1)

$$\hat{y} = \phi(\omega^{\top} x) \tag{1}$$

where ω is a weight vector, x the input vector, $\hat{y}$ is the output of the neuron, and ϕ is an activation function. Training can be perform using different paradigms please see [7,36].

The proposed adaptive single neuron PID controller is illustrated in Figure 4, where e is the error (2) of the system under consideration, this error is the difference between the reference (y_r) and the system output (y). The error e' is the output of the anti-windup scheme (8). x_1, x_2, and x_3 are the inputs to the neuron and they are define as the proportional error (3), the integral of the error (5), and the derivative of the error (4) [37]. The weights ω_1, ω_2, and ω_3, are adapted online using the extended Kalman filter algorithm described in Section 2.1, and they represent the K_P, K_I, and K_D gains, respectively. v is the weighted sum (6), u is the output of the neuron (7), where as activation function $\tanh(\cdot)$ is selected and α scales the amplitude of tanh.

$$e(k) = y_r(k) - y(k) \tag{2}$$

$$x_1(k) = e(k) \tag{3}$$

$$x_2(k) = e(k) - e(k-1) \tag{4}$$

$$x_3(k) = \sum_{j=1}^{k} e'(k) \tag{5}$$

$$v(k) = \sum_{i=1}^{3} \omega_i(k)\, x_i(k) \tag{6}$$

$$u(k) = \alpha \tanh(v(k)) \tag{7}$$

the activation function $\tanh(\cdot)$ reacts in the range $[-1, 1]$. Moreover, the parameter α needs to be selected depend on the control action requires for control tasks.

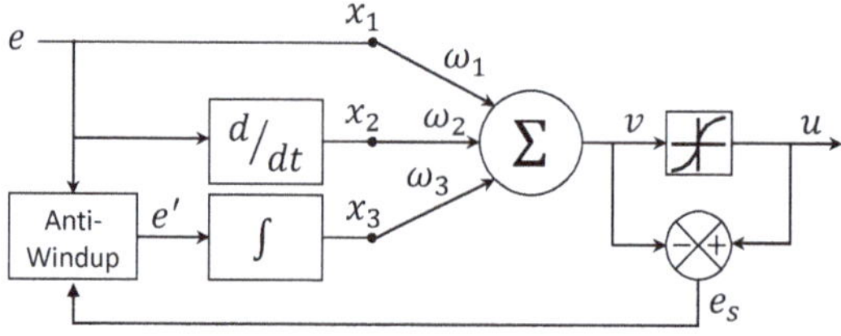

Figure 4. Adaptive single neuron PID controller.

In the presence of actuator saturation, the cumulative error in the integral action produces windup effects. Back-calculation methods are often used to overcome these inconveniences. In general, back-calculation methods consist of reducing the integral action by feeding back the difference between saturated and unsaturated control signals. In order to prevent the windup effects, the feed-back difference e_s is computed (9). The error e_s defines the differences between the outputs u and v, and it is only considered when the value of v exceeds the saturated control signal u scaled by the parameter α, see (7). The anti-windup scheme consists of computing the error e', which handle the integral action. This scheme is defined as

$$e'(k) = \begin{cases} e(k) + e_s(k) & \textit{if } |v(k)| > \alpha \\ e(k) & \textit{otherwise} \end{cases} \tag{8}$$

where

$$e_s(k) = u(k) - v(k) \tag{9}$$

The anti-windup scheme is conducted as follows: when the control signal u is saturated, then the error e' is considered to be the actual error e plus the feed-back difference e_s to reduce the cumulative error. When the control signal u is not saturated, the error e' is the same as the error e. In this case, the cumulative error is not reduced. Typically, limitations in actuators provoke saturation. However, the presented anti-windup scheme prevents the actuator saturation by the selection of α.

It is important to remark that most of the reported adaptive neural PID controllers present training strategies mainly based on the backpropagation algorithm. In this work, the proposed adaptive neuron is trained online using an Extended Kalman Filter (EKF) algorithm that is described in the next section. An important advantage of online training is that the network can adapt itself to changes in the nature of the problem under consideration.

2.1. Adaptive Neuron Training Based on the EKF Algorithm

Several algorithms were reported for the training of neural networks, most of them with the problem of having a slow learning rate and high sensitivity to initial conditions. On the other hand, algorithms based on the Kalman filter are an alternative [34,38], as they reduce the number of neurons and epochs. Moreover, they improve learning convergence [39], and they proved to be reliable for online and offline training for several applications for both feedforward and recurrent neural networks [34,38,40–42]. The Kalman filter is used to estimate the state of a linear system with additive white noise, both in the state and in the output using a recursive solution, which uses previous state and the current input [34]. In the case of neural networks, due to the nonlinear mapping EKF is required, and the weights of the neural network become the state variables to be estimated. The error between the measured output and the network output is considered additive white noise. The training goal is to find an optimal weight vector that minimizes the prediction error [34,38]. The EKF is used due to the mapping of the neural network is nonlinear [34,38,43]. The idea behind the use of the EKF to train the ANN is that other training algorithms, such as gradient descent, recursive least squares and backpropagation, are particular cases of the Kalman filter; for this reason, the EKF is suitable for training [44,45]. Moreover, Neural networks trained with the EKF demonstrated faster learning speeds and convergence times than networks trained with algorithms based on backpropagation [34,43].

The training algorithm for the proposed scheme is an EKF algorithm as proposed in [38], which was proven to be reliable for many applications using multilayer and recurrent neural networks [35,38,41]. Control scheme of the training is depicted in Figure 5. The training of this neuron is performed by an online supervised scheme in which the desired response is the desired trajectory, this is the reason such an error is the one considered for the training, which is going to train the neuron to give control signal such as the trajectory error is minimized.

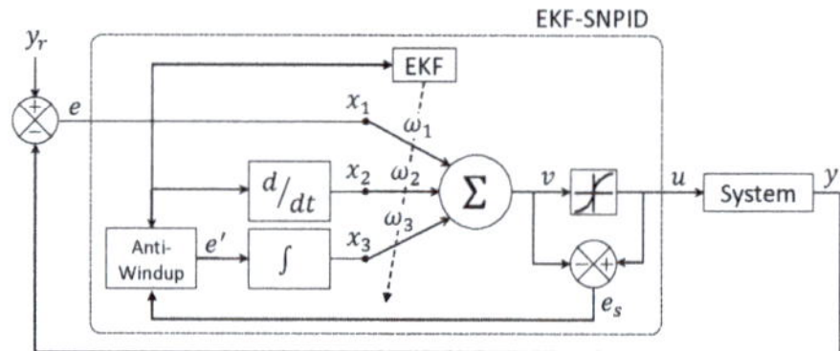

Figure 5. Adaptive single neuron anti-windup PID control scheme. EKF-SNPID: the proposed adaptive neuron PID controller trained with EKF algorithm.

The EKF training algorithm is given by Equations (10)– (12) [43].

$$\mathbf{K}(k) = \mathbf{P}(k)\,\mathbf{H}(k)\left[\mathbf{R}(k) + \mathbf{H}^\top(k)\,\mathbf{P}(k)\,\mathbf{H}(k)\right]^{-1} \tag{10}$$

$$\omega(k+1) = \omega(k) + \eta\mathbf{K}(k)\,\mathbf{e}(k) \tag{11}$$

$$\mathbf{P}(k+1) = \mathbf{P}(k) - \mathbf{K}(k)\,\mathbf{H}^\top(k)\,\mathbf{P}(k) + \mathbf{Q}(k) \tag{12}$$

$$\mathbf{H}_j(k) = \left[\frac{\partial u(k)}{\partial \omega_j(k)}\right] = \left[\frac{\partial u(k)}{\partial v(k)}\frac{\partial v(k)}{\partial \omega_j(k)}\right] = \alpha\,\mathrm{sech}^2(v(k))\,x_j(k) \tag{13}$$

where $\omega \in \mathbb{R}^3$ is the weight vector, $\mathbf{K} \in \mathbb{R}^{3\times1}$ is the Kalman gain vector, $\mathbf{P} \in \mathbb{R}^{3\times3}$, $\mathbf{Q} \in \mathbb{R}^{3\times3}$, and $\mathbf{R} \in \mathbb{R}^{1\times1}$ are covariance matrices of weight estimation error, estimation noise, and error noise, respectively. $\eta \in \mathbb{R}$ is the Kalman filter learning rate, and $\mathbf{H} \in \mathbb{R}^{3\times1}$ is a matrix whose entries $\mathbf{H}_j$ are the derivative of the neural network output with respect to each weight Equation (13), $u \in \mathbb{R}$ is the neuron output, the error $\mathbf{e} \in \mathbb{R}$ is defined as the difference between the desired output and the neuron output (2) [43].

The training objective is to have a set of weight ω such as the signal u minimizes the error between the desired reference and the system output in (2). In the EKF algorithm, the learning rate η is a scalar. However, it is proposed to use $\eta \in \mathbb{R}^{3\times1}$ to provide a learning rate to adjust each weight ω_i independently with η_i, respectively. Finally, the proposed neuron PID controller trained with the Extended Kalman Filter algorithm is called EKF-SNPID. A brief description of the proposed approach is given in the flowchart of Figure 6.

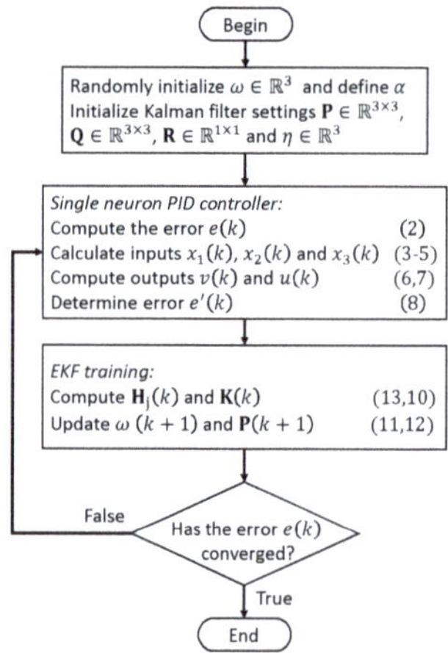

Figure 6. Flowchart of the proposed adaptive neuron PID controller trained with EKF algorithm.

3. Implementation of the Adaptive Neuron PID Controller for an Omnidirectional Mobile Robot

Omnidirectional mobile robots were used in many robotic applications due to their movement capabilities, which allow them simultaneously to move towards any position and reach any desired orientation [46–49]. In contrast to the limit movement capabilities of a conventional mobile robot with two or four wheels due to their nonholonomic kinematics constraints [50]. In this work, the proposed single neuron PID controller is to make an omnidirectional mobile robot reach a desired position and orientation; for this case, the control computes velocities.

3.1. Omnidirectional Mobile Robot Kinematics

The considered omnidirectional mobile robot in this work is made up of four mecanum wheels placed similarly to a conventional vehicle (Figure 7). The pose of the robot with respect to the world frame is given by three degrees of freedom (DOF), which are the positions x, and y, and the orientation θ. Velocities of the mobile robot in the base frame are given by v_x, v_y, and v_θ. Velocity v_i correspond to the velocity of each wheels with $i = 1, 2, 3, 4$, where $v_i = r_i \times w_i$, where r_i is the radius of the wheels, and w_i is angular velocity. The parameter L is half of the distance between the front and the rear wheels, and l is half of the distance between the left and right wheels.

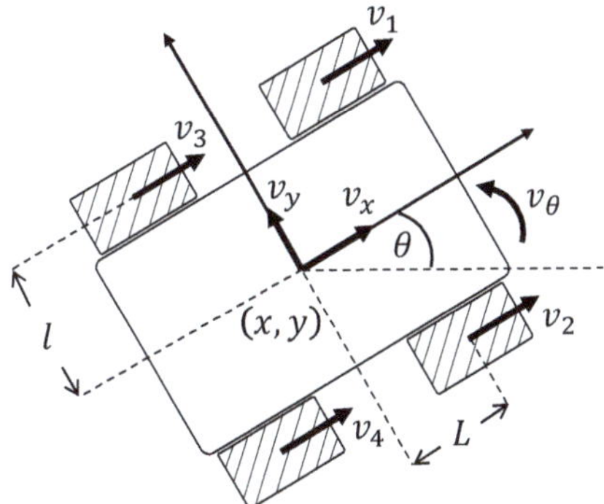

Figure 7. Schematic of an omnidirectional mobile robot conformed of four mecanum wheels. See Figure 18 for a real mobile robot.

The inverse kinematics of the omnidirectional mobile robot in the base frame are given by (14) [51].

$$\begin{bmatrix} v_1(t) \\ v_2(t) \\ v_3(t) \\ v_4(t) \end{bmatrix} = \begin{bmatrix} 1 & 1 & (L+l) \\ 1 & 1 & (L+l) \\ 1 & 1 & -(L+l) \\ 1 & -1 & (L+l) \end{bmatrix} \begin{bmatrix} v_x(t) \\ v_y(t) \\ v_\theta(t) \end{bmatrix} \tag{14}$$

Velocities $\begin{bmatrix} v_x & v_y & v_\theta \end{bmatrix}^\top$ are mapped into the velocities $\begin{bmatrix} \dot{x} & \dot{y} & \dot{\theta} \end{bmatrix}^\top$ of the world frame using the transformation (15).

$$\begin{bmatrix} \dot{x}(t) \\ \dot{y}(t) \\ \dot{\theta}(t) \end{bmatrix} = \begin{bmatrix} \cos(\theta(t)) & -\sin(\theta(t)) & 0 \\ \sin(\theta(t)) & \cos(\theta(t)) & 0 \\ 0 & 0 & 1 \end{bmatrix} \begin{bmatrix} v_x(t) \\ v_y(t) \\ v_\theta(t) \end{bmatrix} \tag{15}$$

Equation (16) is obtained from Equations (14) and (15).

$$\begin{bmatrix} v_1(t) \\ v_2(t) \\ v_3(t) \\ v_4(t) \end{bmatrix} = \begin{bmatrix} \sqrt{2}\sin\left(\theta+\frac{\pi}{4}\right) & -\sqrt{2}\cos\left(\theta+\frac{\pi}{4}\right) & -(L+l) \\ \sqrt{2}\cos\left(\theta+\frac{\pi}{4}\right) & \sqrt{2}\sin\left(\theta+\frac{\pi}{4}\right) & (L+l) \\ \sqrt{2}\cos\left(\theta+\frac{\pi}{4}\right) & \sqrt{2}\sin\left(\theta+\frac{\pi}{4}\right) & -(L+l) \\ \sqrt{2}\sin\left(\theta+\frac{\pi}{4}\right) & -\sqrt{2}\cos\left(\theta+\frac{\pi}{4}\right) & (L+l) \end{bmatrix} \begin{bmatrix} \dot{x}(t) \\ \dot{y}(t) \\ \dot{\theta}(t) \end{bmatrix} = \mathbf{J}(\theta(t)) \begin{bmatrix} \dot{x}(t) \\ \dot{y}(t) \\ \dot{\theta}(t) \end{bmatrix} \tag{16}$$

where $\mathbf{J}(\theta)$ is a transformation matrix, which maps the mobile robot velocities $\begin{bmatrix} \dot{x} & \dot{y} & \dot{\theta} \end{bmatrix}^\top$ into the velocities for each of the wheels.

For further information about the kinematics of the considered mecanum omnidirectional mobile robot, please go to references [52,53].

3.2. Conventional Velocity Control Design

To achieve the position tracking, proper velocities have to be computed, then, the controller has to find velocities $u_x(k)$, $u_y(k)$, and $u_\theta(k)$ at step time k, to drive the mobile robot, from the current pose (17) to the desired pose (18). The error between the desired pose and current pose is defined as (19):

$$\begin{bmatrix} x(k) & y(k) & \theta(k) \end{bmatrix}^\top \tag{17}$$

$$\begin{bmatrix} x_r(k) & y_r(k) & \theta_r(k) \end{bmatrix}^\top \tag{18}$$

$$\begin{bmatrix} x_e(k) \\ y_e(k) \\ \theta_e(k) \end{bmatrix} = \begin{bmatrix} x_r(k) - x(k) \\ y_r(k) - y(k) \\ \theta_r(k) - \theta(k) \end{bmatrix} \tag{19}$$

A PID control can be use to compute the velocities $v_i(k)$ of the mobile robot in order to asymptotically stabilize the system, for $i = 1,2,3,4$ [51]. A discrete PID control laws [37] at step k are given in Equations (20)–(22), respectively. For this control scheme, there is one PID module for each error $x_e(k)$, $y_e(k)$, and $\theta_e(k)$. Parameters K_P^x, K_I^x and K_D^x are the proportional, integrative and derivative gains for error x_e, respectively. Similarly, the parameters K_P^y, K_I^y and K_D^y are the gains for error y_e, and K_P^θ, K_I^θ and K_D^θ are the gains for error θ_e.

$$u_x(k) = K_P^x x_e(k) + K_I^x \sum_{j=1}^{k} x_e(j) + K_D^x \left[x_e(k) - x_e(k-1)\right] \tag{20}$$

$$u_y(k) = K_P^y y_e(k) + K_I^y \sum_{j=1}^{k} y_e(j) + K_D^y \left[y_e(k) - y_e(k-1)\right] \tag{21}$$

$$u_\theta(k) = K_P^\theta \theta_e(k) + K_I^\theta \sum_{j=1}^{k} \theta_e(j) + K_D^\theta \left[\theta_e(k) - \theta_e(k-1)\right] \tag{22}$$

Based on (16), the control outputs $u_x(k)$, $u_y(k)$ and $u_\theta(k)$ are mapped to each wheel control velocity $u_j(k)$ with $j = 1,2,3,4$, using transformation matrix $\mathbf{J}(\theta(k))$ as follows

$$\begin{bmatrix} u_1(k) \\ u_2(k) \\ u_3(k) \\ u_4(k) \end{bmatrix} = \mathbf{J}(\theta(k)) \begin{bmatrix} u_x(k) \\ u_y(k) \\ u_\theta(k) \end{bmatrix} \tag{23}$$

The conventional PID is widely used to control mobile robots due to its simplicity and performance [51,54–56]. It is well known that the main problem of the conventional PID is the manual tuning of the proportional, integrative, and derivative gains. However, this paper presents an adaptive single neuron PID approach to overcome this inconvenience. The proposed approach can adjust itself online during the operation of the system.

3.3. Control Scheme Using the Adaptive Single Neuron PID Controller

The proposed control scheme based on the adaptive single neuron PID controller trained with EKF algorithm (EKF-SNPID) of an omnidirectional mobile robot is shown in Figure 8. In order to minimize the errors x_e, y_e and θ_e, an adaptive PID controller module is designed for each DOF. Then, each output control signal (24) corresponding to each single neuron PID control module which is mapped into the control velocities signals (25) for each wheel, respectively. The system output is given by the pose of the mobile robot (26).

$$\begin{bmatrix} u_x(k) & u_y(k) & u_\theta(k) \end{bmatrix}^\top \tag{24}$$

$$\begin{bmatrix} u_1(k) & u_2(k) & u_3(k) & u_4(k) \end{bmatrix}^\top \tag{25}$$

$$\begin{bmatrix} x(k) & y(k) & \theta(k) \end{bmatrix}^\top \tag{26}$$

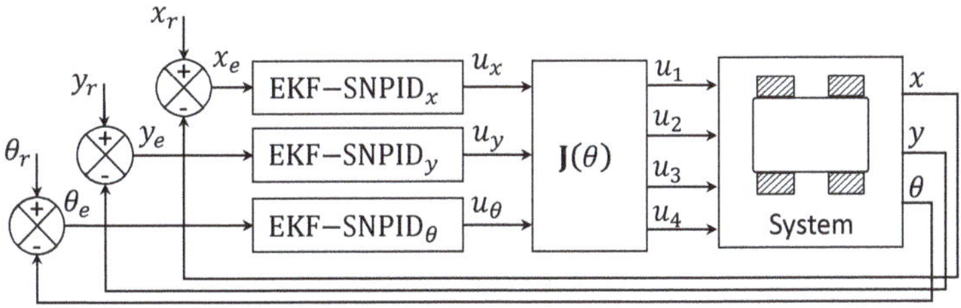

Figure 8. Adaptive single neuron anti-windup PID control scheme for a omnidirectional mobile robot.

It is important to note that even with the inclusion of the anti-windup block to the single neuron architecture (25), the computed control signal is bounded, then such control signal designed for a dynamic system, such as the one for the mobile robot (15) is a feed forward neural control law, which composes a stable system [57–59].

4. Results

In Section 4.1 results of both simulation and experimental tests of the proposed adaptive single neuron PID controller trained with extended Kalman filter (EKF-SNPID) compared to conventional PID, backpropagation trained adaptive neural PID controller (BP-PIDNN) [4], and an adaptive neuron PID controller trained with Hebbian learning rule (HR-PIDNN) [19]. Then, in Section 4.1.1, the proposed EKF single neuron PID controller is compared with a conventional PID controller with a back-calculation anti-windup method [29].

The presented kinematic model of the mobile robot is given in continuous time domain. This model is not used by the proposed EKF single neuron controller scheme. For simulation tests, the kinematic model is programmed on Matlab® where it is discretize using zero-order hold Matlab® function method. For the experimental tests, the information is sent a received on discrete instants of time.

4.1. Simulation Results

For this section, it is important to clear out that in the literature, the reported adaptive neural PID controllers do not integrate an anti-windup scheme. In this way, the proposed scheme is compared against different adaptive neural PID controllers as they where reported. Moreover, limits for control signals are not set due to lose of controllability of the techniques the proposal is compared with.

Simulation experiments consist of the tracking of different trajectories. Test are implemented in Matlab®. Parameters for the omnidirectional mobile robot are set as: $L = 0.2355$ m, $l = 0.15$ m, $r = 0.0475$ m, and sampling time $k = 0.05$ s.

For conventional PID controller, proportional gains are set as: $K_P^x = 0.3$, $K_P^y = 0.3$ and $K_P^\theta = 0.2$, integral gains $K_I^x = 0.2$, $K_I^y = 0.2$ and $K_I^\theta = 0.1$, and derivative gains $K_D^x = 0.1$, $K_D^y = 0.1$ and $K_D^\theta = 0.05$. In addition, for adaptive neural PID controllers, weights are initialized randomly.

Parameter setting for proposed EKF-SNPID controller are: matrices$\mathbf{P}$ and $\mathbf{Q}$ are initialized as diagonal matrices with $\mathbf{P}_{11} = \mathbf{P}_{22} = \mathbf{P}_{33} = 1$ and $\mathbf{Q}_{11} = \mathbf{Q}_{22} = \mathbf{Q}_{33} = 0.1$, the parameter $\mathbf{R} = 0.0001$, Kalman filter learning rates $\eta_1 = 0.1$, $\eta_3 = 0.1$ and $\eta_2 = 0.01$ and $\alpha = 0.3$. The selection of these parameters was chosen experimentally. Considered trajectories at each step time k are generated as follows:

- a sinusoidal trajectory is computed as

$$
\begin{aligned}
x_r(k) &= 0.05\,k \\
y_r(k) &= 0.2\sin\left(\left(\frac{1}{0.5}\right) x_r(k)\,\pi\right) \\
\theta_r(k) &= \frac{\pi}{8}
\end{aligned}
$$

- a rose curve trajectory generated as

$$
\begin{aligned}
a &= 0.2 + 0.05\cos\left(3\,(0.05)\,k\pi\right) \\
x_r(k) &= a\cos\left(0.05k\pi\right) \\
y_r(k) &= a\sin\left(0.05k\pi\right) \\
\theta_r(k) &= \frac{\pi}{4}
\end{aligned}
$$

To compare the performance of the considered controllers, resulting tracking and control signals are shown in graphs. Additionally, the root mean square (RMS) and mean absolute deviation (MAD) measures of the obtained errors are reported.

4.1.1. Simulation Test: Sinusoidal Reference

Trajectory tracking results for the sinusoidal trajectory reference are shown in Figure 9. As with the circular trajectory case, the proposed adaptive PID controller achieves the best tracking among the other compared controllers.

Velocity control signals for the sinusoidal trajectory reference are provided in Figure 10. The proposed EKF-SNPID controller reports faster performance than the others.

The system response for the sinusoidal trajectory reference is shown in Figure 11 where Figure 11a shows the system response for x. In this case, it is shown that all compared controllers achieve the reference successfully. Respect to system response for y, the proposed adaptive PID controller outperformed the others, see Figure 11b. The proposal reported a less steady-state error. In contrast, the other controllers have bigger errors. In Figure 11c, the system response for θ is presented. In this case, there is not overshoot provided by the proposed EKF-SNPID.

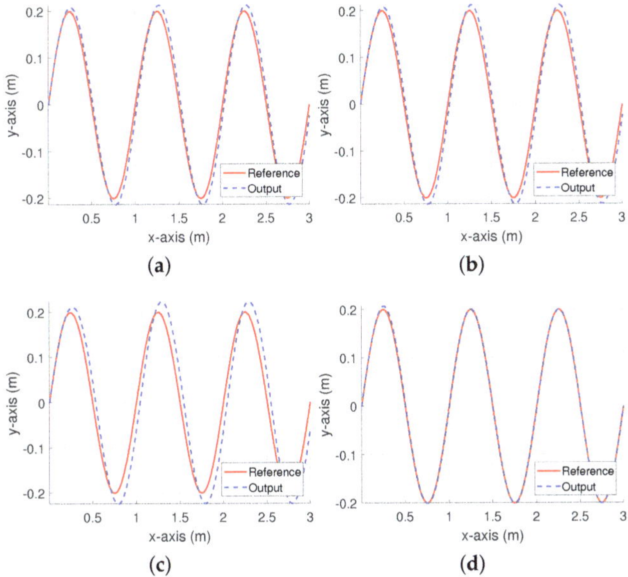

Figure 9. Trajectory following results for the sinusoidal trajectory. (**a**) Trajectory following for PID. (**b**) Trajectory following for BP-PIDNN. (**c**) Trajectory following for HR-PIDNN. (**d**) Trajectory following for EKF-SNPID.

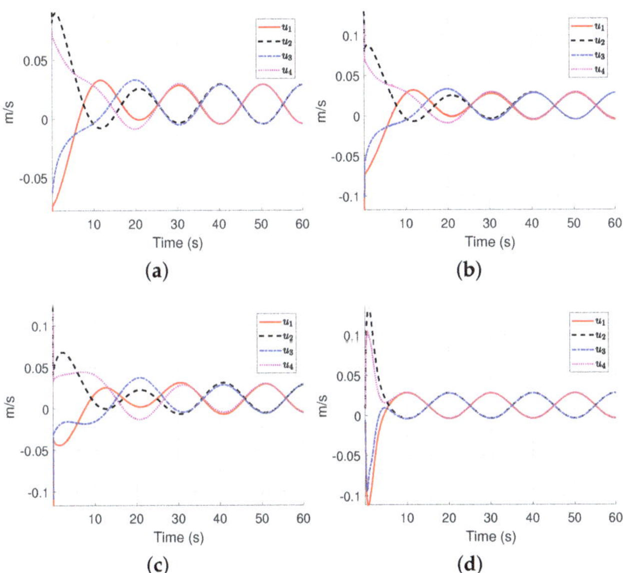

Figure 10. Velocity control signal results for the sinusoidal trajectory. (**a**) Velocity control signal of PID. (**b**) Velocity control signal of BP-PIDNN. (**c**) Velocity control signal of HR-PIDNN. (**d**) Velocity control signal of EKF-SNPID.

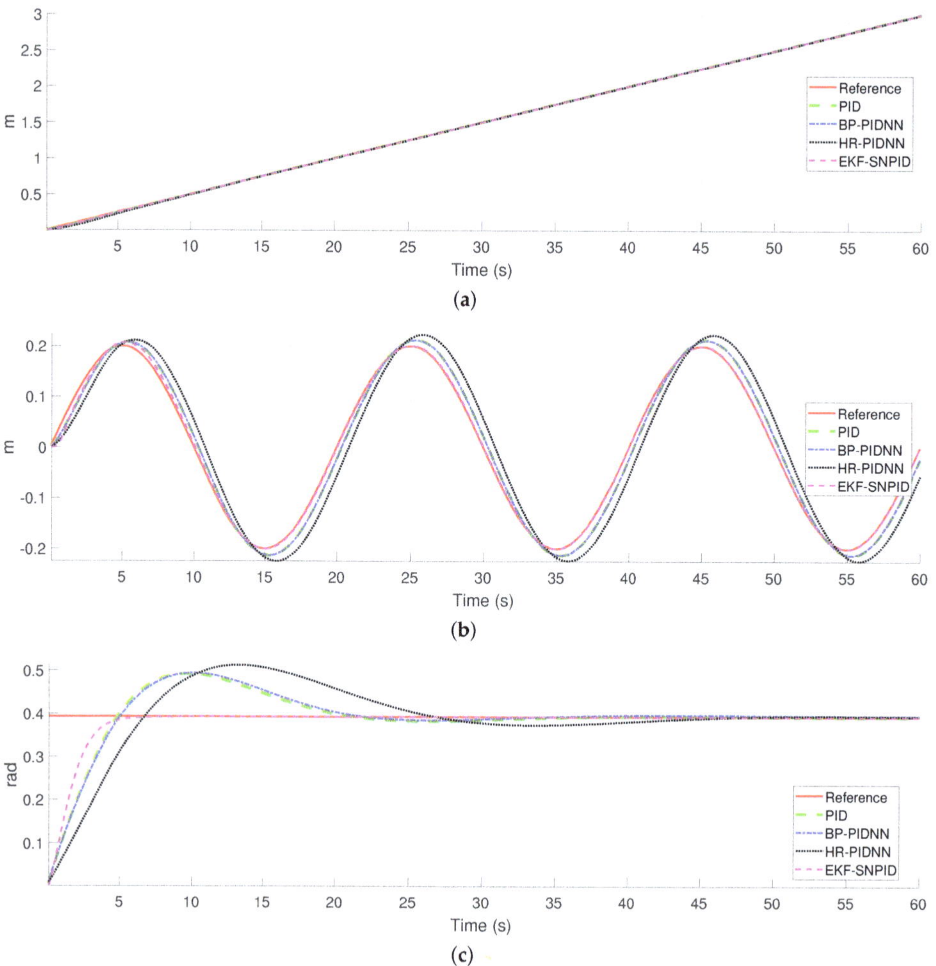

Figure 11. System response for the sinusoidal trajectory. (**a**) System response for x. (**b**) System response for y. (**c**) System response for θ.

The RMS and MAD results for the sinusoidal trajectory reference are presented in Table 1. The proposed adaptive neuron PID controller outperforms the other compared controllers.

Table 1. Simulation results for the sinusoidal trajectory. The best results are highlighted in bold.

Measure	Method	x_e	y_e	θ_e
RMS	PID	4.6702×10^{-3}	1.8181×10^{-2}	6.6922×10^{-2}
	BP-PIDNN	5.0489×10^{-3}	1.7526×10^{-2}	6.7741×10^{-2}
	HR-PIDNN	9.4373×10^{-3}	3.9668×10^{-2}	8.5651×10^{-2}
	EKF-SNPID	$\mathbf{4.1447}\times10^{-3}$	$\mathbf{5.0465}\times10^{-3}$	$\mathbf{5.0092}\times10^{-2}$
MAD	PID	3.0401×10^{-3}	1.6364×10^{-2}	3.0477×10^{-2}
	BP-PIDNN	3.1118×10^{-3}	1.5772×10^{-2}	3.1253×10^{-2}
	HR-PIDNN	6.2210×10^{-3}	3.5599×10^{-2}	4.7817×10^{-2}
	EKF-SNPID	$\mathbf{1.2864}\times10^{-3}$	$\mathbf{2.6786}\times10^{-3}$	$\mathbf{1.9203}\times10^{-2}$

4.1.2. Simulation Test: Rose Curve Reference

Figure 12 shows the trajectory following results for the rose curve trajectory. Also for this case, the proposed adaptive PID controller outperformed the other controllers.

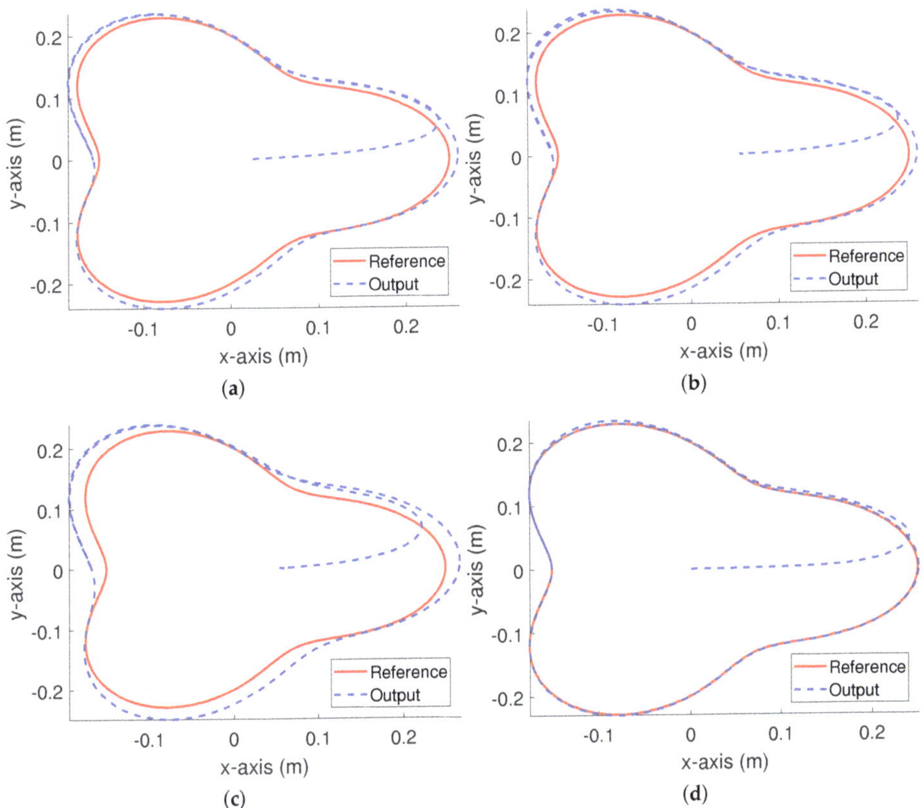

Figure 12. Trajectory following results for the rose curve trajectory. (**a**) Trajectory following for PID) . (**b**) Trajectory following for BP-PIDNN. (**c**) Trajectory following for HR-PIDNN. (**d**) Trajectory following for EKF-SNPID.

Figure 13 provides the velocity control signal for the rose curve trajectory. Velocity control signals for EKF-SNPID are lower than the others.

Figure 14 shows the system response for the rose curve trajectory test. The system response for x, y, and θ are shown in Figure 14. The proposed adaptive PID controller follows in a better way the desired references than the other controllers. In this case, the proposed approach has the lowest steady-state errors. Additionally, the proposed controller reported a fast system response without overshoot respect to θ, see Figure 14c.

Table 2 reports the RMS and MAD results for the rose curve trajectory. In this case, the BP-PIDNN reported the best RMS respect to the x_e error. On the other hand, the proposed EKF-SNPID controller performed better than the other controllers for the y_e and θ_e error. Additionally, the proposal achieved the best results with the lowest MAD results.

Based on the reported results so far, the BP-PIDNN controller performed better than the conventional PID controller. However, the performance of the proposed EKF-SNPID controller is superior to the compared controllers, respect to the best transient response, the smaller steady-state errors, and lower overshoot results. Additionally, the performance of the PID controller for tracking is acceptable. However, it is necessary to adjust the PID gains manually to improve its performance. In contrast, the proposed adaptive EKF-SNPID controller does not require any adjustment.

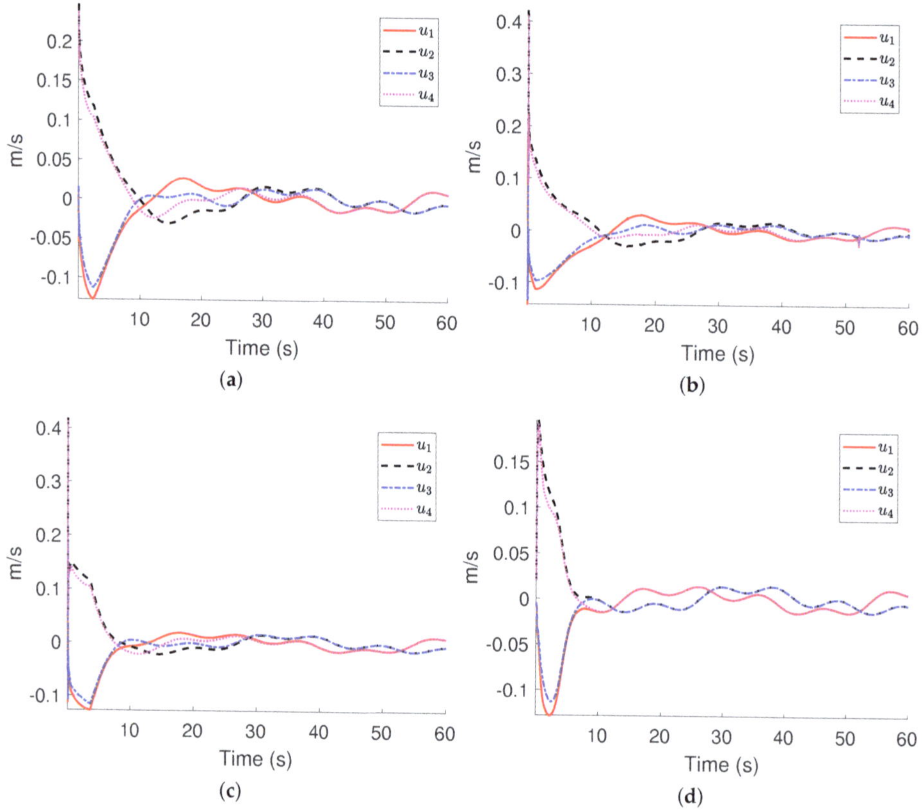

Figure 13. Velocity control signal results for the rose curve trajectory. (**a**) Velocity control signal of PID. (**b**) Velocity control signal of BP-PIDNN. (**c**) Velocity control signal of HR-PIDNN. (**d**) Velocity control signal of EKF-SNPID.

Table 2. Simulation results for the rose curve trajectory. The best results are highlighted in bold.

Measure	Method	x_e	y_e	θ_e
RMS	PID	2.0057×10^{-2}	8.5850×10^{-3}	1.4285×10^{-1}
	BP-PIDNN	$\mathbf{1.8210}\times10^{-2}$	9.6169×10^{-3}	1.5274×10^{-1}
	HR-PIDNN	2.7587×10^{-2}	1.8114×10^{-2}	1.3674×10^{-1}
	EKF-SNPID	1.8662×10^{-2}	$\mathbf{3.7495}\times10^{-3}$	$\mathbf{1.2650}\times10^{-1}$
MAD	PID	1.0300×10^{-2}	6.6289×10^{-3}	6.6001×10^{-2}
	BP-PIDNN	9.6779×10^{-3}	7.4597×10^{-3}	7.7155×10^{-2}
	HR-PIDNN	1.8695×10^{-2}	1.4109×10^{-2}	5.9445×10^{-2}
	EKF-SNPID	$\mathbf{4.9995}\times10^{-3}$	$\mathbf{5.7139}\times10^{-3}$	$\mathbf{3.4104}\times10^{-2}$

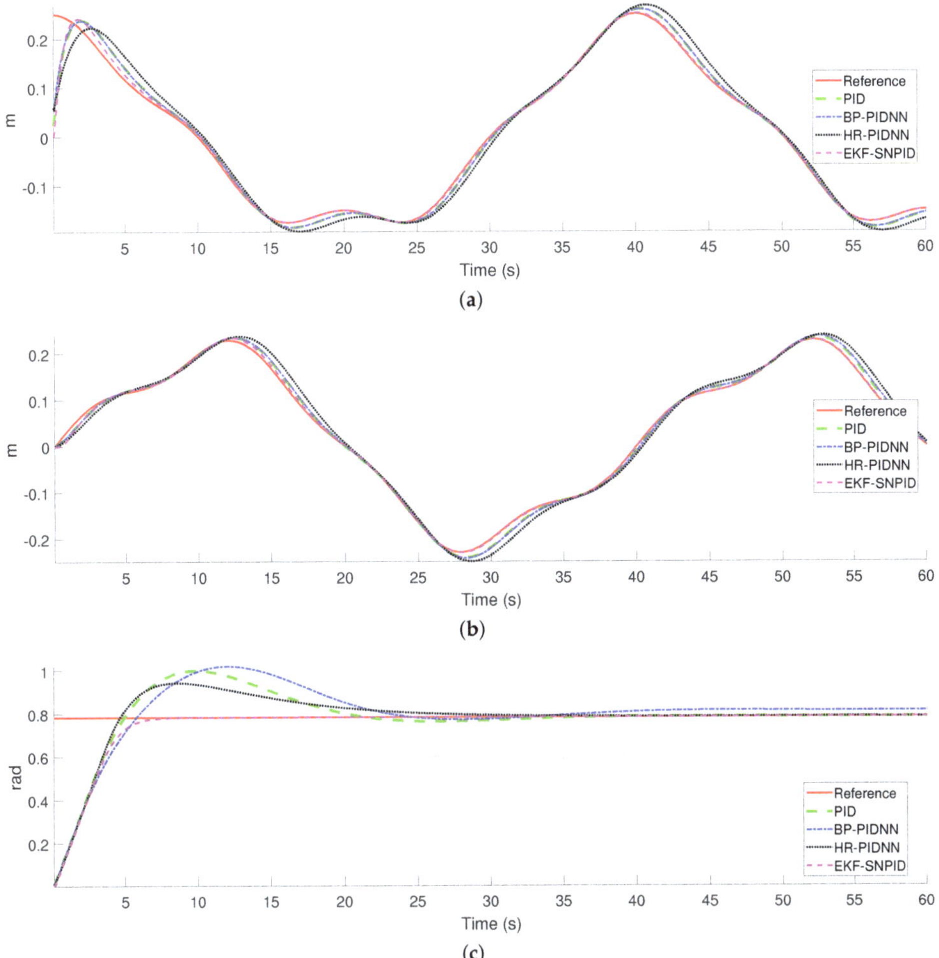

Figure 14. System response for the rose curve trajectory. BP-PIDNN: Neural PID controller trained with Back Propagation algorithm. HR-PIDNN: Neuron PID controller trained with Hebbian learning rule. EKF-SNPID: the proposed neural PID controller trained with Extended Kalman filter. Label s: seconds. Label m: meters. Label rad: radian. (**a**) System response for x. (**b**) System response for y. (**c**) System response for θ.

4.2. Anti-Windup Tests

The following simulations results show the performance of the proposed neuron PID controller with the integration of the back-calculation anti-windup method. Simulations consist of the tracking of a trajectory with offsets. These offsets provoke velocity saturation in the mobile robot when trying to reach the necessary speeds to get to the trajectory. The performance of the proposed adaptive single neuron controller is compared against a PID with back-calculation anti-windup [29]. The conventional PID is also considered for simulations to contrast the windup effects. These tests are implemented in Matlab®.

The omnidirectional mobile robot of the previous simulation is also used for the following tests. Similarly, parameter settings of the considered controllers are chosen. Moreover, the back-calculation anti-windup on PID requires a parameter for the feeding back information [29], which is defined as T_s, and its value is selected as 0.1. This tuning is not required by the proposed EKF-SNPID controller.

The considered trajectory at each step time k are given as follows:

- An approximated trapezoidal trajectory which is calculated as

$$\begin{aligned} x_r(k) &= 0.1k \\ b &= 3.0 + 0.3\sin\left(\left(\frac{1}{1.5}\right)x_r(k)\,\pi\right) \\ y_r(k) &= \begin{cases} 3.2 & \textbf{if } b \geq 3.2 \\ -3.2 & \textbf{if } b < -3.2 \\ b & \textbf{otherwise} \end{cases} \\ \theta_r(k) &= \frac{\pi}{3} \end{aligned}$$

To compare the performance of the considered controllers in this test, results are shown in graphs. Moreover, the RMS and MAD results are presented in terms of percent error. To compare the performance of the proposed approach against the PID controller, we proposed to measure the percent error as

$$\%\,\text{error} = \frac{\text{PID}_{RMS} - \text{EKF-SNPID}_{RMS}}{\text{PID}_{RMS}} \times 100 \tag{27}$$

where a positive percent error indicates that the proposed approach performs better than PID, and a negative percent error indicates that PID overcomes the proposed approach. The same equation is used to compare the proposed controller against the PID with anti-windup. Similarly, the percent error of MAD results are also measured.

Simulation Test: Trapezoidal Reference

Trajectory tracking and Velocity control signal for the trapezoidal reference are give in Figure 15. The conventional PID presents a high tracking error caused by the windup effects, see Figure 15a. The Figure 15c shows that the PID with anti-windup reports better tracking results, but the tracking response is slow compared to the proposed EKF-SNPID with anti-windup that shows the fastest tracking response without windup effects, see Figure 15e. The Figure 15b,d,f reports that the saturation limits are reached by all controllers. The control signals are saturated at $u_x = 0.3$ m/s, $u_y = 0.3$ m/s and $u_x = 0.3$ rad/s.

The Figure 16 reports the system response for the trapezoidal trajectory tracking. The conventional PID reports the highest overshoot. The PID reduces the overshoot with anti-windup. Moreover, the proposed EKF-SNPID suppressed the overshoot and it reaches the reference faster than the others.

The percent errors for this test are give in Figure 17. As can be seen, a percent error is shown for each error of the mobile robot pose x_e, y_e and θ_e. A percent error for RMS and MAD results is also computed. Figure 17a reports the comparison of the proposed EKF-SNPID against the conventional PID and Figure 17b illustrates the of EKF-SNPID against the PID with anti-windup. The reported results for both comparison are positive percent errors. This indicates that the proposed adaptive EKF-SNPID performs betters than the compared controllers.

This test indicates that the conventional PID controller suffers from overshot, which is provoked by saturation velocity. The use of the back-calculation method in the PID highly reduces this inconvenience. However, its response time and tracking results are poor. The conventional back-calculation requires the tune of a gain parameter to improve this performance. In contrast, the proposed EKF-SNPID with anti-windup outperformed the conventional PID and the PID with anti-windup. The proposal reduces overshoot effectively. Moreover, it has a better response time and tracking performance, even in the presence of saturation limits. Additionally, it does not require any adjustment or tuning.

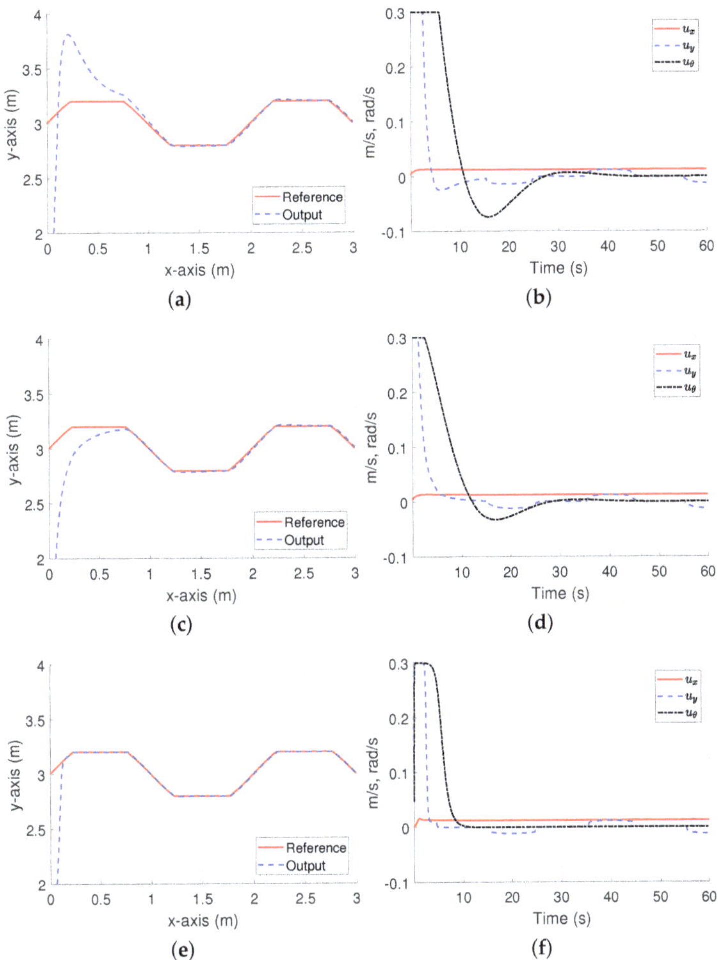

Figure 15. Trajectory following and Velocity control signal results for the trapezoidal trajectory with offsets. (**a**). Trajectory following for PID. (**b**) Velocity control signal of PID. (**c**) Trajectory following for PID with anti-windup. (**d**) Velocity control signal of PID with anti-windup. (**e**) Trajectory following for EKF-SNPID. (**f**) Velocity control signal of EKF-SNPID.

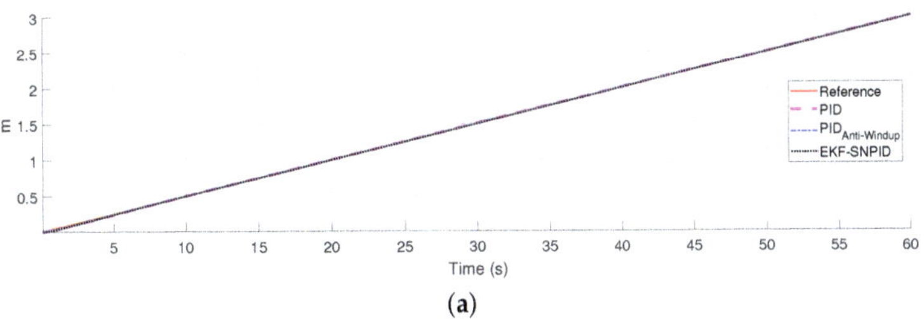

Figure 16. *Cont.*

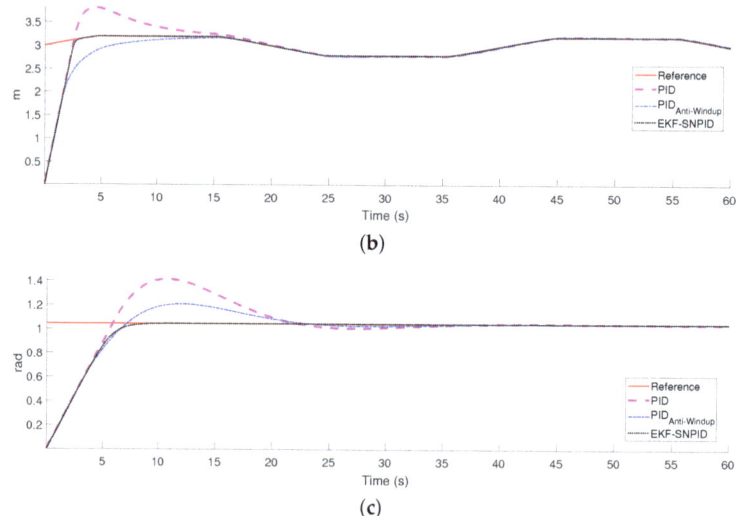

Figure 16. System response for the trapezoidal trajectory with offsets. (**a**) System response for x. (**b**) System response for y. (**c**) System response for θ.

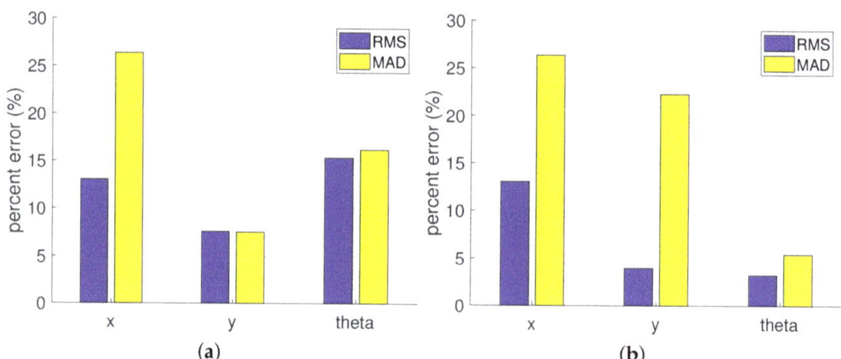

Figure 17. RMS and MAD percent errors results for the trapezoidal trajectory with offsets. The percent errors x, y and theta were computed based on the RMS and MAD results of the mobile robot error pose x_e, y_e and θ_e, respectively. (**a**) Percent error for EKF-SNPID vs PID. (**b**) Percent error for EKF-SNPID vs. PID with anti-windup.

4.3. Experimental Results

Experiment tests are performed using a four-wheeled omnidirectional mobile robot. The considered omnidirectional mobile robot is a KUKA® Youbot® platform (Figure 18). Its parameters are $L = 0.2355$ m and $l = 0.1$ m, and the wheel radius $r = 0.0475$ m. For the experimental test, the performance of the proposed EKF-SNPID controller, which includes the back-calculation anti-windup, is compared to the PID controller with anti-windup. The PID and the proposed approach were implemented in C++ using the KUKA® Youbot® API. The sampling time for the system is set to 0.05 s.

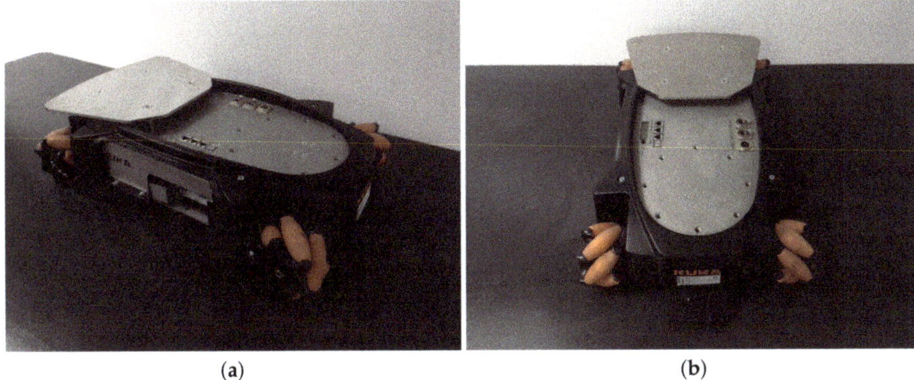

(a) (b)

Figure 18. Omnidirectional KUKA® Youbot® platform. (**a**) side and (**b**) front views.

In the case of the PID controller, proportional gains are set as $K_P^x = 1.5$, $K_P^y = 1.5$ and $K_P^\theta = 0.5$, integral gains are set as $K_I^x = 0.5$, $K_I^y = 0.5$ and $K_I^\theta = 0.3$, and derivative gains are set as $K_D^x = 0.01$, $K_D^y = 0.01$ and $K_D^\theta = 0.01$. For the back-calculation method, the parameter tuning T_s was selected as 0.1.

For the proposed adaptive EKF-SNPID controller, Kalman filter settings are given below. The matrices **P** and **Q** are initialized as diagonal matrices with $\mathbf{P}_{11} = \mathbf{P}_{22} = \mathbf{P}_{33} = 1$ and $\mathbf{Q}_{11} = \mathbf{Q}_{22} = \mathbf{Q}_{33} = 0.1$. The parameter r is set as 0.0001. The Kalman filter learning rates are set as $\eta_1 = 0.2$, $\eta_3 = 0.2$ and $\eta_2 = 0.01$ and $\alpha = 0.3$. Moreover, initial weights are set randomly. All these parameters were chosen experimentally.

The experiments consist of some trajectory tracking tasks. The use of two trajectories with different degrees of difficulty. was considered. The considered trajectories at step time k are generated as follows:

- a sinusoidal trajectory is computed as

$$\begin{aligned} x_r(k) &= 0.1k \\ y_r(k) &= 0.2\sin\left(x_r(k)\,\pi\right) \\ \theta_r(k) &= -\frac{\pi}{4} \end{aligned}$$

- a rose curve trajectory with offsets is generated as

$$\begin{aligned} a &= 0.2 + 0.05\cos\left(5\,(0.1)\,k\pi\right) \\ x_r(k) &= 2.5 + a\cos\left(0.1k\pi\right) \\ y_r(k) &= 2.5 + a\sin\left(0.1k\pi\right) \\ \theta_r(k) &= \frac{\pi}{4} \end{aligned}$$

For the simulation results, to compare the performance of the proposed EKF-SNPID and the PID controller, tracking and control signals results are illustrated in graphs. Moreover, the RMS and MAD measures of the achieved errors are reported in tables. Additionally, the RMS and MAD results are reported in terms of percent error as well.

4.3.1. Experimental test: sinusoidal reference

Figure 19 reports trajectory tracking and the velocity control signal of the test for a sinusoidal desired reference. Figure 19a,b show that the proposed EKF-SNPID controller outperformed PID.

In this case, the proposal showed the best tracking result compared to PID, which performs poorly. From Figure 19c,d, it is seen that proposed EKF single neuron PID controller is adjusting itself to reject perturbation and changes during experimental test.

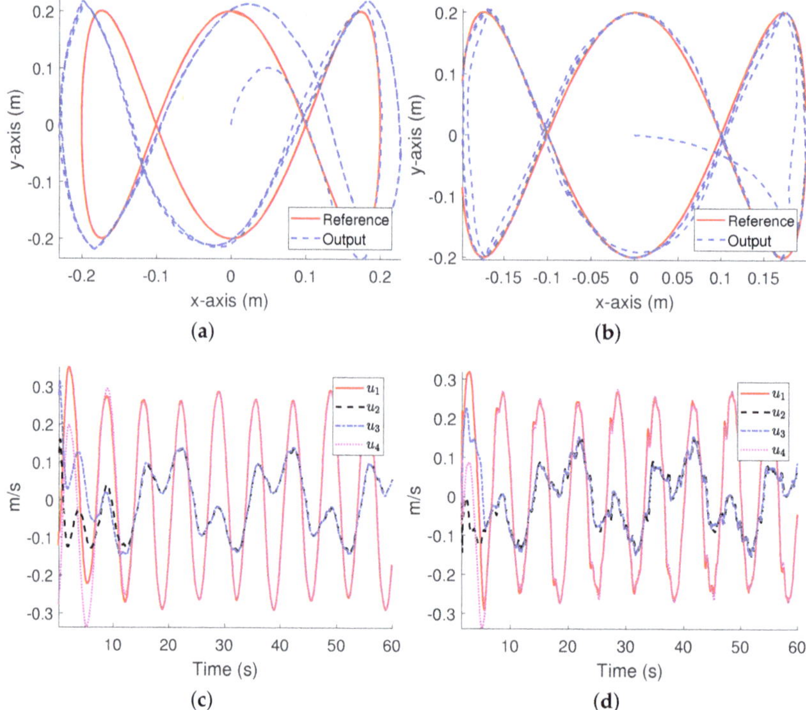

Figure 19. Experimental results for the sinusoidal trajectory. (**a**) Trajectory following for PID. (**b**) Trajectory following for EKF-SNPID. (**c**) Velocity control signal of PID. (**d**) Velocity control signal of EFK-PIDNN.

Figure 20 shows system response results for the sinusoidal trajectory. System response for x, y and θ are reported in Figure 20a–c, respectively. In this test, PID controller shows bigger steady-state errors. In contrast, the adaptive EKF-SNPID controller performed better. In the case of the system response for θ, PID controller provides a bigger overshoot and transient response, see Figure 20c.

Table 3 reports the RMS and MAD results for the sinusoidal trajectory. In this case, the RMS and MAD results of the proposed EKF-SNPID controller are better than PID.

Table 3. Experimental results for the sinusoidal trajectory. The best results are highlighted in bold.

Measure	Method	x_e	y_e	θ_e
RMS	PID	1.0232×10^{-2}	2.4106×10^{-2}	1.3638×10^{-1}
	EKF-SNPID	**8.6468** $\times 10^{-3}$	**7.8939** $\times 10^{-3}$	**1.1402** $\times 10^{-1}$
MAD	PID	4.8462×10^{-3}	2.1612×10^{-2}	5.3743×10^{-2}
	EKF-SNPID	**2.9027** $\times 10^{-3}$	**5.6546** $\times 10^{-3}$	**3.7291** $\times 10^{-2}$

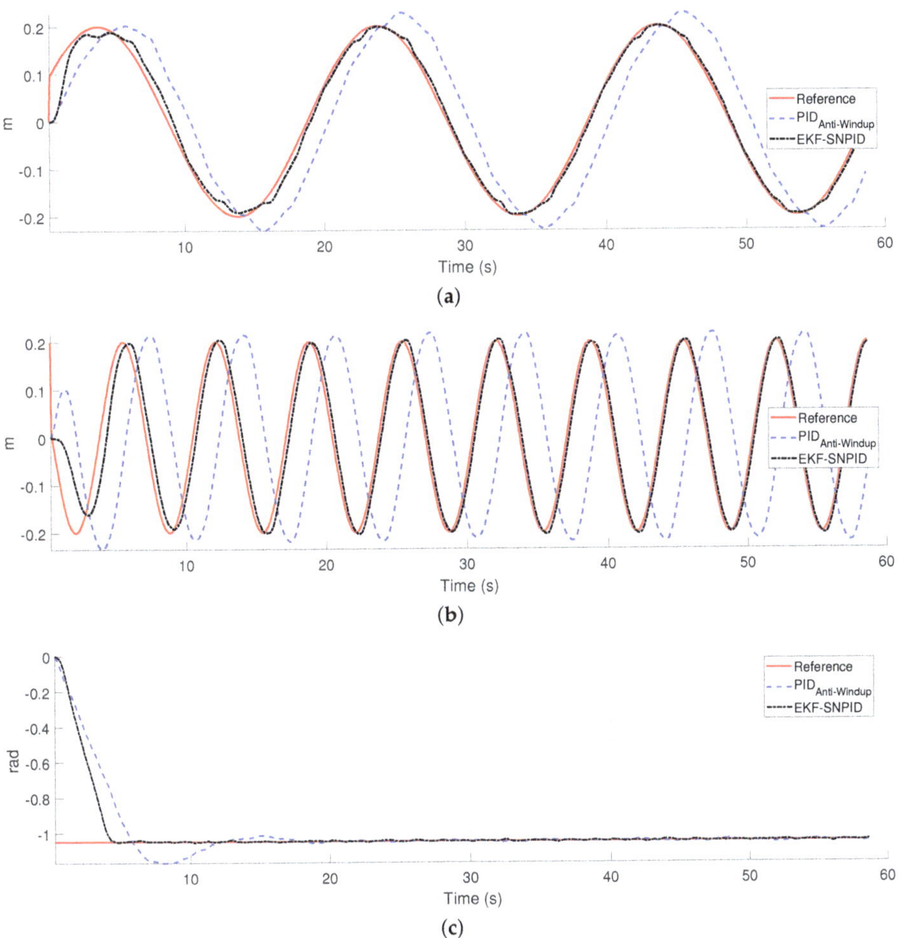

Figure 20. System response for the sinusoidal trajectory. (**a**) System response for x. (**b**) System response for y. (**c**) System response for θ.

4.3.2. Experimental Test: Rose Curved Reference

Trajectory tracking and velocity control signals of test for rose curve trajectory are shown in Figure 21. This test includes velocity saturation since the trajectory includes offsets. Figure 21a,b show the trajectory results for PID and for proposed EKF-SNPID controllers, respectively. In this test, the EKF-SNPID performed better than PID. The velocities control signal for both controllers reached the saturation limits in the first 4 seconds, see Figure 21c,d.

Figure 22 gives the system response for rose curve trajectory. The system response for x is shown in Figure 22a and the system response for y is shown in Figure 22b. For both cases, the adaptive EKF-SNPID controller outperformed PID with a better tracking results and faster system response. Additionally, the proposal reported the smallest steady-state errors. Both controller suppressed overshot, but PID controller showed slow response. The Figure 22c shows the system response for θ. In this case, there is no presence of overshoot in the system response of the proposed adaptive controller.

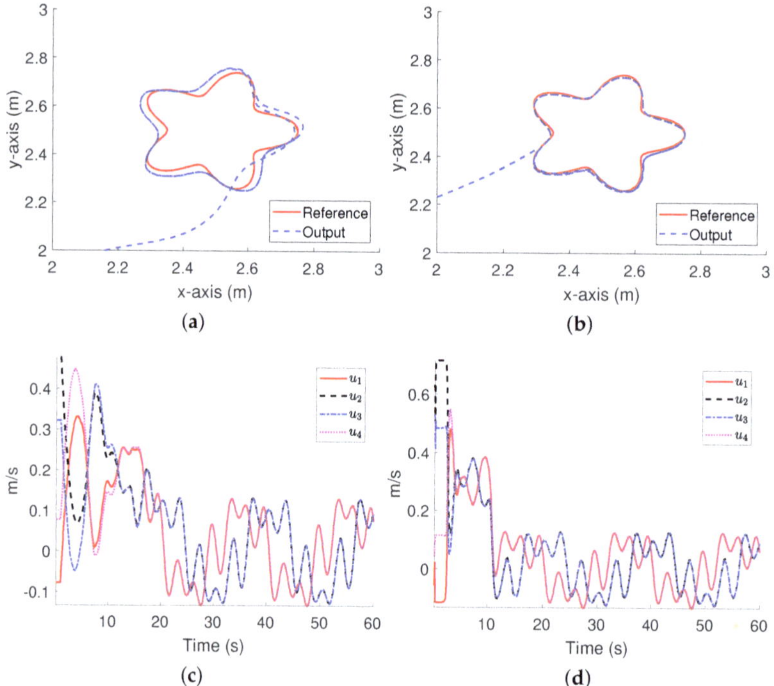

Figure 21. Experimental results for the rose curve trajectory. (**a**) Trajectory following for PID. (**b**) Trajectory following for EKF-SNPID. (**c**) Velocity control signal of PID. (**d**) Velocity control signal of EKF-SNPID.

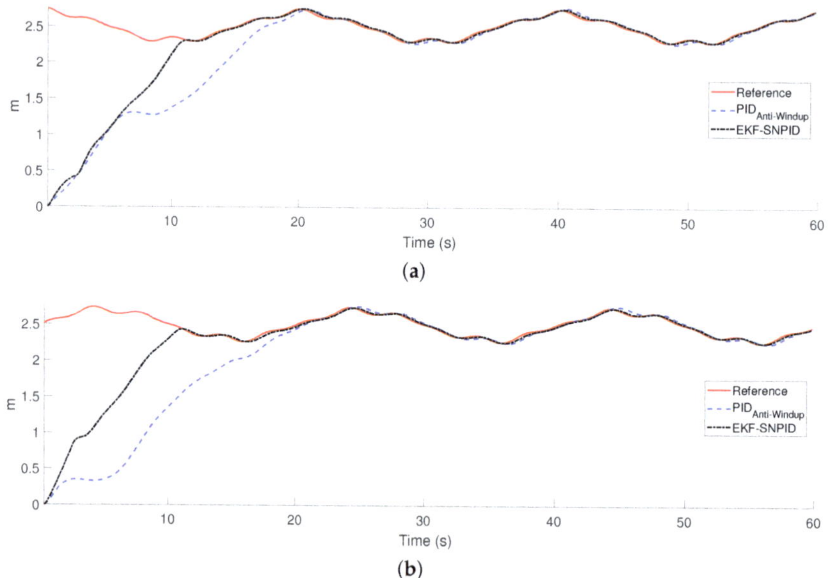

Figure 22. *Cont.*

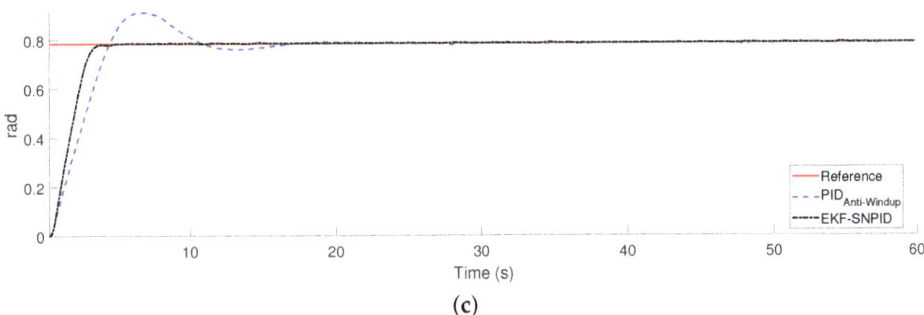

(c)

Figure 22. System response for the rose curve trajectory. (**a**) System response for x. (**b**) System response for y. (**c**) System response for θ.

The RMS and MAD results for rose curve trajectory tests are reported in Table 4. The adaptive EKF-SNPID controller performed better than PID. In this case, better RMS and MAD results are provided by the proposal.

Table 4. Experimental results for the rose curve trajectory. The best results are highlighted in bold.

Measure	Method	x_e	y_e	θ_e
RMS	PID	3.8207×10^{-2}	3.4422×10^{-2}	1.3833×10^{-1}
	EKF-SNPID	$\mathbf{2.6344}\times10^{-2}$	$\mathbf{2.0574}\times10^{-2}$	$\mathbf{1.1510}\times10^{-1}$
MAD	PID	2.9415×10^{-2}	2.7517×10^{-2}	5.4825×10^{-2}
	EKF-SNPID	$\mathbf{1.2798}\times10^{-2}$	$\mathbf{1.3251}\times10^{-2}$	$\mathbf{3.8303}\times10^{-2}$

Finally, the percent errors for experimental results are give in Figure 23. This Figure reported the comparison among the adaptive EKF-SPID and PID with anti-windup. RMS and MAD percent errors for each error of the mobile robot pose x_e, y_e and θ_e are shown in Table 4. Figure 23a presents the percent results for the sinusoidal tracking results and the Figure 23b shows the percent results for the rose curve tracking results. The percent error for both tracking results are positive percentages, which indicates that the proposed adaptive EKF-SNPID performs betters than PID.

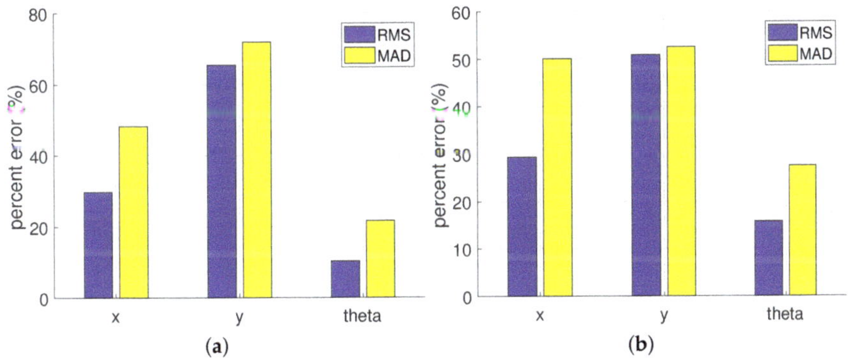

(a) (b)

Figure 23. RMS and MAD percent errors results for EKF-SNPID vs PID with anti-windup. The percent errors x, y and theta were computed based on the RMS and MAD results of the errors x_e, y_e and θ_e from Table 4, respectively. (**a**) Percent error results for sinusoidal trajectory. (**b**) Percent error results for rose curve trajectory.

Based on the results of the experiments, the performance of the adaptive EKF-SNPID controller is superior to PID controller with anti-windup. The results of PID controller present more significant steady-state errors for the required references. In contrast, the proposed approach reported smaller steady-state errors. PID controller needs to adjust its parameter setting to improve the performance. On the other hand, these experiments proved that the proposed EKF-SNPID controller can adjust its parameters to improve its performance. Moreover, the performance of the PID under perturbations and changes during the operation of the experiment is poor. In contrast, the proposed EKF-SNPID can adjust itself to overcomes these inconveniences. Additionally, the performance of EKF-SNPID to handle with windup effects is superior to PID controller with conventional back-calculation anti-windup method. Both controllers suppressed the overshoot provoked by saturation limits. However, the system time response of the proposed EKF-SNIP is faster.

The parameter setting for conventional PID controller was heuristically determined. Simulations and experiments results indicates that it is required to adjust the parameters of PID control to improve its performance. Even if the model is known and the gains are properly adjusted, unmodelled dynamics and uncertainties are present, moreover, the model changes when adding or changing sensors and it should be tuned again regardless of the method used. In contrast, the proposed single neuron PID adjusts dynamically these parameters to overcome this inconvenient.

5. Conclusions

In this paper, an adaptive single neuron PID controller trained with the EKF algorithm was proposed. Simulations and experiments test was implemented to show the performance of the proposed adaptive PID controller. Tests were designed for position tracking control of an omnidirectional mobile robot. A comparative study against a conventional PID controller and other adaptive neural PID controllers were included. The simulations report that the performance of the proposed adaptive neuron PID controller is superior to the compared controllers, respect to best transient response, smaller steady-state errors, and lower overshoot results. The use of the EKF algorithm improves faster learning and convergence speeds compared to backpropagation. An anti-windup test was also included to test the performance of the proposed controller under saturation limits. The proposal showed a better time and tracking response than the PID controller with a back-calculation anti-windup method. The experimental results indicate that the proposed adaptive controller outperformed the performance of PID with better system response under perturbations and changes during the operation of the experimental. Moreover, the conventional PID controller requires the tuning of its gains to improve the performance, which is not necessary for the proposed adaptive single neuron PID controller.

Based on the obtained results, the proposed scheme offers good performance compared to more computationally demanding techniques. This makes it an excellent option to be implemented in similar systems or as part of more complex systems as autonomous vehicles, where fast, reliable responses are needed. Also, it can be seen as future work and implementation of a multilayer approach of the proposal.

As future work, the proposed adaptive PID control can be used to control nonholonomic mobile robots by including the nonholonomic constraints in the control scheme. Moreover, the parameter setting of the EKF algorithm was heuristically determined. However, to improve the performance of the proposed adaptive PID, the use of metaheuristic techniques can be used to optimize the EKF parameters. Additionally, the proposed control method can be extended to handle time-delay systems. Finally, the application of the proposed controller is online based on the Youbot API through ethernet cable connection on the onboard computer. A study of the proposed scheme considering latency is left as future work.

Author Contributions: All of the experiments in this paper have been designed and performed by J.H.-B. and J.D.R. The reported results on this work were analyzed and validated by A.Y.A., C.L.-F., N.A.-D. and J.G.-A. All authors are credited for their contribution on the writing and edition of the presented manuscript. All authors have read and approved the final version of the manuscript.

Funding: The authors thank the support of CONACYT Mexico, through Projects CB256769 and CB258068 ("Project supported by *Fondo Sectorial de Investigación para la Educación*"), and PN4107 ("Problemas Nacionales").

Conflicts of Interest: The authors declare no conflict of interest.

Abbreviations

The following abbreviations are used in this manuscript:

ANN	Artificial Neural Networks
PID	Proporional Integrative Derivative
P	Proportional error
I	Integral of the error
D	Derivative of the error
DOF	Degrees of freedom
BP-PIDNN	Neuron PID controller trained with Back Propagation algorithm
HR-PIDNN	Neuron PID controller trained with Hebbian learning rule
EKF-SNPID	Single neuron PID controller trained with Extended Kalman filter
m	meters
m/s	meters per second
s	seconds
rad	radian
rad/s	radian per second
RMS	Root Mean Square
MAD	Mean Absolute Deviation

References

1. Åström, K.; Hägglund, T. *Pid Controllers*; Setting the standard for automation; International Society for Measurement and Control: Research Triangle Park, NC, USA, 1995.
2. Tian, Y.C.; Tadé, M.O.; Tang, J. A nonlinear PID controller with applications. In Proceedings of the 14th IFAC World Congress 1999, Beijing, Chia, 5–9 July 1999; Volume 32, pp. 2657–2661. [CrossRef]
3. Ogata, K. *Modern Control Engineering*; Instrumentation and controls series; Prentice Hall: Upper Saddle River, NJ, USA, 2010.
4. Chen, Y.m.; He, Y.l.; Zhou, M.f. Decentralized PID neural network control for a quadrotor helicopter subjected to wind disturbance. *J. Cent. South Univ.* **2015**, *22*, 168–179. [CrossRef]
5. Hernández-Alvarado, R.; García-Valdovinos, L.; Salgado-Jiménez, T.; Gómez-Espinosa, A.; Fonseca-Navarro, F. Neural network-based self-tuning PID control for underwater vehicles. *Sensors* **2016**, *16*, 1429. [CrossRef] [PubMed]
6. Zhang, M.; Li, W. Single Neuron PID Model Reference Adaptive Control Based on RBF Neural Network. In Proceedings of the International Conference on Machine Learning and Cybernetics, Dalian, China, 18–21 August 20006; pp. 3021–3025. [CrossRef]
7. Haykin, S. *Neural Networks and Learning Machines*; Pearson Education: London, UK, 2011.
8. Baek, J.; Jin, M.; Han, S. A New Adaptive Sliding-Mode Control Scheme for Application to Robot Manipulators. *IEEE Trans. Ind. Electron.* **2016**, *63*, 3628–3637. [CrossRef]
9. Rahmani, M.; Ghanbari, A.; Ettefagh, M.M. Robust adaptive control of a bio-inspired robot manipulator using bat algorithm. *Expert Syst. Appl.* **2016**, *56*, 164 – 176. [CrossRef]
10. Chen, Y.; Kong, H.; Li, Z.; Ke, F. Trajectory-Tracking for a Mobile Robot Using Robust Predictive Control and Adaptive Control. In Proceedings of the 3rd International Conference on Advanced Robotics and Mechatronics (ICARM), Singapore, 18–20 July 2018; pp. 30–35.
11. Pirník, R.; Hruboš, M.; Nemec, D.; Mravec, T.; Božek, P. Integration of Inertial Sensor Data into Control of the Mobile Platform. In Proceedings of the 2015 Federated Conference on Software Development and Object Technologies, Zilina, Slovakia, 19 November 2015; Janech, J., Kostolny, J., Gratkowski, T., Eds.; Springer International Publishing: Cham, Switzerland, 2017; pp. 271–282.
12. Wang, H.; Guo, D.; Liang, X.; Chen, W.; Hu, G.; Leang, K.K. Adaptive Vision-Based Leader–Follower Formation Control of Mobile Robots. *IEEE Trans. Ind. Electron.* **2017**, *64*, 2893–2902. [CrossRef]

13. Carlucho, I.; Paula, M.D.; Wang, S.; Petillot, Y.; Acosta, G.G. Adaptive low-level control of autonomous underwater vehicles using deep reinforcement learning. *Robot. Auton. Syst.* **2018**, *107*, 71–86. [CrossRef]
14. Makavita, C.D.; Jayasinghe, S.G.; Nguyen, H.D.; Ranmuthugala, D. Experimental Study of Command Governor Adaptive Control for Unmanned Underwater Vehicles. *IEEE Trans. Control. Syst. Technol.* **2019**, *27*, 332–345. [CrossRef]
15. Kebria, P.M.; Khosravi, A.; Nahavandi, S.; Shi, P.; Alizadehsani, R. Robust Adaptive Control Scheme for Teleoperation Systems With Delay and Uncertainties. *IEEE Trans. Cybern.* **2019**, 1–11. [CrossRef]
16. Bozek, P.; Ivandic, Z.; Lozhkin, A.; Lyalin, V.; Tarasov, V. Solutions to the characteristic equation for industrial robot's elliptic trajectories. *Tehnicki vjesnik Technical Gazette* **2016**, *23*, 1017–1023.
17. Bako, B.; Božek, P. Trends in Simulation and Planning of Manufacturing Companies. In Proceedings of the International Conference on Manufacturing Engineering and Materials, ICMEM, Nový Smokovec, Slovakia, 6–10 June 2016; Volume 149, pp. 571–575.
18. Liu, T.; Juang, J. A single neuron PID control for twin rotor MIMO system. In Proceedings of the IEEE/ASME International Conference on Advanced Intelligent Mechatronics, Singapore, 14–17 July 2009; pp. 186–191. [CrossRef]
19. Rivera-Mejía, J.; Léon-Rubio, A.; Arzabala-Contreras, E. PID based on a single artificial neural network algorithm for intelligent sensors. *J. Appl. Res. Technol.* **2012**, *10*, 262–282. [CrossRef]
20. Lopez-Franco, C.; Gomez-Avila, J.; Alanis, A.; Arana-Daniel, N.; Villaseñor, C. Visual servoing for an autonomous hexarotor using a neural network based PID controller. *Sensors* **2017**, *17*, 1865. [CrossRef] [PubMed]
21. Jiao, J.; Chen, J.; Qiao, Y.; Wang, W.; Wang, C.; Gu, L. Single Neuron PID Control of Agricultural Robot Steering System Based on Online Identification. In Proceedings of the IEEE Fourth International Conference on Big Data Computing Service and Applications (BigDataService), Bamberg, Germany, 26–29 March 2018; pp. 193–199. [CrossRef]
22. Shu, H.; Pi, Y. PID neural networks for time-delay systems. *Comput. Chem. Eng.* **2000**, *24*, 859–862. [CrossRef]
23. Huailin Shu.; Xiucai Guo.; Hua Shu. PID neural networks in multivariable systems. In Proceedings of the IEEE Internatinal Symposium on Intelligent Control, Vancouver, BC, Canada, 30–30 October 2002; pp. 440–444. [CrossRef]
24. Sento, A.; Kitjaidure, Y. Neural network controller based on PID using an extended Kalman filter algorithm for multi-variable non-linear control system. In Proceedings of the Eighth International Conference on Advanced Computational Intelligence (ICACI), Chiang Mai, Thailand, 14–16 February 2016; pp. 302–309. [CrossRef]
25. Zeng, G.Q.; Xie, X.Q.; Chen, M.R.; Weng, J. Adaptive population extremal optimization-based PID neural network for multivariable nonlinear control systems. *Swarm Evol. Comput.* **2019**, *44*, 320–334, [CrossRef]
26. Gomez-Avila, J. Chapter 5—Adaptive PID Controller Using a Multilayer Perceptron Trained With the Extended Kalman Filter for an Unmanned Aerial Vehicle. In *Artificial Neural Networks for Engineering Applications*; Alanis, A.Y., Arana-Daniel, N., López-Franco, C., Eds.; Academic Press: Cambridge, MA, USA, 2019; pp. 55–63. [CrossRef]
27. Zhang, Y.; Yu, X.; Bi, M.; Xu, S. An adaptive neural PID controller for torque control of hybrid electric vehicle. In Proceedings of the 6th International Conference on Computer Science Education (ICCSE), Singapore, 3–5 August 2011; pp. 901–903. [CrossRef]
28. Bobtsov, A.; Guirik, A.; Budko, M.; Budko, M. Hybrid parallel neuro-controller for multirotor unmanned aerial vehicle. In Proceedings of the 8th International Congress on Ultra Modern Telecommunications and Control Systems and Workshops (ICUMT), Lisbon, Portugal, 18–20 October 2016; pp. 1–4. [CrossRef]
29. Åström, K.; Murray, R. Feedback Systems: An Introduction for Scientists and Engineers. In *Feedback Systems: An Introduction for Scientists and Engineers*; Princeton University Press: Princeton, NJ, USA, 2008.
30. Kumar, S.; Negi, R. A comparative study of PID tuning methods using anti-windup controller. In Proceedings of the 2nd International Conference on Power, Control and Embedded Systems, Allahabad, India, 17–19 December 2012; pp. 1–4. [CrossRef]
31. Angel, L.; Viola, J.; Paez, M. Evaluation of the windup effect in a practical PID controller for the speed control of a DC-motor system. In Proceedings of the IEEE 4th Colombian Conference on Automatic Control (CCAC), Medellín, Colombia, 15–18 October 2019; pp. 1–6. [CrossRef]

32. Bohn, C.; Atherton, D.P. An analysis package comparing PID anti-windup strategies. *IEEE Control. Syst. Mag.* **1995**, *15*, 34–40.
33. Kheirkhahan, P. Robust anti-windup control design for PID controllers. In Proceedings of the 17th International Conference on Control, Automation and Systems (ICCAS), Jeju, South Korea, 18–21 October 2017; pp. 1622–1627. [CrossRef]
34. Haykin, S. Kalman Filtering and Neural Networks. In *Adaptive and Cognitive Dynamic Systems: Signal Processing, Learning, Communications and Control*; Wiley: Hoboken, NJ, USA, 2004.
35. Alanis, A.; Sanchez, E. *Discrete-Time Neural Observers: Analysis and Applications*; Elsevier Science: Amsterdam, The Netherlands, 2017.
36. Goodfellow, I.; Bengio, Y.; Courville, A. *Deep Learning*; MIT Press: Cambridge, MA, USA, 2016.
37. Moradi, M.H.; Katebi, M.R.; Johnson, M.A. Predictive PID control: A new algorithm. In Proceedings of the IECON'01, 27th Annual Conference of the IEEE Industrial Electronics Society (Cat. No.37243), Denver, CO, USA, 29 November–2 December 2001; Volume 1, pp. 764–769. [CrossRef]
38. Sanchez, E.; Alanis, A.; Loukianov, A. *Discrete-Time High Order Neural Control: Trained with Kalman Filtering*; Studies in Computational Intelligence; Springer: Berlin/Heidelberg, Germany, 2008.
39. Feldkamp, L.A.; Prokhorov, D.V.; Feldkamp, T.M. Conditioned adaptive behavior from Kalman filter trained recurrent networks. In Proceedings of the International Joint Conference on Neural Networks, Portland, OR, USA, 20–24 July 2003; Volume 4, pp. 3017–3021.
40. Camacho, J.D.; Villaseñor, C.; Alanis, A.Y.; Lopez-Franco, C.; Arana-Daniel, N. KAdam: Using the Kalman Filter to Improve Adam algorithm. In *Progress in Pattern Recognition, Image Analysis, Computer Vision, and Applications*; Nyström, I., Hernández Heredia, Y., Milián Núñez, V., Eds.; Springer International Publishing: Cham, Switzerland, 2019; pp. 429–438.
41. Alanis, A.; Arana-Daniel, N.; Lopez-Franco, C. *Artificial Neural Networks for Engineering Applications*; Elsevier Science: Amsterdam, The Netherlands, 2019.
42. Rios, J.; Alanis, A.; Lopez-Franco, C.; Arana-Daniel, N.; Sanchez, E. *Neural Networks Modelling and Control: Applications for Unknown Nonlinear Delayed Systems in Discrete-Time*; Elsevier Science & Technology: Amsterdam, The Netherlands, 2020.
43. Sanchez, E.; Alanis, A. *Redes Neuronales: Conceptos Fundamentales y Aplicaciones a Control Automático*; Automática Robótica, Pearson Educación: Madrid, Spain, 2006.
44. Leung, H.; Haykin, S. The complex backpropagation algorithm. *IEEE Trans. Signal Process.* **1991**, *39*, 2101–2104. [CrossRef]
45. Ruck, D.W.; Rogers, S.K.; Kabrisky, M.; Maybeck, P.S.; Oxley, M.E. Comparative analysis of backpropagation and the extended Kalman filter for training multilayer perceptrons. *IEEE Trans. Pattern Anal. Mach. Intell.* **1992**, 686–691. [CrossRef]
46. Kundu, A.S.; Mazumder, O.; Dhar, A.; Lenka, P.K.; Bhaumik, S. Scanning Camera and Augmented Reality Based Localization of Omnidirectional Robot for Indoor Application. *Procedia Comput. Sci.* **2017**, *105*, 27–33. [CrossRef]
47. Wu, J.; Lv, C.; Zhao, L.; Li, R.; Wang, G. Design and implementation of an omnidirectional mobile robot platform with unified I/O interfaces. In Proceedings of the IEEE International Conference on Mechatronics and Automation (ICMA), Changchun, China, 5–8 August 2017; pp. 410–415. [CrossRef]
48. Zhang, G.; Qin, W.; Qin, Q.; He, B.; Liu, G. Varying gain MPC for consensus tracking with application to formation control of omnidirectional mobile robots. In Proceedings of the 12th World Congress on Intelligent Control and Automation (WCICA), Guilin, China, 12–15 June 2016; pp. 2957–2962. [CrossRef]
49. Sharifi, M.; Young, M.; Chen, X.; Clucas, D.; Pretty, C. Mechatronic design and development of a non-holonomic omnidirectional mobile robot for automation of primary production. *Cogent Eng.* **2016**, *3*. [CrossRef]
50. Li, Z.; Yang, C.; Su, C.; Deng, J.; Zhang, W. Vision-Based Model Predictive Control for Steering of a Nonholonomic Mobile Robot. *IEEE Trans. Control. Syst. Technol.* **2016**, *24*, 553–564. [CrossRef]
51. Tsai, C.; Tai, F.; Lee, Y. Motion controller design and embedded realization for Mecanum wheeled omnidirectional robots. In Proceedings of the 9th World Congress on Intelligent Control and Automation, Taipei, Taiwan, 21–25 June 2011; pp. 546–551. [CrossRef]

52. Viboonchaicheep, P.; Shimada, A.; Kosaka, Y. Position rectification control for Mecanum wheeled omni-directional vehicles. In Proceedings of the IECON'03, 29th Annual Conference of the IEEE Industrial Electronics Society (IEEE Cat. No.03CH37468), Roanoke, VA, USA, 2–6 November 2003; Volume 1, pp. 854–859. [CrossRef]
53. Taheri, H.; Qiao, B.; Ghaeminezhad, N. Kinematic Model of a Four Mecanum Wheeled Mobile Robot. *Int. J. Comput. Appl.* **2015**, *113*, 6–9. [CrossRef]
54. Ardiyanto, I. Task Oriented Behavior-Based State-Adaptive PID (Proportional Integral Derivative) Control for Low-Cost Mobile Robot. In Proceedings of the Second International Conference on Computer Engineering and Applications, Bali Island, Indonesia, 26–29 March 2010; Volume 1, pp. 103–107. [CrossRef]
55. Normey-Rico, J.E.; Alcalá, I.; Gómez-Ortega, J.; Camacho, E.F. Mobile robot path tracking using a robust PID controller. *Control. Eng. Pract.* **2001**, *9*, 1209–1214. [CrossRef]
56. Carlucho, I.; Paula, M.D.; Villar, S.A.; Acosta, G.G. Incremental Q-learning strategy for adaptive PID control of mobile robots. *Expert Syst. Appl.* **2017**, *80*, 183–199. [CrossRef]
57. Khalil, H.K. *Nonlinear Systems*, 3rd ed.; Prentice-Hall: Upper Saddle River, NJ, 2002; The book can be consulted by contacting: PH-AID: Wallet, Lionel.
58. Ren, C.; Sun, Y.; Ma, S. Passivity-based control of an omnidirectional mobile robot. *Robot. Biomimetics* **2016**, *3*. [CrossRef] [PubMed]
59. Shukla, S.; Singh, P.; Behera, L. A passivity based system design for non-holonomic mobile robot in presence of delay and traffic disturbances. In Proceedings of the 12th IEEE Conference on Industrial Electronics and Applications (ICIEA), Siem Reap, Cambodia, 18–20 June 2017; pp. 733–738. [CrossRef]

Article

Facilitating Autonomous Systems with AI-Based Fault Tolerance and Computational Resource Economy

Kyriakos M. Deliparaschos [1,2,*,†], Konstantinos Michail [1,3,†] and Argyrios C. Zolotas [2,†]

1 Electrical and Computer Engineering and Informatics, Cyprus University of Technology, 3036 Limassol, Cyprus; kon_michael@ieee.org
2 Centre for Autonomous and Cyber-Physical Systems, SATM, Cranfield University, Bedford MK43 0AL, UK; a.zolotas@cranfield.ac.uk
3 SignalGeneriX, 23C Gregory Afxentiou str., P.O. Box 59627, 4011 Limassol, Cyprus
* Correspondence: k.deliparaschos@cranfield.ac.uk or k.deliparaschos@cut.ac.cy
† The authors contributed equally to this work.

Received: 14 April 2020; Accepted: 1 May 2020; Published: 11 May 2020

Abstract: Proposed is the facilitation of fault-tolerant capability in autonomous systems with particular consideration of low computational complexity and system interface devices (sensor/actuator) performance. Traditionally model-based fault-tolerant/detection units for multiple sensor faults in automation require a bank of estimators, normally Kalman-based ones. An AI-based control framework enabling low computational power fault tolerance is presented. Contrary to the bank-of-estimators approach, the proposed framework exhibits a single unit for multiple actuator/sensor fault detection. The efficacy of the proposed scheme is shown via rigorous analysis for several sensor fault scenarios for an electro-magnetic suspension testbed.

Keywords: fault tolerance; reconfigurable control; Maglev; neural networks; artificial intelligence

---

1. Introduction

Modern control systems require careful, reliable and economic design with maximum performance, normally imposing several design trade-offs (economic design, reliability, performance). In particular, reliability in control systems is vital especially in safety-critical systems (i.e., where faults must be accommodated before the impaired system becomes unstable). In non-safety-critical system cases, like production lines, reliability supports a normal operation regime avoiding production delays and/or unnecessary maintenance. In areas more aligned to autonomy, such as in Unmanned Area Vehicles (UAVs), the problem of considering control methods for reliability adds to the computational power of the already limited resources [1–6].

Autonomous systems must be trustworthy, and trustworthiness has been a popular topic of discussion in the current autonomous systems literature [7,8]. Reliability facilitates trustworthiness, and in this context fault accommodation can be achieved with a priori design of a controller that has the ability to take remedial actions so that the stability of the control system is maintained even with degraded performance. The stability and performance of a control system depends upon the healthy operation of its interfaces (actuators and sensors) and various approaches to design such capability appear in the literature (both model-based and model-free methods) [9–17]. Fault-tolerant control (FTC) supports reliability [18], the approach normally classified as either Passive (PFTC) or Active (AFTC) [19]. Passive FTC type requires a prior knowledge of the faults, while the Active type (used in this work) does not necessitate such knowledge of the fault rather a Fault Detection and Isolation (FDI) mechanism with reconfigurable control. Reconfigurable FTC control has gained significant attention over recent years given the demand on reliable system design [20–22] and in the area of cyber-physical systems [23].

Referring to sensor fault tolerance, especially after sensors failure, a few methods exist that use the information from the remaining healthy sensors, in order to reconstruct the lost signal of the faulty ones [24]. The latter methods include, use of a bank of Neural Networks (NNs) or use of Kalman Estimators (KE) [1,25]. Both approaches are worth considering when aiming in avoiding sensor redundancy. In contrast to the KEs approach, NNs have increased False Alarm Rates (FAR), mainly because they leave a very small residual after fault estimation [24]. Despite that, they are widely used since they can be designed without having precise knowledge of the model of the system under test [26–29]. In the above approaches, where many actuators/sensors exist, an (in-parallel) bank of estimators for multiple faults detection is employed [30]. For example, if there is one actuator with n_y sensors, then the number of sensor fault combinations that could happen is $2^{n_y} - 1$ (where n_y is the total number of sensors, assuming that not all sensors can fail). Hence, to be able to detect those faults it requires the same number of estimators. However, this increases the complexity of the control design and requires additional computational resources, since the estimators must work in parallel.

NNs have been used to a great extend in many engineering fields including control systems [31,32], as well as for Fault Detection (FD) methods in FTC systems [33,34] and more specifically in Sensor FDI [24,35,36]. We proposed an AI-based FD mechanism, referred to as *i*FD, based on the use of Neural Network approach that performs a similar task to the conventional bank-of-estimators FD albeit offers substantially reduced computational complexity. Visually this is depicted in Figure 1, the bank of the estimators running in parallel (dotted lines). The authors, in their brief paper [19], presented the original concept framework from an automatic fault-tolerant control viewpoint. This paper considerably extends the original results (with a particular emphasis on their interpretation) and present how reliable system autonomy can be facilitated.

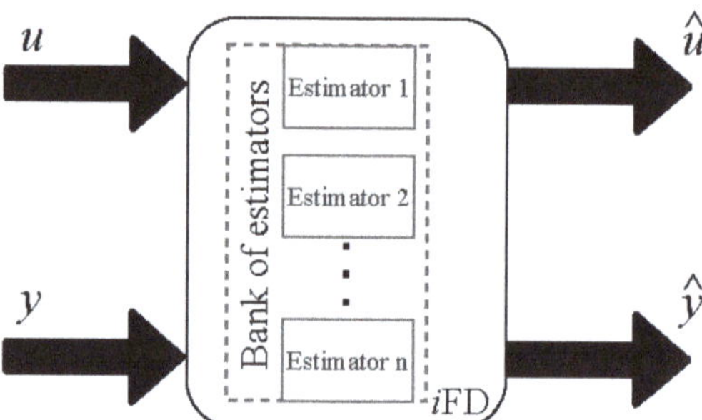

Figure 1. Bank of estimators (dotted lines) vs. proposed *i*FD (bold lines) for actuators/sensors estimates.

We discuss the framework solution with the help of a practical example, i.e., an Electro-Magnetic Suspension (EMS) testbed (typically forms the suspension platform of Maglev train) to support the vehicle compartment and maintain acceptable passenger ride quality. The rationale behind this choice being that EMS (Maglev) is an inherently unstable, non-linear, safety-critical system, subject to non-trivial control performance and reliability requirements (hence offering a challenging application for the validation of the work). Sensor faults are first modelled and then the simulation results using various fault scenarios showcase the proposed method.

The rest of the paper is organized as follows: Section 2 describes the proposed *i*FD approach including a short description of the NN training, and Section 3.1 shows the efficacy of the proposed method based on the analysis of the results (with the help of the practical example of the EMS—the test case details been discussed in Appendix A). Conclusions are discussed in Section 4.

2. Proposed Fault Detection Scheme

We employ the FD unit to detect actuator/sensor faults and activate controller reconfiguration. The proposed FD scheme is NN-based and the concept is depicted in Figure 2. Industrial systems typically exhibit Multiple-Input, Multiple Outputs. Control inputs relate to actuation (i.e., industrial drives, motors, pumps, electro-magnets etc.), in a control setting indicated by variable $\mathcal{U}$. Outputs measure several useful parameters (both for control purposes and monitoring purposes), typically indicated by variable $\mathcal{Y}$. Control design satisfies a desired performance relevant to the application.

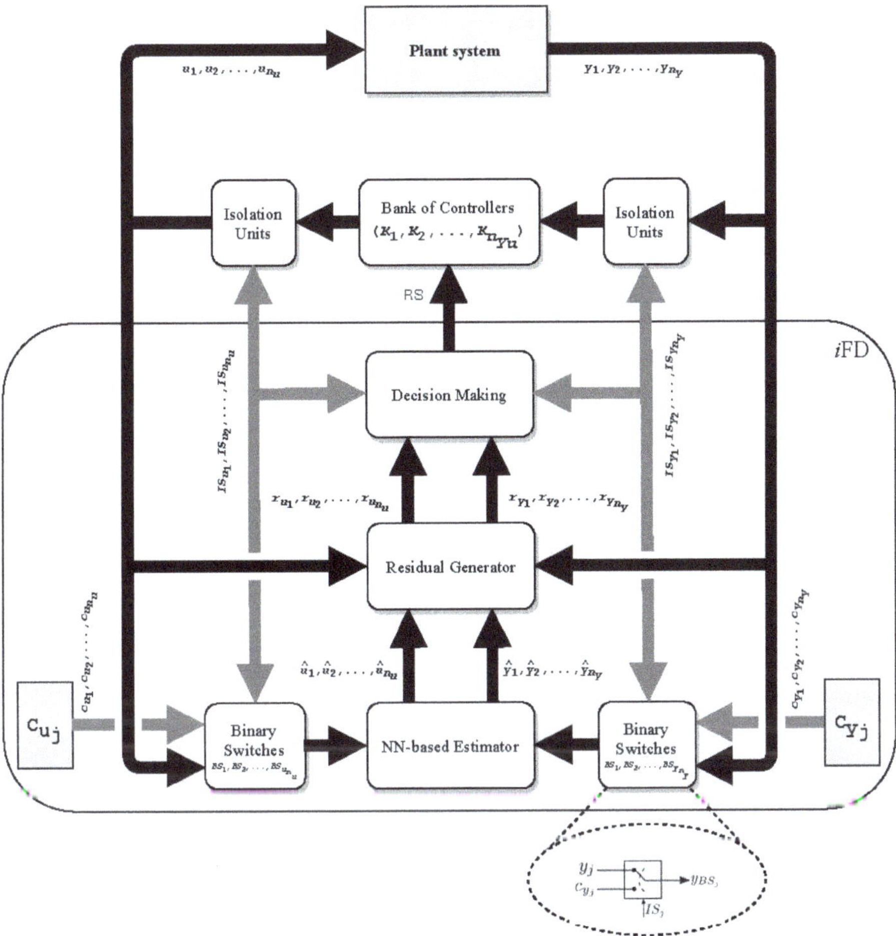

Figure 2. FDI general diagram with proposed *i*FD and including the binary switch operator.

When one or more actuators and/or sensors are impaired, those signals are distorted leading to performance degradation or possibly instability of the closed loop. The sets of actuators and sensors are defined as $\mathcal{U} = [u_1, u_2, \dots u_{n_u}]$ and $\mathcal{Y} = [y_1, y_2, \dots y_{n_y}]$, where u_j is the jth actuator, y_j is the jth sensor, n_u is the total number of actuators and n_y is the total number of sensors.

The control loop features a bank of controllers $[K_1, K_2, \dots K_{n_{yu}}]$ and two isolation units for isolating faulty actuator and sensor signals when these happen. The *i*FD mechanism is employed to detect the faults and is comprised of a NN-based estimator, a Residual Generator (RG) and a Decision Mechanism (DM). The NN estimator is trained in such a way that the actuators and sensors signals are estimated and fed into the RG which in its turn compares the real signals with the estimated ones. Immediately after the RG has completed the comparison, it advances the residuals to the DM, leading to a decision whether a component is faulty or not.

The key point in the proposed *i*FD method is the estimator training approach to observe $\mathcal{U}$ and $\mathcal{Y}$ (the training process discussed in Section 2.1).

The inputs to the estimator are obtained from the Binary Switches (BS) depicted in Figure 2. The BS have three inputs; one represents the real measured values of the $\mathcal{U}$ and $\mathcal{Y}$ and the other comes from the functions $\mathcal{C}_{u_j}$ and $\mathcal{C}_{y_j}$, defined as $\mathcal{C}_{u_j} = [c_{u_1}, c_{u_2} \ldots c_{u_{nu}}]$ and $\mathcal{C}_{y_j} = [c_{y_1}, c_{y_2} \ldots c_{y_{ny}}]$. $\mathcal{C}_{u_j}$ and $\mathcal{C}_{y_j}$ represent two arrays that contains predefined functions, used during the training and operation of the *i*FD. Typically, calculating these values is benefited by designer experience as they tend to be application dependent. The third input (IS_{y_j}) is a binary input which controls the switching operation between the inputs e.g., from y_1 to c_{y_1}. The output y_{BS_j} of the BS is given by (1)

$$y_{BS_j} = \{ \begin{matrix} y_j, & if & IS_{y_j}=1 \\ c_{y_j}, & if & IS_{y_j}=0 \end{matrix} \tag{1}$$

The residual generation is a vital task although is out of scope in this paper. The moving average filter defined in (2), manages to accommodate the noise coming from the sensors to reduce the FAR [37].

$$r_{y_j} = \sum_{j-(N-1)}^{j} \frac{(y_j - \hat{y}_j)^2}{N} \tag{2}$$

where r_{y_j} is the residual, y_j and $\hat{y}_j$ are the jth real and estimated signals (for the actuators the y is replaced by u), and N is the total number of past samples.

The DM decides whether one or more actuating and/or sensing components are impaired or not by comparing the relative residual r_j with a predefined threshold (engineer must define the threshold). Threshold selection is a non-trivial task to perform, since it impacts both fault detection sensitivity of the DM and the FAR. A very sensitive DM to faults (thresholds too low) causes increase of FAR, while a less sensitive DM (thresholds too high) may cause instability due to delayed reconfiguration. Threshold selection is normally done as trial-and-error process with designer experience in the application beneficial. Within the AI remit of fault estimation, threshold selection has received attention [38] but the details of this aspect are beyond the scope of this paper. In the proposed *i*FD, the Reconfiguration Signal (RS) at the output of DM enables controller reconfiguration. The proposed *i*FD setup works as follows:

- Given normal operation, the actuator(s)/sensors signals are estimated and then fed into the residual generator. The generator calculates the residual (which would correspond to a very small value under normal operation) for each actuator/sensor and feeds it to the DM. The DM outputs the IS_j and RS signals.
- With one or more actuator(s)/sensors failing the corresponding residuals r_j increase and the DM will detect the change by comparison with the relevant threshold levels. Next, the ISs will switch in order to activate the corresponding BS to modify its output to c_j; while the RS will feed the appropriate data value to enable controller reconfiguration. The BSs will also isolate the signals of the faulty devices from the estimator so that the latter 'sees' some 'known' data based on its training/knowledge (this is described in the next section).
- Finally, the isolation units will remove faulty devices from the loop and the controller reconfigures to maintain performance and stability.

2.1. Offline Training of the iFD Unit: Obtaining the Learning Set

As in all AI-based solutions, training of the algorithm is the key point. The NN *i*FD unit is trained based on data accumulated from an extensive set of scenaria on subsets of the main sensor set, $\mathcal{Y}$.

The collected data are then packed in a structure shown on Table 1. The first column presents the sensor set number from 1 to n_{yn}, defined as,

$$n_{yu} = 2^{(n_u+n_y)} - 2^{n_u} - 2^{n_y} + 1 \tag{3}$$

The second column shows the status of the sensor set, i.e., all possible sensor/actuator fault scenarios are covered, and the next two columns show the measured sensors and actuators signals. The last two columns reflect the estimated sensors and actuators signals. The data set $\mathcal{D}$ with dimensions $d_r \times d_c$ is given by,

$$d_r = n_{yu} \times k \tag{4}$$

$$d_c = n_y + n_u + n_{\hat{y}} + n_{\hat{u}} \tag{5}$$

where $n_{\hat{u}}$ and $n_{\hat{y}}$ are the number of estimated signals for actuators and sensors, respectively. $\mathcal{D}$ is constructed with data from numerical simulations with each sensor/actuator fault scenario. Where the sensor(s) and/or actuator(s) is (are) assumed to be faulty, a known function c_{u_j}, c_{y_j} replaces the k data points. In an automatic setup the design engineer is required to select a set of functions $\mathcal{C}_{u_j}$ and $\mathcal{C}_{y_j}$ to replace the non-predicted outputs from the faulty actuators and sensors respectively. In an autonomous setting, experience from automation and use of reinforced learning design will enable the aforementioned choice (this part is studied as future research and is not the main aspect of the work presented here).

When an actuator and/or sensor fault occurs, the corresponding function c_{u_j} and/or c_{y_j} is/are connected to the *i*FD. This is a result of the *i*FD learning capability to respond to sensor/actuator faults in a way that the *i*FD unit itself continually checks for faults on the full sensor set and its subsets.

An electro-magnetic suspension (EMS) system testbed (details in Appendix A) is used as the practical platform to illustrate the proposed solution.

2.2. Visualizing the iFD Applied on the EMS Example

The EMS system is excited by a single control input (i.e., $n_u = 1$), $\mathcal{U} = \{u_c\}$, and provides four output measurements (i.e., $n_y = 4$), namely $\mathcal{Y} = \{i, (z_t - z), \dot{z}, \ddot{z}\}$. The outputs are employed for controller design. The $\mathcal{H}_\infty$ Loop-Shaping Design Procedure (LSDP) robust control approach was used to enable the necessary closed-loop performance. Please note that the airgap $(z_t - z)$ is required by default as a standard input [39]. With $(z_t - z)$ a default measurement, the total number of actuator/sensor sets are $n_{yn} = 8$. Hence, eight LSDP controllers (i.e., $K_{(z_t-z)}$, $K_{i,(z_t-z)}$, $K_{(z_t-z),\dot{z}}$, $K_{(z_t-z),\ddot{z}}, \ldots$, etc.) are used to cover the spectrum of (seven) possible sensor combinations that could occur as indicated in Figure 3. Details on the design of LSDP controllers is discussed in [40]. The *i*FD concept presented here can be considered for the detection and isolation of sensor faults in a typical FTC system arrangement for other engineering test platforms. In the case of one or more sensors failures, the relevant fault is detected and isolated (note that the isolation switches and the binary switches are merged together as shown in the figure) and the system switches to the alternative controller $K_\bullet$ (for closed-loop purposes).

Table 1. Structure of the data used for the *i*FD training (basis from [19]).

Actuator/ Sensor Set Number	Actuator/ Sensor Status	Measured										Estimated									
		y_1	y_2	y_3	…	y_{n_y}	u_1	u_2	u_3	…	u_{n_u}	$\hat{y}_1$	$\hat{y}_2$	$\hat{y}_3$	…	$\hat{y}_{n_{\hat{y}}}$	$\hat{u}_1$	$\hat{u}_2$	$\hat{u}_3$	…	$\hat{u}_{n_{\hat{u}}}$
	healthy set	$D^1_{y_1^1}$	$D^1_{y_2^1}$	$D^1_{y_3^1}$	…	$D^1_{y_{n_y}^1}$	$D^1_{u_1^1}$	$D^1_{u_2^1}$	$D^1_{u_3^1}$	…	$D^1_{u_{n_u}^1}$	$D^1_{\hat{y}_1^1}$	$D^1_{\hat{y}_2^1}$	$D^1_{\hat{y}_3^1}$	…	$D^1_{\hat{y}_{n_{\hat{y}}}^1}$	$D^1_{\hat{u}_1^1}$	$D^1_{\hat{u}_2^1}$	$D^1_{\hat{u}_3^1}$	…	$D^1_{\hat{u}_{n_{\hat{u}}}^1}$
1	$\mathcal{Y} = y_1, y_2 \dots y_{n_y}$	⋮	⋮	⋮		⋮	⋮	⋮	⋮		⋮	⋮	⋮	⋮		⋮	⋮	⋮	⋮		⋮
	$\mathcal{U} = u_1, u_2, \dots u_{n_u}$	$D^1_{y_1^k}$	$D^1_{y_2^k}$	$D^1_{y_3^k}$	…	$D^1_{y_{n_y}^k}$	$D^1_{u_1^k}$	$D^1_{u_2^k}$	$D^1_{u_3^k}$	…	$D^1_{u_{n_u}^k}$	$D^1_{\hat{y}_1^k}$	$D^1_{\hat{y}_2^k}$	$D^1_{\hat{y}_3^k}$	…	$D^1_{\hat{y}_{n_{\hat{y}}}^k}$	$D^1_{\hat{u}_1^k}$	$D^1_{\hat{u}_2^k}$	$D^1_{\hat{u}_3^k}$	…	$D^1_{\hat{u}_{n_{\hat{u}}}^k}$
	Faulty	$c^2_{y_1^1}$	$D^2_{y_2^1}$	$D^2_{y_3^1}$	…	$D^2_{y_{n_y}^1}$	$D^2_{u_1^1}$	$D^2_{u_2^1}$	$D^2_{u_3^1}$	…	$D^2_{u_{n_u}^1}$	$c^2_{\hat{y}_1^1}$	$D^2_{\hat{y}_2^1}$	$D^2_{\hat{y}_3^1}$	…	$D^2_{\hat{y}_{n_{\hat{y}}}^1}$	$D^2_{\hat{u}_1^1}$	$D^2_{\hat{u}_2^1}$	$D^2_{\hat{u}_3^1}$	…	$D^2_{\hat{u}_{n_{\hat{u}}}^1}$
2	y_1	⋮	⋮	⋮		⋮	⋮	⋮	⋮		⋮	⋮	⋮	⋮		⋮	⋮	⋮	⋮		⋮
		$c^2_{y_1^k}$	$D^2_{y_2^k}$	$D^2_{y_3^k}$	…	$D^2_{y_{n_y}^k}$	$D^2_{u_1^k}$	$D^2_{u_2^k}$	$D^2_{u_3^k}$	…	$D^2_{u_{n_u}^k}$	$c^2_{\hat{y}_1^k}$	$D^2_{\hat{y}_2^k}$	$D^2_{\hat{y}_3^k}$	…	$D^2_{\hat{y}_{n_{\hat{y}}}^k}$	$D^2_{\hat{u}_1^k}$	$D^2_{\hat{u}_2^k}$	$D^2_{\hat{u}_3^k}$	…	$D^2_{\hat{u}_{n_{\hat{u}}}^k}$
	Faulty	$c^3_{y_1^1}$	$c^3_{y_2^1}$	$D^3_{y_3^1}$	…	$D^3_{y_{n_y}^1}$	$D^3_{u_1^1}$	$D^3_{u_2^1}$	$D^3_{u_3^1}$	…	$D^3_{u_{n_u}^1}$	$c^3_{\hat{y}_1^1}$	$c^3_{\hat{y}_2^1}$	$D^3_{\hat{y}_3^1}$	…	$D^3_{\hat{y}_{n_{\hat{y}}}^1}$	$D^3_{\hat{u}_1^1}$	$D^3_{\hat{u}_2^1}$	$D^3_{\hat{u}_3^1}$	…	$D^3_{\hat{u}_{n_{\hat{u}}}^1}$
3	y_1, y_2	⋮	⋮	⋮		⋮	⋮	⋮	⋮		⋮	⋮	⋮	⋮		⋮	⋮	⋮	⋮		⋮
		$c^3_{y_1^k}$	$c^3_{y_2^k}$	$D^3_{y_3^k}$	…	$D^3_{y_{n_y}^k}$	$D^3_{u_1^k}$	$D^3_{u_2^k}$	$D^3_{u_3^k}$	…	$D^3_{u_{n_u}^k}$	$c^3_{\hat{y}_1^k}$	$c^3_{\hat{y}_2^k}$	$D^3_{\hat{y}_3^k}$	…	$D^3_{\hat{y}_{n_{\hat{y}}}^k}$	$D^3_{\hat{u}_1^k}$	$D^3_{\hat{u}_2^k}$	$D^3_{\hat{u}_3^k}$	…	$D^3_{\hat{u}_{n_{\hat{u}}}^k}$
	Faulty	$c^4_{y_1^1}$	$c^4_{y_2^1}$	$D^4_{y_3^1}$	…	$D^4_{y_{n_y}^1}$	$c^4_{u_1^1}$	$D^4_{u_2^1}$	$D^4_{u_3^1}$	…	$D^4_{u_{n_u}^1}$	$c^4_{\hat{y}_1^1}$	$c^4_{\hat{y}_2^1}$	$D^4_{\hat{y}_3^1}$	…	$D^4_{\hat{y}_{n_{\hat{y}}}^1}$	$c^4_{\hat{u}_1^1}$	$D^4_{\hat{u}_2^1}$	$D^4_{\hat{u}_3^1}$	…	$D^4_{\hat{u}_{n_{\hat{u}}}^1}$
4	y_1, y_2, u_1	⋮	⋮	⋮		⋮	⋮	⋮	⋮		⋮	⋮	⋮	⋮		⋮	⋮	⋮	⋮		⋮
		$c^4_{y_1^k}$	$c^4_{y_2^k}$	$D^4_{y_3^k}$	…	$D^4_{y_{n_y}^k}$	$c^4_{u_1^k}$	$D^4_{u_2^k}$	$D^4_{u_3^k}$	…	$D^4_{u_{n_u}^k}$	$c^4_{\hat{y}_1^k}$	$c^4_{\hat{y}_2^k}$	$D^4_{\hat{y}_3^k}$	…	$D^4_{\hat{y}_{n_{\hat{y}}}^k}$	$c^4_{\hat{u}_1^k}$	$D^4_{\hat{u}_2^k}$	$D^4_{\hat{u}_3^k}$	…	$D^4_{\hat{u}_{n_{\hat{u}}}^k}$
	Faulty	$c^5_{y_1^1}$	$D^5_{y_2^1}$	$D^5_{y_3^1}$	…	$D^5_{y_{n_y}^1}$	$c^5_{u_1^1}$	$c^5_{u_2^1}$	$D^5_{u_3^1}$	…	$D^5_{u_{n_u}^1}$	$c^5_{\hat{y}_1^1}$	$D^5_{\hat{y}_2^1}$	$D^5_{\hat{y}_3^1}$	…	$D^5_{\hat{y}_{n_{\hat{y}}}^1}$	$c^5_{\hat{u}_1^1}$	$c^5_{\hat{u}_2^1}$	$D^5_{\hat{u}_3^1}$	…	$D^5_{\hat{u}_{n_{\hat{u}}}^1}$
5	y_1, u_1, u_2	⋮	⋮	⋮		⋮	⋮	⋮	⋮		⋮	⋮	⋮	⋮		⋮	⋮	⋮	⋮		⋮
		$c^5_{y_1^k}$	$D^5_{y_2^k}$	$D^5_{y_3^k}$	…	$D^5_{y_{n_y}^k}$	$c^5_{u_1^k}$	$c^5_{u_2^k}$	$D^5_{u_3^k}$	…	$D^5_{u_{n_u}^k}$	$c^5_{\hat{y}_1^k}$	$D^5_{\hat{y}_2^k}$	$D^5_{\hat{y}_3^k}$	…	$D^5_{\hat{y}_{n_{\hat{y}}}^k}$	$c^5_{\hat{u}_1^k}$	$c^5_{\hat{u}_2^k}$	$D^5_{\hat{u}_3^k}$	…	$D^5_{\hat{u}_{n_{\hat{u}}}^k}$
⋮	⋮	⋮	⋮	⋮		⋮	⋮	⋮	⋮		⋮	⋮	⋮	⋮		⋮	⋮	⋮	⋮		⋮
	Faulty	$c^{n_{yu}}_{y_1^1}$	$c^{n_{yu}}_{y_2^1}$	…	$c^{n_{yu}}_{y_{n_y-1}^1}$	$D^{n_{yu}}_{y_{n_y}^1}$	$c^{n_{yu}}_{u_1^1}$	$c^{n_{yu}}_{u_2^1}$	…	$c^{n_{yu}}_{u_{n_u-1}^1}$	$D^{n_{yu}}_{u_{n_u}^1}$	$c^{n_{yu}}_{\hat{y}_1^1}$	$c^{n_{yu}}_{\hat{y}_2^1}$	…	$c^{n_{yu}}_{\hat{y}_{n_y-1}^1}$	$D^{n_{yu}}_{\hat{y}_{n_{\hat{y}}}^1}$	$c^{n_{yu}}_{\hat{u}_1^1}$	$c^{n_{yu}}_{\hat{u}_2^1}$	…	$c^{n_{yu}}_{\hat{u}_{n_u-1}^1}$	$D^{n_{yu}}_{\hat{u}_{n_{\hat{u}}}^1}$
n_{yu}	$y_1 \dots y_{n_y-1}$	⋮	⋮		⋮	⋮	⋮	⋮		⋮	⋮	⋮	⋮		⋮	⋮	⋮	⋮		⋮	⋮
	$u_1, \dots u_{n_u-1}$	$c^{n_{yu}}_{y_1^k}$	$c^{n_{yu}}_{y_2^k}$	…	$c^{n_{yu}}_{y_{n_y-1}^k}$	$D^{n_{yu}}_{y_{n_y}^k}$	$c^{n_{yu}}_{u_1^k}$	$c^{n_{yu}}_{u_2^k}$	…	$c^{n_{yu}}_{u_{n_u-1}^k}$	$D^{n_{yu}}_{y_{n_u}^k}$	$c^{n_{yu}}_{\hat{y}_1^k}$	$c^{n_{yu}}_{\hat{y}_2^k}$	…	$c^{n_{yu}}_{\hat{y}_{n_y-1}^k}$	$D^{n_{yu}}_{\hat{y}_{n_{\hat{y}}}^k}$	$c^{n_{yu}}_{\hat{u}_1^k}$	$c^{n_{yu}}_{\hat{u}_2^k}$	…	$c^{n_{yu}}_{\hat{u}_{n_u-1}^k}$	$D^{n_{yu}}_{\hat{y}_{n_{\hat{u}}}^k}$

k is the total number of samples at each actuator/sensor set.

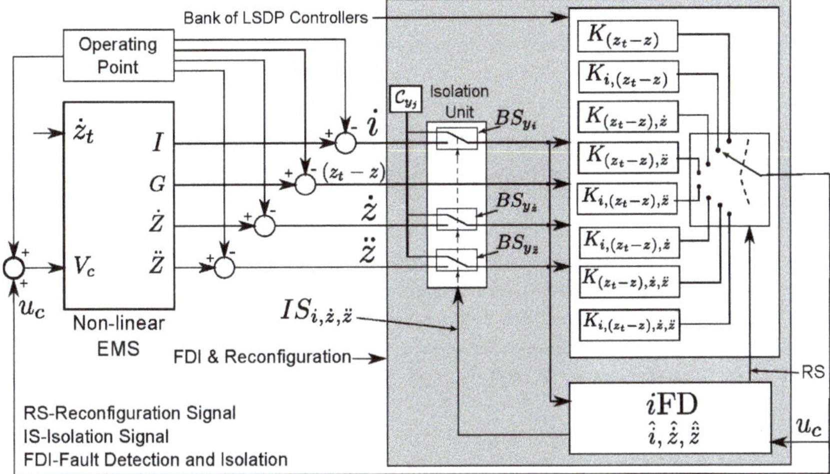

Figure 3. The proposed *i*FD solution using the EMS application example (basis from [19]).

Sensor faults can be categorized into additive and multiplicative categories. As clearly shown in Figure 4, there exist three types of faults in each category, namely: (i) abrupt or step-type fault, (ii) incipient or soft fault and (iii) indeterminate fault. The faults accounted for in this work fall into the additive and multiplicative categories and are both abrupt and incipient types of faults. In the first case, the output of the sensor is added or multiplied with a function f_a and f_m respectively. Indeterminate faults could also belong to any combinations of the aforementioned faults, but they are assumed to occur in random width time windows and amplitudes. These faults are treated like permanent by the *i*FD, and they are 'captured' once they appear.

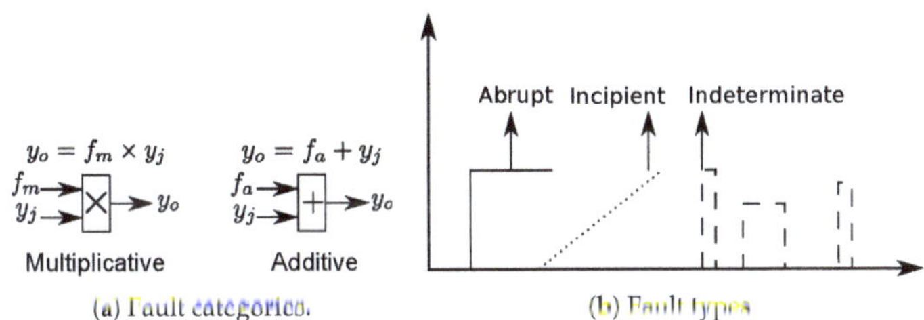

(**a**) Fault categories. (**b**) Fault types.

Figure 4. Sensor fault categories (**a**) and types (**b**) (basis from [19]).

2.3. Neural Network Algorithm

A plethora of NN algorithm types exist in the research literature; however in this paper a dynamic non-linear input-output Neural Network model with tapped delay lines at the input is employed for time-series prediction. The NN's task is to perform similar to that of the—more conventional—bank of Kalman estimators (KE) in the feedback loop, as well as to predict future values based on past values of one or multiple time-series. In particular, to predict $\phi(t)$ series based on n_c past values of $\theta(t)$ series so that $\phi(t) = \lambda(\theta(t-1), \ldots, \theta(t-n_c))$.

This type of NN algorithm is adapted to fit the inputs and targets of the suspension for the *i*FD realization. It has a total of five inputs (u_c and $i, (z_t - z), \dot{z}, \ddot{z}$) and three estimated outputs ($\hat{i}, \hat{\dot{z}}, \hat{\ddot{z}}$). Its internal architecture is shown in Figure 5, and is realized as a hidden layer (with one delay and

20 hidden neurons) and an output layer with sigmoid and linear functions, respectively. Each neuron's output is generally described by (6),

$$o_n = \sum_{j=1}^{n_n} \Delta_j w_j + \psi \tag{6}$$

where o_n is the neuron's output, n_n is the number of neuron's inputs, w are the weights (is a vector equal to the size of the neuron's inputs), Δ are the delay lines and ψ is the bias point (considered to be one for all neurons).

A fast convergence method for training moderate sized feed-forward neural networks is the Levenberg-Marquardt backpropagation algorithm (for details see [41,42] (Chapters 11–12)).

The training data that were (later used to train the NN), were successively collected in equal time windows T for all failing set combinations with a sampling time τ_s, on an online working state model with the $\mathcal{H}_\infty$ LSDP controllers in the feedback loop. The stopping criteria was set to a Mean Square Error (MSE $\leq 10^{-5}$) or a maximum of 1000 epochs.

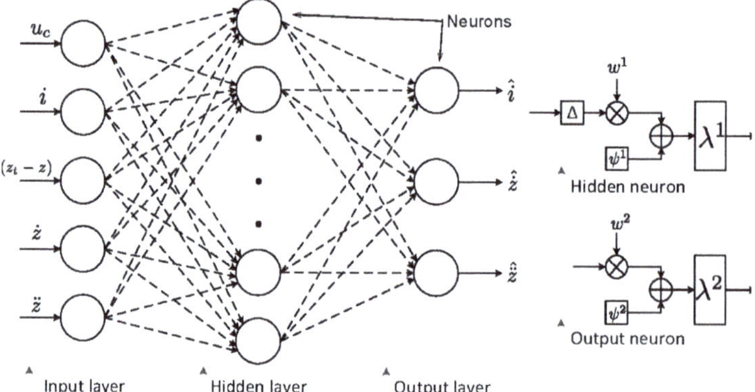

Figure 5. The Neural network architecture for the *i*FD as applied on the EMS.

3. Results Discussion

3.1. Efficacy and Assessment of the Proposed iFD

A rigorous analysis follows next, to show the effectiveness of the proposed *i*FD unit with results from realistic simulations (MATLAB/Simulink) on the EMS system.

An offline training with a typical NN-based estimator, described in Section 2.3, was performed prior to the FTC design of the suspension. The total training process (1000 epochs) required c.7 min on a mobile workstation laptop (equipped with Intel® Core™ i7-9750H @ 2.6 GHz, 32 GB RAM), with the parameters described in the previous section. The total number of sensor and actuators, their estimated signals and the number of actuator/sensor sets are, $n_y = 4$, $n_u = 1$, $n_{\hat{y}} = 3$, $n_{\hat{u}} = 1$ and $n_{yu} = 8$ respectively.

The training data from each sensor set comprising $\mathcal{D}$, were collected using sample rate $\tau_s = 1$ kHz under total simulation time $T = 6.6$ s. The data set $\mathcal{D}$ consists of data subsets $\mathcal{D}_d$ and $\mathcal{D}_s$, drawn from the deterministic and stochastic suspension responses (Subscripts d and s indicate the deterministic and stochastic cases respectively). The data set used for training is as follows,

$$\mathcal{D}^{d_r \times d_c} = \begin{bmatrix} \mathcal{D}_d^{d_{r_d} \times d_{c_d}} \\ \mathcal{D}_s^{d_{r_s} \times d_{c_s}} \end{bmatrix} \tag{7}$$

where $d_r = d_{r_d} + d_{r_s}$ and $d_c = d_{c_d} = d_{c_s}$ are both found from (4). The total number of columns is calculated as $d_c = 9$, while the total samples per set is $k = 6.6 \times 1000$, hence $d_{r_d} = d_{r_s} = 52{,}800$ and

$d_r = 105{,}600$. The functions used for the training of the NN are $\mathcal{C}_{y_j} = \{c_i = 0, c_{\dot{z}} = 0, c_{\ddot{z}} = 0\}$ with dimensions $k \times 3$. These functions are both used in the stochastic and deterministic responses of the suspension where a fault is assumed, as explained in Section 2.1.

Overall, there are 4 sensors in the full sensor set ($\mathcal{Y}$), while it is taken for granted that the actuator, u_c, and the airgap sensor $(z_t - z)$ cannot fail (in case where fault tolerance is required for the airgap then redundant components can be used under a voting scheme). Three sensor fault possibilities are considered: the current i, the vertical velocity $\dot{z}$ and the acceleration $\ddot{z}$. Sensor faults are generally classified into abrupt and incipient fault types. We also use additive and multiplicative faults with bias for each one of the sensors $i, \dot{z}$ and $\ddot{z}$. Figure 6 illustrates current sensor measurement, i, fault profile (similar pattern fault profiles are used for other sensors in this work). For all cases, faults start developing at time $t_f = 1$ s (this is marked at point A in all relevant figures).

Figure 6a for the impaired sensor, illustrates the normal current value superimposed with a low frequency band-limited random signal, $v(t)$, at frequencies of 10 rad/s and zero mean, white noise characteristics and power spectral density of $S_i = 5$ limited to $\omega_i = 1.6$ Hz. Consequently, the additive/abrupt fault profile for the current sensor (f_{aa_i}) is given by,

$$f_{aa_i}(t) = \{ {}^{v(t)\ if \quad t_f \le t < \infty}_{0 \quad if \quad t < t_f} \tag{8}$$

Sensor output $\dot{z}$ and $\ddot{z}$ follow a similar pattern (same bandwidth) and $S_{\dot{z}} = 0.03$ and $S_{\ddot{z}} = 2$ respectively. Next, Figure 6b depicts the multiplicative/abrupt case (f_{ma_i}) where the current sensor is suddenly damaged at $t = 1$ s and as a result its output becomes five times larger than normal (9).

$$f_{ma_i}(t) = \{ {}^{5 \quad if \quad t_f \le t < \infty}_{1 \quad if \quad 0 \le t < t_f} \tag{9}$$

The same fault profile is used for the other two sensors, $\dot{z}$ and $\ddot{z}$. For the current measurement, a bias (abrupt) type of failure is shown in Figure 6e. Clearly the sensor output abruptly increases (in a step manner) to its maximum value, i.e., in this case max $y_{o_i} = 10$ A.

The incipient types of faults are illustrated in Figure 6c and Figure 6d respectively. In the former figure the additive/incipient fault on the current measurement (f_{ai_i}) are described by (10). The latter is a ramp type signal with σ_i slope superimposed with a low frequency random signal with band-limited white noise characteristics as previously explained. The aforementioned figure, depicts the multiplicative/incipient (f_{mi_i}) fault described by (11), where the fault starts developing at $t_f = 1$ s and then falls to zero (due to multiplication by zero).

$$f_{ai_i}(t) = \{ {}^{\sigma_i(t-t_f)+v(t) \quad if \quad t_f \le t < \infty}_{0 \qquad\qquad if \quad 0 < t < t_f} \tag{10}$$

$$f_{mi_i}(t) = \{ {}^{\sigma_i(t-t_f)+1+v(t) \quad if \quad t_f \le t < \infty}_{1 \qquad\qquad if \quad 0 < t < t_f} \tag{11}$$

The fault profiles for $\dot{z}$ and $\ddot{z}$ follow the same behavior as above with effective slopes, $\sigma_i = 20$, $\sigma_{\dot{z}} = 6$ and $\sigma_{\ddot{z}} = 20$ and for the PSD, $S_i = 0.5$, $S_{\dot{z}} = 0.03$ and $S_{\ddot{z}} = 0.5$ respectively. The following scenario supports explaining the *i*FD working principle:

- Three sensors are subsequently impaired with a time difference as follows: accelerometer at 0.5 s, velocity at 1.5 s and current at 2 s),
- the deterministic disturbance to the suspension is used and,
- a multiplicative/abrupt fault profile is injected for each sensor at each time instant mentioned above (e.g., for the current sensor see Figure 6b).

The airgap sensor output with fault-free case and with the aforementioned fault scenario is depicted in Figure 7. The figure illustrates the airgap with a fault-free case (i.e., healthy sensor set, $\mathcal{Y}$, with $K_{i,(z_t-z),\dot{z},\ddot{z}}$) and under the fault scenario mentioned. The acceleration sensor is impaired at 0.5 s (point A) and immediately after a controller reconfiguration follows (a new controller, $K_{i,(z_t-z),\dot{z}}$, is introduced in the loop) in order to maintain the stability and performance of the EMS. The EMS response with both fault-free and fault scenario comply with the control performance requirements described in Appendix A. Following the acceleration fault, the velocity one fails at $t = 1.5$ s (designated at point B) and the current sensor follows at $t = 2$ s (marked at point C). The subsequent faults are successfully detected and accommodated via appropriate switching on $K_{i,(z_t-z)}$ and $K_{(z_t-z)}$ respectively.

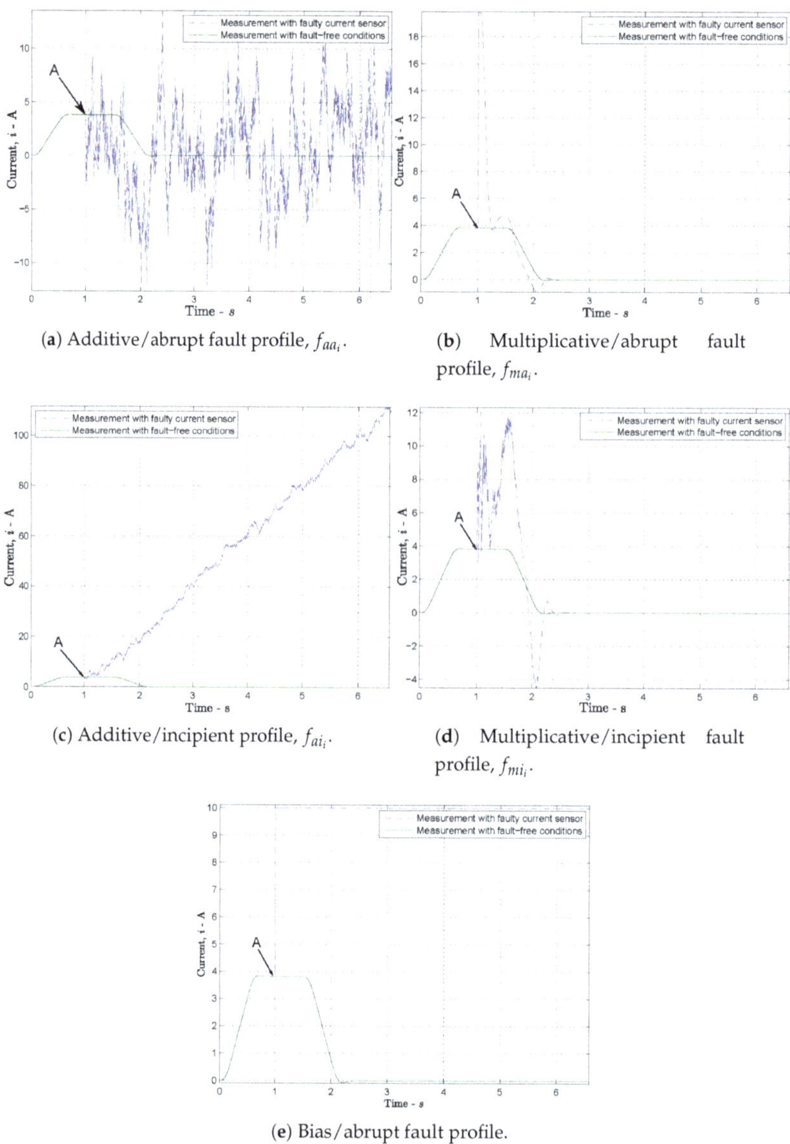

(**a**) Additive/abrupt fault profile, f_{aa_i}.

(**b**) Multiplicative/abrupt fault profile, f_{ma_i}.

(**c**) Additive/incipient profile, f_{ai_i}.

(**d**) Multiplicative/incipient fault profile, f_{mi_i}.

(**e**) Bias/abrupt fault profile.

Figure 6. Coil's current sensor, i, fault profiles.

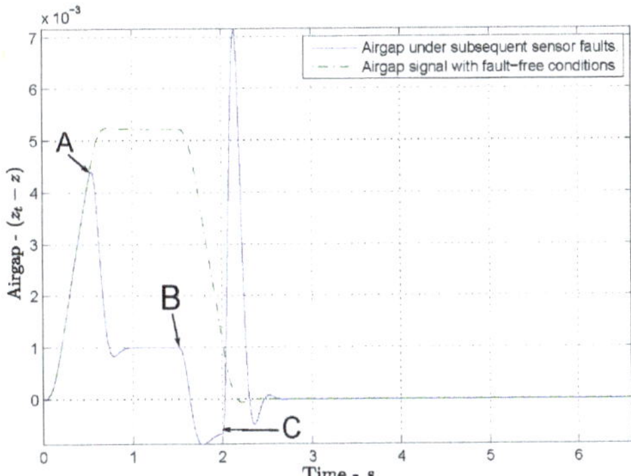

Figure 7. Airgap sensor signal with fault-free and with the fault scenario, *id*: 8 of Table 2.

The sensor fault accommodation for each sensor failure is integrated in three steps. To assist in explaining the steps of the procedure, the current sensor fault will be interpreted: (i) *Sensor FD*: when the fault occurs at $t = 1$ s, the residual of the current measurement, r_{y_i} starts increasing and as soon as it passes the threshold (see Figure 8) the fault is detected. (ii) *Fault Isolation*: at this stage the faulty sensor is removed from the loop using a BS, while a 'known' function $c_{y_i} = 0$ is connected at the input of the *i*FD. Figure 9 clearly shows the signal at the input and output of the BS, as well as the signal at the output of the *i*FD. (iii) *Controller reconfiguration*: after the faulty sensor isolation, a reconfiguration signal is generated and the new controller, $K_{(z_t-z)}$, is introduced in the loop.

Careful investigation of Figure 9 (after the fault occurs at point C) shows that the unit detect the fault after a few time steps and that one time step is required for the BS_{y_i} portion to permanently change its output to $c_{y_i} = 0$. Hence, the residual remains large which justifies the reason the BS output will never return to its previous stage when/if the fault vanishes. The same figure also shows the input to BS_{y_i} with the two previous sensor faults, i.e., acceleration and velocity (at point A and B respectively).

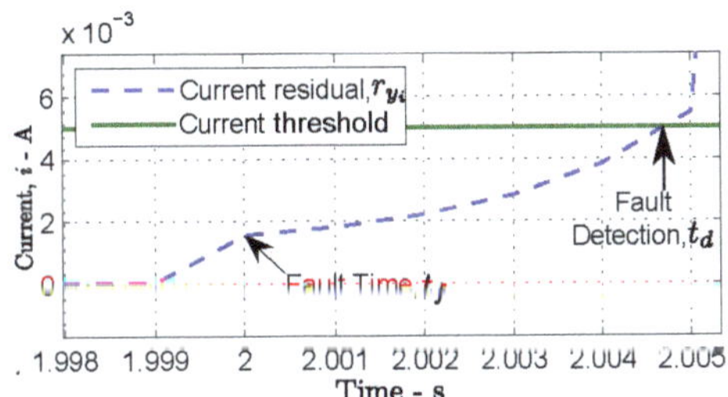

Figure 8. Current sensor residual, r_{y_i}, when fault occurs at 2 s.

Table 2 indicates the resulted performance of the suspension and the false alarm (FA) after 70 tests, analytically 35 for each deterministic and stochastic responses of the suspension. The first column of the table describes the sensor fault scenarios used to test the proposed *i*FD. Typically, rows 2–4 show the results for single sensor faults that occur at $t = 1$ s, while the rest rows show the results with subsequence faults starting from 0.5 s with a time difference of 1 s. The first six columns present the performance with abrupt faults while the rest four, show the performance with incipient faults. The track inputs exciting the EMS were discussed in Appendix A. Multiplicative (Mult.) and Additive (Add.) faults are used, as well as a bias (Bis.) fault that occurs abruptly. Per the scenario case, entry ✓ indicates that EMS performance is successfully maintained, while entry "x" indicates the opposite. In addition, if a FA arises is marked with a red color ▽.

Close investigation of the aforementioned table of results indicates that the *i*FD successfully detects and reconfigures the controller under all scenaria (maintaining the appropriate performance levels). In some cases, i.e., in *id*: 7–8, an FA appears meaning that a sensor is 'shown' impaired although truly is healthy. This is an important finding towards facilitating reliable system autonomy. The threshold setting for the residual plays a substantial role and this needs to be addressed in a reliable autonomous system (as an FA could hinder perception and hence impact system stability).

Given that the residual threshold of such sensor cases is increased to avoid the FA, and the sensors are impaired themselves, then the FD could delay long enough to cause instability. Two particular issues that have been noted and looked further as future research are: (i) in the NN-trained FD unit, in general a small residual remains after a fault occurs in some scenaria, (ii) the coupled nature of the closed-loop (FD unit, reconfiguration mechanisms, decision making). A deep learning approach is currently investigated to address that uncertainty envelope.

Table 2. Performance with various sensor fault scenarios for the EMS.

id	Faulty Sensor(s)	Abrupt Fault						Incipient Fault			
		Mult./FA		Add./FA		Bis./FA		Mult./FA		Add./FA	
		Sth.	Dtm.	Sth.	Dtm.	Sth.	Dtm.	Sth.	Dtm.	Sth.	Dtm.
1	Fault-free	√/x	√/x	√/x	√/x	√/x	√/x	√/x	√/x	√/x	√/x
2	i	√/x	√/x	√/x	√/x	√/x	√/x	√/x	√/x	√/x	√/x
3	$\dot{z}$	√/x	√/x	√/x	√/x	√/x	√/x	√/x	√/x	√/x	√/x
4	$\ddot{z}$	√/x	√/x	√/x	√/x	√/x	√/x	√/x	√/x	√/x	√/x
5	$i \rightarrow \dot{z}$	√/x	√/x	√/x	√/x	√/x	√/x	√/x	√/x	√/x	√/x
6	$i \rightarrow \ddot{z}$	√/x	√/x	√/x	√/x	√/x	√/x	√/x	√/x	√/x	√/x
7	$\dot{z} \rightarrow \ddot{z}$	√/x	√/▽	√/x	√/▽	√/x	√/▽	√/x	√/▽	√/▽	√/▽
8	$\ddot{z} \rightarrow \dot{z} \rightarrow i$	√/x	√/▽	√/x	√/▽	√/x	√/▽	√/x	√/▽	√/x	√/▽

Mult.—Multiplicative, Add.—Additive, FA—False Alarm, Sth.—Stochastic, Dtm.—Deterministic, Bis.—Bias.

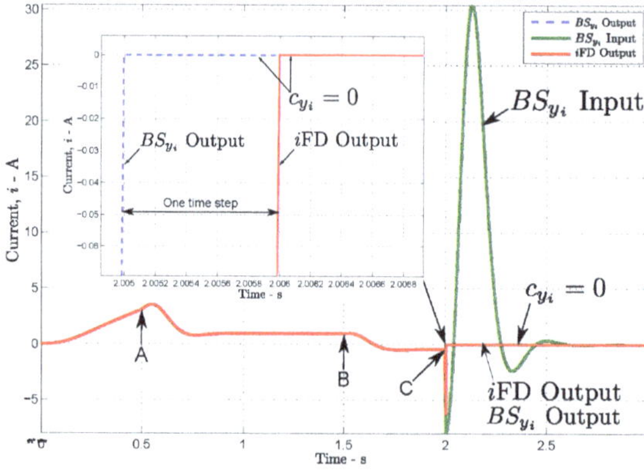

Figure 9. Input-output of the BS_{y_i} of the current sensor, i, and the output of the *i*FD.

Sensor FD time investigation is seen on Table 3. Column-wise: column 1 maps the identifier number for the chosen scenario, column 2 lists the sensors (remark: underlined channels indicate faulty ones with incident occurring at t_f which is shown column 3). The rest of the columns refer to FD time, t_d, in particular the ones marked using boldface font indicated fault occurrence is detected (while boldface entries with superscript "*" are the false alarms).

Careful investigation of the results shows that with abrupt single sensor failures (*id*: 1–3) the fault detection succeeds at the instance the sensor fails. In the incipient fault cases, a short delay within 0.030–0.399 s is noted. However, this delay does not hinder performance due to the robust controller. Although delays are observed in FD throughout these three and the next two scenaria, none of these cause FAs. In the last two scenarios (i.e., in scenario 6: two subsequent faults on velocity and acceleration outputs are considered and in scenario 7: acceleration, velocity and current sensors are impaired sequentially) FAs appear mainly in the current sensor case. Please note that other observed fault delays in *i*FD are successfully accommodated via the control reconfiguration.

Table 3. Sensor (-Snr.) fault detection time, t_d, for the various fault scenarios.

id	Snr.	t_f (s)	**Fault Detection Time, $t_d(s)$**									
			Abrupt Fault						**Incipient Fault**			
			Mult.		**Add.**		**Bis.**		**Mult.**		**Add.**	
			Sth.	**Dtm.**	**Sth.**	**Dtm.**	**Sth.**	**Dtm.**	**Sth.**	**Dtm.**	**Sth.**	**Dtm.**
1	$\underline{i}$	1	1.000	1.000	1.000	1.000	1.000	1.000	**1.052**	1.000	1.000	1.000
	$\dot{z}$	-	-	-	-	-	-	-	-	-	-	-
	$\ddot{z}$	-	-	-	-	-	-	-	-	-	-	-
2	i	-	-	-	-	-	-	-	-	-	-	-
	$\underline{\dot{z}}$	1	1.000	1.000	1.000	1.000	1.000	1.000	1.000	1.000	**2.399**	1.000
	$\ddot{z}$	-	-	-	-	-	-	-	-	-	-	-
3	i	-	-	-	-	-	-	-	-	-	-	-
	$\dot{z}$	-	-	-	-	-	-	-	-	-	-	-
	$\underline{\ddot{z}}$	1	1.000	1.000	1.000	1.000	1.000	1.000	**1.030**	1.000	**2.075**	1.000
4	$\underline{i}$	0.5	0.500	0.500	0.500	0.500	0.500	0.500	**0.527**	0.500	0.500	0.500
	$\underline{\dot{z}}$	1.5	1.500	**1.502**	1.500	1.500	1.500	1.500	**1.517**	1.500	**2.335**	**1.511**
	$\ddot{z}$	-	-	-	-	-	-	-	-	-	-	-
5	$\underline{i}$	0.5	0.500	0.500	0.500	0.500	0.500	0.500	**0.527**	0.500	0.500	0.500
	$\dot{z}$	-	-	-	-	-	-	-	-	-	-	-
	$\underline{\ddot{z}}$	1.5	1.500	1.500	1.500	1.500	1.500	1.500	1.500	**1.510**	**1.514**	1.500
6	i	-	-	**1.700 ***	-	**1.700 ***	-	**1.700 ***	-	**1.700 ***	**2.400 ***	**1.700 ***
	$\underline{\dot{z}}$	0.5	0.500	0.500	**2.070**	0.500	0.500	0.500	0.500	0.500	**2.423**	**0.534**
	$\underline{\ddot{z}}$	1.5	1.500	1.500	1.500	1.500	1.500	**1.700**	1.500	1.500	**2.403**	1.500
7	$\underline{i}$	2.0	2.000	**0.654 ***	2.000	**0.654 ***	2.000	**0.650 ***	**2.614**	**0.654 ***	2.000	**0.654 ***
	$\underline{\dot{z}}$	1.5	**1.517**	**1.502**	1.500	1.500	1.500	1.500	**1.517**	**1.502**	**2.336**	**1.511**
	$\underline{\ddot{z}}$	0.5	**0.506**	0.500	0.500	0.500	0.500	0.500	**0.508**	0.500	0.500	0.500

3.2. Comparison of the Execution Time

Using the full sensor set, $\mathcal{Y}$ is possible to compare the time taken for a simulation to complete i.e., the execution time (t_e) using the *i*FD with that for a bank of eight in-parallel KEs as shown in Figure 10a. The execution time is measured at a high-level simulation in Simulink platform iteratively ($\times 100$) with each estimator. The execution time for each simulation is illustrated in Figure 10b. The mean simulation time for the *i*FD is 0.5 s, while that of the bank-of-estimators setup 7.9 s. Clearly the *i*FD is c. $\times 16$ faster, in terms of mean exec. time, compared to the bank of KEs (in addition the proposed scheme offers c. $\times 13$ lesser standard deviation i.e., 0.15 s). The comparison showcases a two-fold aspect: (i) the efficacy of the proposed fault detection approach in terms of fast actuators/sensors fault detection, (ii) a level of re-assurance as the proposed approach performs within the same envelope of performance of the conventional bank-of-estimators approach.

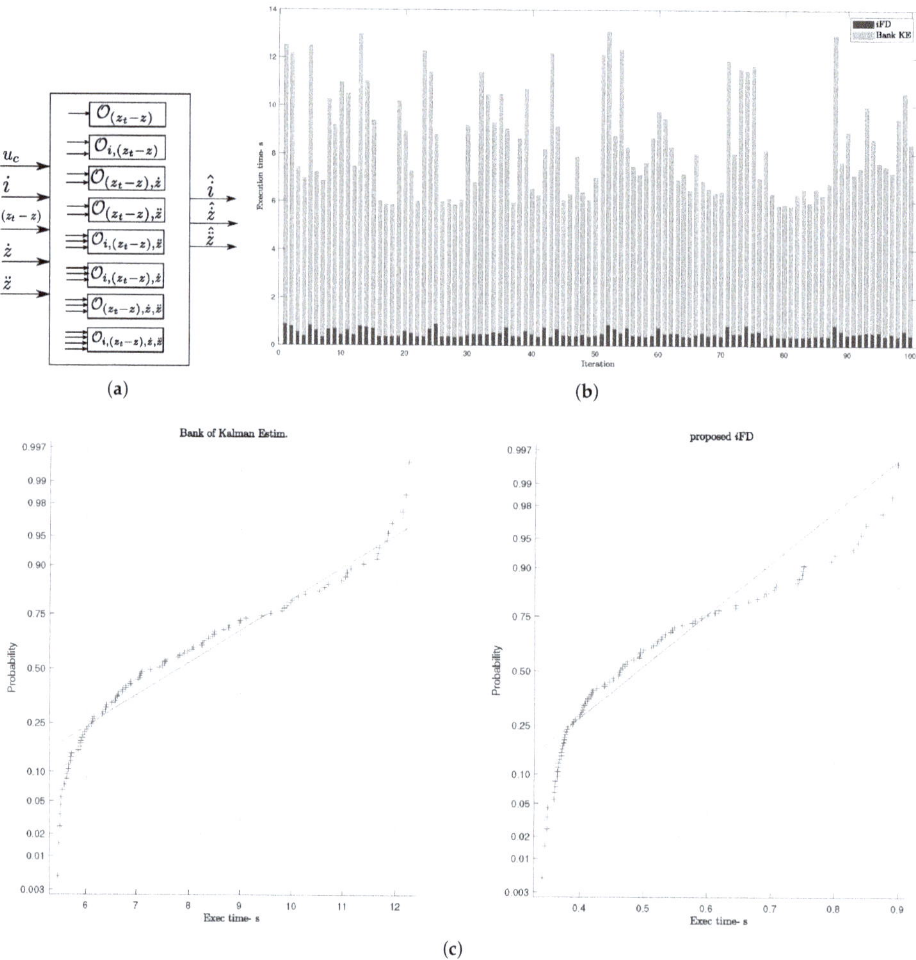

Figure 10. Computational complexity comparison bank of KE vs. *i*FD (AI-based) for multiple sensors failure detection on the EMS system: (**a**) bank of KE (inner box) vs. *i*FD architecture (outer box), (**b**) execution time performance, (**c**) normal probability plot of execution time data.

Figure 10c shows the normal probability plots for the conventional bank of estimators and proposed AI-based schemes. Clearly with very different execution time average, it is seen that the former setup is closer to a normal distribution, the latter NN-based solution favors faster execution time (while providing comparable performance). The importance of this result is two-fold, which we study further, (i) maintaining similar level of performance/reliability to an acceptable conventional solution enables certification [43] and (ii) correlating the process with descriptive statistics supports faster training for the autonomous system solution. To summarize, the following specific comments are highlighted for the proposed scheme:

1. a single AI-based estimator unit can be used in the FDI instead of the conventional bank of estimators benefiting substantial computational resource reduction;
2. the AI-based unit maintaining similar level of performance/reliability to an acceptable conventional solution enables certification [43];

3. faults of multiple characteristics can be handled appropriately by the single AI-based unit as in the more conventional cases;
4. correlating the process with descriptive statistics will support faster autonomous system training.

4. Conclusions

We presented an AI-based, using neural networks, method referred to as *i*FD, towards more reliable system autonomy via detecting actuator/sensor faults. The methods were supported by a detailed analysis on multiple (70) fault scenarios using an electro-magnetic suspension system testbed. Various sensor fault types were studied i.e., abrupt/multiplicative, abrupt/additive, abrupt/bias, incipient/multiplicative and incipient/additive characteristics. The results clearly show that: a single AI-based estimator can be used in the FDI instead of the conventional bank of estimators, which benefits computational resource reduction. The paper indicates important point towards reliable AI-based system autonomy from the FD and threshold discussion. Currently the scheme follows a conservative approach that considers indeterminate faults as permanent ones which are removed as soon as they are detected (to enforce a level of reliability in performance). It is worth mentioning that training (and availability of data) plays an important role in the development.

Author Contributions: The authors contributed to the article as follows: Conceptualization, K.M.D., A.C.Z.; Methodology, K.M.D. and K.M.; Software, K.M.D.; Validation, K.M.D., K.M., A.C.Z.; Resources, K.M.D., A.C.Z.; Data curation, K.M.D., A.C.Z.; Writing–original draft preparation, K.M., K.M.D.; Writing–review and editing, K.M.D., A.C.Z.; Visualization, K.M.D., K.M., A.C.Z.; Supervision, A.C.Z., K.M.D. All authors have read and agreed to the published version of the manuscript.

Funding: This research received no external funding.

Acknowledgments: The authors would like to thank Roger Goodall (now at University of Huddersfield, UK) for his advice during the early stages of this work on the electro-magnetic suspension model considerations and testbed configuration setup. The authors would also like to acknowledge Spyros Tzafestas' (Deceased 13 April 2019) advice and contributions on AI and Neural Networks development in the earlier developments of this work; this paper is dedicated to his memory.

Conflicts of Interest: The authors declare no conflict of interest.

Appendix A. The EMS System—A Test Case

We use a single-stage EMS system platform representation (typically representing a quarter of a Maglev vehicle [44]. The model is actually extensively discussed in [45]. It is the nature of the EMS system that makes it appealing as a test system [46], i.e., being inherently unstable, safety-critical system, under non-trivial control requirements. Figure A1a presents the EMS schematic diagram, the input excitation profile and Figure A1c lists the performance constraints.

The characteristics of the system are discussed in [45]. In particular, the vehicle mass (M_c) is supported on the electromagnet with the distance between them being the so-called airgap $(z_t - z)$. Normally the airgap does not exceed 15 mm. The rest of the variables of interest are: flux B, force F, current I, airgap $(z_t - z)$, input voltage to the EM V_c, vehicle position Z. Please note that the designed operating condition for the EMS system is at airgap of 15 mm, with nominal force 9810 N, nominal flux 1 T, nominal current $I_o = 10$ A, and operating voltage 100 V. For the linearized time invariant state space model of the EMS and the electromagnet parameter calculations the reader is referred to [19,45].

In addition, for simulation purposes the operational scenario uses deterministic rail track information due to the intended changes of the rail's inclination. This transition, vertical direction only, onto the rail's gradient is simulated by the signal shown in Figure A1b. We also consider stochastic track irregularities as random variations of the rail's vertical layout typically caused by the installation process i.e., accumulated inaccuracies and unevenness. Such track elements normally are difficult to measure (some information can be provided from a monitored track database, or a specialist track condition measurement vehicle), while attempts to estimate irregularities from vehicle-based sensor exist in the current literature [47]. Considering the vertical direction, the velocity variations can be approximated by a double-sided power spectrum density [19].

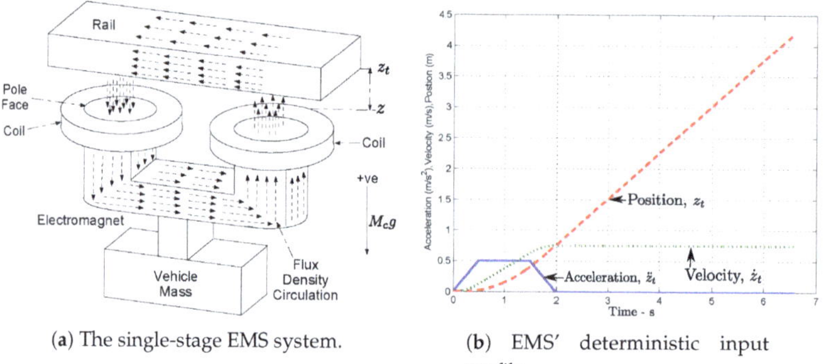

(**a**) The single-stage EMS system.

(**b**) EMS' deterministic input profile.

EMS limitations	Value
Stochastic rail profile	
RMS of acceleration, $\ddot{z}_{rms}$	$\leq 1ms^{-2}$
RMS of airgap variation, $(z_t - z)_{rms}$	$\leq 5mm$
RMS of control effort, $u_{c_{rms}}$	$\leq 300V$
Deterministic rail profile	
Maximum of airgap deviation, $(z_t - z)_p$	$\leq 7.5mm$
Maximum of control effort, u_{c_p}	$\leq 300V$
Airgap settling time, t_s	$\leq 3s$
Airgap steady state error, $(z_t - z)_{e_{ss}}$	$= 0$

(**c**) EMS control constraints.

Figure A1. EMS systems and its constraints.

Moreover, the control design requirements of an EMS system are dependent on the train type and its speed [48]. The EMS system should follow the gradient onto the rail (deterministic) and remain insensitive to track irregularities. Figure A1c summarizes the control performance requirements.

References

1. Napolitano, M.; Windon, D.; Casanova, J.; Innocenti, M.; Silvestri, G. Kalman filters and neural-network schemes for sensor validation in flight control systems. *IEEE Trans. Control Syst. Technol.* **1998**, *6*, 596–611. [CrossRef]
2. Rago, C.; Prasanth, R.; Mehra, R.K.; Fortenbaugh, R. Failure detection and identification and fault tolerant control using the IMM-KF with applications to the Eagle-Eye UAV. In Proceedings of the 37th IEEE Conference on Decision and Control, Tampa, FL, USA, 18 December 1998; Volume 4, pp. 4208–4213.
3. Ranjbaran, M.; Khorasani, K. Fault recovery of an under-actuated quadrotor Aerial Vehicle. In Proceedings of the 49th IEEE Conference on Decision and Control, Atlanta, GA, USA, 15–17 December 2010; pp. 4385–4392.
4. Petritoli, E.; Leccese, F.; Ciani, L. Reliability and Maintenance Analysis of Unmanned Aerial Vehicles. *Sensors* **2018**, *18*, 3171. [CrossRef] [PubMed]
5. Nguyen, N.P.; Hong, S.K. Fault Diagnosis and Fault-Tolerant Control Scheme for Quadcopter UAVs with a Total Loss of Actuator. *Energies* **2019**, *12*, 1139. [CrossRef]
6. Chen, X.M.; Wu, C.X.; Wu, Y.; Xiong, N.X.; Han, R.; Ju, B.B.; Zhang, S. Design and Analysis for Early Warning of Rotor UAV Based on Data-Driven DBN. *Electronics* **2019**, *8*, 1350. [CrossRef]
7. Lussier, B.; Chatila, R.; Guiochet, J.; Ingrand, F.; Lampe, A.; Olivier Killijian, M.; Powell, D. Fault Tolerance in Autonomous Systems: How and How Much? In Proceedings of the 4th IARP/IEEE-RAS/EURON Joint Workshop on Technical Challenge for Dependable Robots in Human Environments, Nagoya, Japan, 16–18 June 2005; 10p.
8. Kate Devitt, S. Trustworthiness of Autonomous Systems. In *Foundations of Trusted Autonomy*; Abbass, H.A., Scholz, J., Reid, D.J., Eds.; Studies in Systems, Decision and Control; Springer International Publishing: Cham, Switzerland, 2018; pp. 161–184. [CrossRef]

9. Napolitano, M.; An, Y.; Seanor, B. A fault tolerant flight control system for sensor and actuator failures using neural networks. *Aircr. Des.* **2000**, *3*, 103–128. [CrossRef]
10. Campa, G.; Fravolini, M.L.; Seanor, B.; Napolitano, M.R.; Gobbo, D.D.; Yu, G.; Gururajan, S. On-line learning neural networks for sensor validation for the flight control system of a B777 research scale model. *Int. J. Robust Nonlinear Control* **2002**, *12*, 987–1007. [CrossRef]
11. Lunze, J.; Schroder, J. Sensor and actuator fault diagnosis of systems with discrete inputs and outputs. *IEEE Trans. Syst. Man Cybern. Part B* **2004**, *34*, 1096–1107. [CrossRef]
12. Edwards, C.; Tan, C. Sensor fault tolerant control using sliding mode observers. *Control Eng. Pract.* **2006**, *14*, 897–908. [CrossRef]
13. Talebi, H.; Khorasani, K. An intelligent sensor and actuator fault detection and isolation scheme for nonlinear systems. In Proceedings of the 46th IEEE Conference on Decision and Control, New Orleans, LA, USA, 12–14 December 2007; pp. 2620–2625.
14. Talebi, H.A.; Khorasani, K. A neural network-based actuator gain fault detection and isolation strategy for nonlinear systems. In Proceedings of the 46th IEEE Conference on Decision and Control, New Orleans, LA, USA, 12–14 December 2007; pp. 2614–2619.
15. Heredia, G.; Ollero, A.; Bejar, M.; Mahtani, R. Sensor and actuator fault detection in small autonomous helicopters. *Mechatronics* **2008**, *18*, 90–99. [CrossRef]
16. Yetendje, A.; Seron, M.M.; Doná, J.A.D.; Martínez, J.J. Sensor fault-tolerant control of a magnetic levitation system. *Int. J. Robust Nonlinear Control* **2010**, *20*, 2108–2121. [CrossRef]
17. Realpe, M.; Vintimilla, B.X.; Vlacic, L. A Fault Tolerant Perception system for autonomous vehicles. In Proceedings of the 2016 35th Chinese Control Conference (CCC), Chengdu, China, 27–29 July 2016; pp. 6531–6536.
18. Ding, S.X. *Data-Driven Design of Fault Diagnosis and Fault-Tolerant Control Systems*; Springer: London, UK, 2014.
19. Michail, K.; Deliparaschos, K.M.; Tzafestas, S.G.; Zolotas, A.C. AI-Based Actuator/Sensor Fault Detection With Low Computational Cost for Industrial Applications. *IEEE Trans. Control Syst. Technol.* **2016**, *24*, 293–301. [CrossRef]
20. Hwang, I.; Kim, S.; Kim, Y.; Seah, C.E. A survey of fault detection, isolation, and reconfiguration methods. *IEEE Trans. Control Syst. Technol.* **2010**, *18*, 636–653. [CrossRef]
21. Zhang, Y.; Jiang, J. Bibliographical review on reconfigurable fault-tolerant control systems. *Annu. Rev. Control* **2008**, *32*, 229–252. [CrossRef]
22. Chibani, A.; Chadli, M.; Ding, S.X.; Braiek, N.B. Design of robust fuzzy fault detection filter for polynomial fuzzy systems with new finite frequency specifications. *Automatica* **2018**, *93*, 42–54. [CrossRef]
23. Ruiz-Arenas, S.; Rusák, Z.; Horváth, I.; Mejí-Gutierrez, R. Systematic exploration of signal-based indicators for failure diagnosis in the context of cyber-physical systems. *Front. Inf. Technol. Electron. Eng.* **2019**, *20*, 152–175. [CrossRef]
24. Samy, I.; Postlethwaite, I.; Gu, D.W. Survey and application of sensor fault detection and isolation schemes. *Control Eng. Pract.* **2011**, *19*, 658–674. [CrossRef]
25. Samy, I.; Postlethwaite, I.; Gu, D.W. A comparative study of NN- and EKF-based SFDA schemes with application to a nonlinear UAV model. *Int. J. Control* **2010**, *83*, 1025–1043. [CrossRef]
26. Tzafestas, S. System Fault Diagnosis Using the Knowledge-Based Methodology. In *Fault Diagmosis in Dynamic Systems*; Patton, R., Frank, P., Clark, R., Eds.; Prentice Hall: Upper Saddle River, NJ, USA, 1989; pp. 502–552.
27. Mok, H.; Chan, C. Online fault detection and isolation of nonlinear systems based on neurofuzzy networks. *Eng. Appl. Artif. Intell.* **2008**, *21*, 171–181. [CrossRef]
28. Skoundrianos, E.N.; Tzafestas, S.G. Fault diagnosis via local neural networks. *Math. Comput. Simul.* **2002**, *60*, 169–180. [CrossRef]
29. Skoundrianos, E.N.; Tzafestas, S.G. Modelling and FDI of Dynamic Discrete Time Systems Using a MLP with a New Sigmoidal Activation Function. *J. Intell. Robot. Syst.* **2004**, *41*, 19–36. [CrossRef]
30. Isermann, R. Supervision, fault-detection and fault-diagnosis methods–an introduction. *Control Eng. Pract.* **1997**, *5*, 639–652. [CrossRef]
31. Albu, A.; Precup, R.E.; Teban, T.A. Results and challenges of artificial neural networks used for decision-making and control in medical applications. *Facta Univ. Ser. Mech. Eng.* **2019**, *17*, 285. [CrossRef]

32. Hunt, K.; Sbarbaro, D.; Żbikowski, R.; Gawthrop, P. Neural networks for control systems-a survey. *Automatica* **1992**, *28*, 1083–1112. [CrossRef]
33. Polycarpou, M.; Helmicki, A. Automated fault detection and accommodation: A learning systems approach. *IEEE Trans. Syst. Man Cybern.* **1995**, *25*, 1447–1458. [CrossRef]
34. Polycarpou, M.; Vemuri, A. Learning methodology for failure detection and accommodation. *IEEE Control Syst. Mag.* **1995**, *15*, 16–24.
35. Frank, P.M.; Köppen-Seliger, B. New developments using AI in fault diagnosis. *Eng. Appl. Artif. Intell.* **1997**, *10*, 3–14. [CrossRef]
36. Reppa, V.; Polycarpou, M.M.; Panayiotou, C.G. Adaptive Approximation for Multiple Sensor Fault Detection and Isolation of Nonlinear Uncertain Systems. *IEEE Trans. Neural Netw. Learn. Syst.* **2013**, *25*, 137–153. [CrossRef]
37. Heredia, G.; Ollero, A.; Mahtani, R.; Béjar, M.; Remuss, V.; Musial, M. Detection of sensor faults in autonomous helicopters. In Proceedings of the IEEE International Conference on Robotics and Automation, Barcelona, Spain, 18–22 April 2005; pp. 2229–2234.
38. Wang, Z.; Lu, C. An Adaptive Threshold Based on RBF Neural Network for Fault Detection of a Nonlinear System. In *Advances in Computer, Communication, Control and Automation*; Wu, Y., Ed.; Lecture Notes in Electrical Engineering; Springer: Berlin/Heidelberg, Germany, 2012; Volume 121, pp. 495–502.
39. McFarlane, D.C.; Glover, K. A loop-shaping design procedure using $\mathcal{H}_\infty$ synthesis. *IEEE Trans. Autom. Control* **1992**, *37*, 759–769. [CrossRef]
40. Michail, K.; Zolotas, A.C.; Goodall, R.M.; Halikias, G. Optimal selection for sensor fault tolerant control of an EMS system via loop-shaping robust control. In Proceedings of the 19th Mediterranean Conference on Control and Automation, Corfu, Greece, 20–23 June 2011; pp. 1112–1117.
41. Hagan, M.; Menhaj, M. Training feedforward networks with the Marquardt algorithm. *IEEE Trans. Neural Netw.* **1994**, *5*, 989–993. [CrossRef]
42. Hagan, M.; Demuth, H.; Beale, M. *Neural Network Design*; PWS Pub. Co.: Boston, MA, USA, 1996.
43. Aravena, J.; Zhou, K.; Li, X.R.; Chowdhury, F. Fault tolerant safe flight controller bank 1. *IFAC Proc. Vol.* **2006**, *39*, 807–812. doi:10.3182/20060829-4-CN-2909.00134. [CrossRef]
44. Goodall, R.M. Generalised Design Models for EMS Maglev. In Proceedings of the Maglev 2008—The 20th International Conference on Magnetically Levitated Systems and Linear Drives, San Diego, CA, USA, 15–18 December 2008.
45. Michail, K. Optimised Configuration of Sensing Elements for Control and Fault Tolerance Applied to an Electro-Magnetic Suspension System. Ph.D. Thesis, Loughborough University, Loughborough, UK, 2009.
46. Bojan-Dragos, C.A.; Radac, M.B.; Precup, R.E.; Hedrea, E.L.; Tanasoiu, O.M. Gain-scheduling control solutions for magnetic levitation systems. *Acta Polytech. Hung.* **2018**, *15*, 89–108. [CrossRef]
47. Rosa, A.D.; Alfi, S.; Bruni, S. Estimation of lateral and cross alignment in a railway track based on vehicle dynamics measurements. *Mech. Syst. Signal Process.* **2019**, *116*, 606–623. [CrossRef]
48. Goodall, R.M. Dynamics and control requirements for EMS Maglev suspensions. In Proceedings of the International Conference on Maglev, Shanghai, China, 26–28 October 2004; pp. 926–934.

MDPI

Article

On-Line Learning and Updating Unmanned Tracked Vehicle Dynamics

Natalia Strawa †, Dmitry I. Ignatyev *, Argyrios C. Zolotas and Antonios Tsourdos

Centre for Autonomous and Cyber-Physical Systems, SATM, Cranfield University, Cranfield MK43 0AL, UK; strawa.natalia@gmail.com (N.S.); a.zolotas@cranfield.ac.uk (A.C.Z.); a.tsourdos@cranfield.ac.uk (A.T.)
* Correspondence: d.ignatyev@cranfield.ac.uk
† Current Address: Hyper Poland, 03-828 Mazowieckie, Poland.

Abstract: Increasing levels of autonomy impose more pronounced performance requirements for unmanned ground vehicles (UGV). Presence of model uncertainties significantly reduces a ground vehicle performance when the vehicle is traversing an unknown terrain or the vehicle inertial parameters vary due to a mission schedule or external disturbances. A comprehensive mathematical model of a skid steering tracked vehicle is presented in this paper and used to design a control law. Analysis of the controller under model uncertainties in inertial parameters and in the vehicle-terrain interaction revealed undesirable behavior, such as controller divergence and offset from the desired trajectory. A compound identification scheme utilizing an exponential forgetting recursive least square, generalized Newton–Raphson (NR), and Unscented Kalman Filter methods is proposed to estimate the model parameters, such as the vehicle mass and inertia, as well as parameters of the vehicle-terrain interaction, such as slip, resistance coefficients, cohesion, and shear deformation modulus on-line. The proposed identification scheme facilitates adaptive capability for the control system, improves tracking performance and contributes to an adaptive path and trajectory planning framework, which is essential for future autonomous ground vehicle missions.

Keywords: unmanned tracked vehicle; inertial parameters; vehicle-terrain interaction; identification; recursive least square with exponential forgetting; generalized Newton–Raphson; Unscented Kalman Filter

Citation: Strawa, N.; Ignatyev, D.I.; Zolotas, A.C.; Tsourdos, A. On-Line Learning and Updating Unmanned Tracked Vehicle Dynamics. *Electronics* **2021**, *10*, 187. https://doi.org/10.3390/electronics10020187

Received: 13 October 2020
Accepted: 11 January 2021
Published: 15 January 2021

Publisher's Note: MDPI stays neutral with regard to jurisdictional claims in published maps and institutional affiliations.

1. Introduction

The growing interest in the development of autonomous platforms establishes new research directions that address challenges associated with autonomous systems operations. Control of unmanned ground vehicles is a complex task due to vehicle-ground interactions that can strongly influence dynamics of the system and, at the same time, are very difficult to be measured or estimated analytically. Most of the important parameters may vary because a vehicle can cross different types of terrains during motion, mass and inertia can vary due to mission plan, etc. All these factors tend to influence control performance negatively and, as a result, trajectory tracking capabilities.

Learning and updating vehicle dynamics on the fly can tackle the aforementioned issues and improve the control and trajectory tracking performance, thus providing a framework for adaptive path-planning and control [1–4].

Usage of tracked vehicles in military, rescue, agricultural and recreational missions where terrain conditions are difficult or unpredictable is common. This is because tracked vehicles perform better than wheeled vehicles due to the larger contact area of tracks which provides better flotation and better mobility over unprepared terrain [3,5]. Hence, tracked vehicles are preferred choice for autonomous off-road tasks.

Skid steering can be characterized by an absence of a separate steering system, like, for example, in the Ackermann steering system. Advantages provided by the skid-steered scheme are a simple mechanical structure, robustness and high maneuverability [6].

Skid-steering, although energy inefficient, is commonly used for mobile robots (both wheeled and tracked) with the requirement of good mobility.

Skid-steering is performed by controlling relative velocities of the drives on the left and right sides of the vehicle; hence, track slippage causes turning. In this case, the motion of the tracked vehicle is governed by the two longitudinal track forces and the lateral friction force. Due to the fact that the friction force depends on the linear and angular velocities, the force equilibrium equation perpendicular to the tracks becomes a non-integrable differential equality constraint [3]. Maneuvering capabilities also depend on the complex vehicle—ground interaction [7].

For identification of varying parameters in real-time applications, kinematic models of vehicles seems more attractive as they put a less computational burden on the on-board instrumentation at the expense of unmodeled dynamic effects. Unfortunately, such simplification might be especially inefficient in case of skid-steering due to significant slippage. Some authors addressed this problem in their researches. For example, Wu et al. [7] proposed a method for experimental estimation of the wheeled vehicle kinematics. In their work, they applied approximating function with identifiable parameters to derive the relationship between the instantaneous center of rotation of the vehicle, its speed and path curvature. A different approach was studied by Sutoh et al. [8], who experimentally obtained the relationship between a rover's wheels input and output velocities for various types of loose terrain such as silica or gravel.

The biggest challenge in the ground vehicle dynamics modeling is a description of the vehicle-ground interaction. For the wheeled vehicles, model obtained experimentally by Pacejka [9] and so-called 'Pacejka's magic formula' is most widely used. In the case of tracked vehicles, the majority of authors applies the work of Wong [5]. A workaround for contact modeling problem was demonstrated in Reference [10], where a tyre-model-free integral control method correcting the wheel slip coefficient in real-time was designed. Many researches, however, utilize the classic dynamics model, including contact forces obtained with analytical functions derived from experimental data. Equations of motion for skid-steering based mobile robots are well-known and can be found to name a few in Reference [6,11–13]. Ahmadi et al. [14] slightly simplified the model by excluding the non-holonomic constraint from the dynamics equations, while Tang et al. [15] considered more complex and generalized track-terrain interactions model. Some advanced methods were also used to describe vehicle-terrain interaction. For example, Economou and Colyer [16] utilized the Fuzzy Logic for modeling of the vehicle-ground interactions based on the experimental results obtained for an electric wheeled skid steer vehicle under steady-state conditions and a variety of motions and surfaces. A neural network was employed to model the steering dynamics of an autonomous vehicle in Reference [17].

The path tracking control problem of a skid-steered vehicle is a fairly well-researched field. Hence, a variety of approaches can be found in the existing literature. Due to the highly nonlinear nature of the considered class of ground vehicles, a common starting point for the trajectory tracking control design is the feedback linearization. Thus, a linear control law can be further applied. The feedback linearization is thoroughly described in Reference [18]. A similar approach was used in Reference [11], and an exponentially stabilizing state feedback was further applied to complete the control design. Feed-forward friction compensation is implemented alongside the feedback linearization in Reference [14]. Their controller is based on the simplified model of the tracked vehicle. The elaborated force-slip relationship is linearized to relate the inputs with the states of the system. A robust recursive LQR design was applied for the mobile robot in Reference [19]. A neural network model of the steering dynamics of an autonomous vehicle developed in Reference [17] was integrated with a Nonlinear Model Predictive Controller to generate feed-forward steering commands. Recently, approaches based on dynamic model of vehicle for optimal path planning and tracking control of unmanned ground vehicles (UGV) were proposed in Reference [20,21]. Moreover, a novel approach introduced in Reference [22] employs a backstepping technique to robustly control the instantaneous center of rotation position of the vehicle so that it relates to the path curvature and the desired vehicle speed. The backstepping method was also utilized by Zou et al. [13], who designed a modified PID computed-torque controller for an unmanned tracked vehicle.

From a practical point of view, the interest of the current research lays in identification of the following parameters: mass, inertia, and slip, as well as the soil parameters, as part of the vehicle-terrain interaction model.

Vehicle mass and inertia estimations become increasingly important with the rising popularity of autonomous vehicles. These parameters are especially vital in the heavy-duty vehicles automation, powertrain and economic cruise control [23,24]. Upon the analysis of approaches presented in the existing literature, exponential forgetting recursive least squares (EF RLS) method is the most widely used for this purpose [23–25] since it is a very powerful and robust method when the system dynamics can be represented in a linear form. A different method for vehicle mass estimation was introduced by Rhode and Gauterin [26], who utilized the total least squares (TLS) regression.

Regarding the slip estimation, although it is possible to measure all the quantities essential for obtaining the slip value from analytical expressions, measurement inaccuracies and errors lead to the results highly inconsistent with the real slip values. As the slip estimation problem has a significant impact on the tracked vehicle control, therefore, many researchers sought the solution.

Dar and Longoria [27] utilized an Extended Kalman Filter (EKF) with state noise compensation to estimate slip, trajectory and orientation of a small tracked vehicle. However, some authors question the accuracy of EKF estimation, especially in the case of the fast dynamics. Instead, they suggest employing an Unscented Kalman Filter (UKF) which can provide more accuracy without linearization [28]. In research [29], not only kinematics of the vehicle but also simplified dynamics were included in the UKF design. Alternatively, a sliding mode observer (SMO) was exploited in Reference [30] to obtain an accurate slip estimation. In their comparative studies, the authors proved that both the UKF and SMO yield better results than those obtained with the EKF.

Another important vehicle dynamics parameter is the maximum tractive effort that certain vehicle can develop on certain soil types. This property has a major influence on path planning as it can restrict the maneuverability of the vehicle and has a huge impact on the energy efficiency [5]. For satisfactory estimation of soil parameters the Newton–Raphson (NR) method is widely used [31–33]. In particular, in Reference [31], the authors conducted a comparison between NR and RLS that manifested that the former yields much better estimation accuracy and robustness as RLS tends to diverge quite often as it is more prone to the measurement noise.

UGVs are characterized by nonlinear dynamics and excited by a combination of multiple external and internal (to the system) factors. Even small variation (and/or uncertainties) in parameters of a nonlinear system can cause dramatic departures from a "nominal" case scenario. Although many research works focused on designing algorithms for the identification of a particular parameter of UGV dynamics, there is a lack of studies addressing the whole (integrated) dynamics or, at least, its major part. A rigorous evaluation of the system performance is required, especially, while traversing unknown terrains (as is the case investigated in this paper). Such an integrated estimator-based approach is extremely valuable for the development of advanced autonomous systems utilizing AI tools to estimate and predict vehicle behavior for uncertain in nature terrain environments and generate optimal path [2,34,35].

Thus, the main aim of this research was the development of an integrated system running on-line algorithms estimating different vehicle model parameters. To reach this aim, the simplicity and computational efficiency were prioritized. Influence of uncertainties of the dynamics model used for the control design on the behavior of an autonomous tracked vehicle was studied. On-line system identification algorithms using the EF RLS, UKF, and NR algorithms were implemented for soil and inertial parameters estimation;

their performance was further evaluated. The system considered in this study consists of the tracked vehicle model, trajectory tracking controller, and the identification module.

This paper is organized as follows. Section 2 provides description of the track-terrain interaction model. The kinematic and dynamic models of the tracked vehicle are given in Sections 3 and 4, correspondingly. Overview of the entire system is provided in Section 5. Design of the control system is described in Section 6. The proposed identification framework is demonstrated in Section 7. The obtained results are discussed in Section 8, and, finally, Section 9 concludes the paper.

2. Track-Terrain Interaction Model

The system identification has the following key elements, namely selection of the model structure, experiment design, and parameter estimation. On-line identification is performed automatically, so it is extremely important to have a good understanding of all aspects of the problem [1]. That is why we provide here a detailed description of vehicle dynamics, including vehicle-terrain interaction model.

2.1. Terramechanics

The current section gives a brief introduction into terramechanics to equip with a basic knowledge of the track-terrain interaction, indispensable for the tracked vehicle modeling.

Mobility of the off-road vehicles can be severely limited by properties of the encounter terrain. Terramechanics provides essential knowledge of the mechanical properties of the terrain, as well as its reaction to vehicular loadings.

Certain types of trafficable terrain can be considered as ideal elastoplastic materials [5]. Unless stress level exceeds a given limit, terrain behavior remains in the elastic range (see Figure 1). This assumption is utilized for stress distribution prediction in the soil. When yield stress is reached in the terrain under vehicular load, strain increases and plastic flow is constituted. Transition to plastic flow is known as the failure of soils. Failure criterion can be described with the Mohr-Coulomb theory, which postulates that the following condition must be satisfied at the point of the material to cause its failure [36]:

$$\tau = c + \sigma \tan \phi, \tag{1}$$

where τ is the shear stress, c is cohesion, σ is normal stress on the sheared surface, and ϕ is the angle of internal shearing resistance.

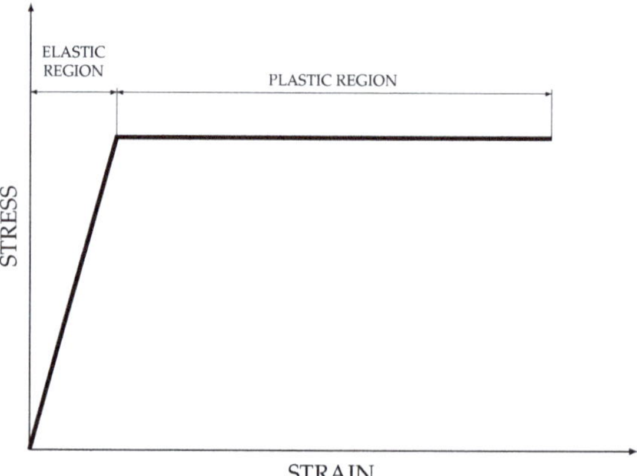

Figure 1. Strain-stress relation of idealized elastoplastic material, adopted from Reference [36].

Cohesive forces bind soil grains together irrespective of normal pressure between the particles, e.g., for saturated clay only cohesion c is presented. However, when grains are not held together by cohesion, they can move upon each other and, while pressed against each other, friction develops, e.g., for dry sand, the shear strength can be expressed by $\sigma \tan\phi$. In practice, the majority of the soils exhibit both plastic and frictional behavior and the shear strength should be hence characterized with Equation (1).

2.2. Tractive Effort and Slip of a Track

For a mathematical description of a ground vehicle motion, a track-terrain interaction should be considered. In the current study, a parametric approach proposed by Bekker [36] is used to do this.

Bekker's method assumes that a track in contact with a terrain is similar to a rigid footing. Moreover, it is assumed in the current research, that the center of gravity (CG) of the tracked vehicle is positioned at the mid-point of the track-terrain contact area. It is also assumed here that a normal pressure produced by the track has an uniform distribution along the track.

During the vehicle motion a torque produced by the motors and applied to the sprocket of the track initiates shearing action on the track-terrain interface, which subsequently results in a development of tractive effort. The maximum tractive effort is bounded by the maximum shear strength of the soil τ_{max} and the track contact area A. Taking into account Equation (1), we can obtain:

$$\begin{aligned} F_{max} &= A\tau_{max} \\ &= A(c + p\tan\phi) \\ &= Ac + W\tan\phi, \end{aligned} \tag{2}$$

where $A = bl$, b is the contact width, l is the contact length, W is the normal load, and c and ϕ are the soil parameters: the cohesion and the angle of internal shearing resistance, respectively. It can be observed that the terrain type critically impacts the maximum shear strength of the soil and, consequently, has great influence on the maximum tractive effort. For instance, as it was mentioned before, dry sand is a frictional soil. Therefore, cohesion is negligible in this case and maximum tractive effort is higher for heavier vehicles. On the other hand, the saturated clay, which is an example of cohesive soil, has a low value of ϕ; hence, mainly the contact area of the track influences the maximum tractive effort value.

It should be pointed out that the tractive effort defined in Equation (2) is a maximum value that the tracked vehicle can develop on a certain terrain. To determine thrust over a full operating range, its relationship with the slip of a track should be examined. The slip of a track is defined as follows [5]:

$$i = 1 - \frac{V}{\omega r} = 1 - \frac{V}{V_t} = \frac{V_t - V}{V_t} = \frac{V_j}{V_t}, \tag{3}$$

where V is the forward speed of a track, and V_t is the theoretical speed defined by the sprocket rotational speed ω and its radius r. Then, V_j is the speed of slip with reference to the ground. Assuming that the track is rigid and cannot stretch, the vehicle is moving along a flat surface with a homogeneous soil property, then every point of the track, which is in contact with the terrain, has the same speed V_j. Therefore, the shear displacement j at a distance x from the front of the track-terrain contact area can be found with the following equation:

$$j = V_j t, \tag{4}$$

with $t = x/V_t$ being the contact time of a considered point and the terrain. Thus, Equation (4) can be rearranged:

$$j = \frac{V_j x}{V_t} = ix. \tag{5}$$

Equation (5) indicates that the shear displacement increases linearly with the distance from the front of the contact area (see Figure 2).

$$i_L = 1 - \frac{v_{t,L}}{\omega_L r} = 1 - \frac{\dot{x} + (b/2)\dot{\theta}}{\omega_L r},$$

$$i_R = 1 - \frac{v_{t,R}}{\omega_L r} = 1 - \frac{\dot{x} - (b/2)\dot{\theta}}{\omega_R r},$$

and the slip angle can be computed as follows:

$$\alpha = \arctan\frac{\dot{y}}{\dot{x}}.$$

Taking into account Equations (8) and (9), the turning radius R, with slip considered, can be obtained:

$$R = \frac{V}{\dot{\theta}} = \frac{b}{2cos\alpha}\frac{\omega_L(1-i_L) + \omega_R(1-i_R)}{\omega_L(1-i_L) - \omega_R(1-i_R)}.$$

Since velocity in the body frame can be expressed through speed and slip angle the following equation can be obtained:

$$\begin{bmatrix}\dot{x}\\ \dot{y}\end{bmatrix} = V\begin{bmatrix}cos\alpha\\ sin\alpha\end{bmatrix}.$$

Therefore, this approach can be further projected to the inertial frame

$$\begin{bmatrix}\dot{X}\\ \dot{Y}\end{bmatrix} = V\begin{bmatrix}c\theta c\alpha - s\theta s\alpha\\ s\theta c\alpha + c\theta s\alpha\end{bmatrix},$$

and, subsequently, one can relate the velocity in the $\mathcal{I}$ frame with the rotational velocities of the sprockets and slips of the tracks, which yields a complete kinematic model of the tracked vehicle [13]:

$$\dot{X} = \frac{r}{2}[\omega_L(1-i_L) + \omega_R(1-i_R)][c\theta - s\theta\tan\alpha], \tag{10}$$

$$\dot{Y} = \frac{r}{2}[\omega_L(1-i_L) + \omega_R(1-i_R)][s\theta + c\theta\tan\alpha], \tag{11}$$

$$\dot{\theta} = \frac{r}{b}[\omega_L(1-i_L) - \omega_R(1-i_R)]. \tag{12}$$

To finalize the mathematical description of the kinematics model of the tracked vehicle, nonholonomic constraint should be imposed [12,14]. The arbitrary planar motion of a body can be represented as a rotation around the instantaneous center of rotation (ICR) [38]. This concept applied to the tracked vehicle is demonstrated in Figure 4. ICR of the tracked vehicle is denoted by O', while the rotation radius vectors, defined in the body-fixed frame and directed from the ICR, are: $\mathbf{d}_{t,i} = \begin{bmatrix}d_{t,ix} & d_{t,iy}\end{bmatrix}^T$, where $i = \{L, R\}$, and $\mathbf{d}_C = \begin{bmatrix}d_{Cx} & d_{Cy}\end{bmatrix}^T$. From the definition of ICR, we can obtain

$$\frac{\|\mathbf{v}_{t,i}\|}{\|\mathbf{d}_{t,i}\|} = \frac{\|\mathbf{V}\|}{\|\mathbf{d}_C\|} = |\omega|, \tag{13}$$

where $\|\cdot\|$ is the Euclidean norm. Equation (13) can be further transformed to the expanded form

$$\frac{v_{t,ix}}{-d_{t,iy}} = \frac{V_x}{-d_{Cy}} = \frac{v_{t,iy}}{d_{t,ix}} = \frac{V_y}{d_{Cx}} = \omega. \tag{14}$$

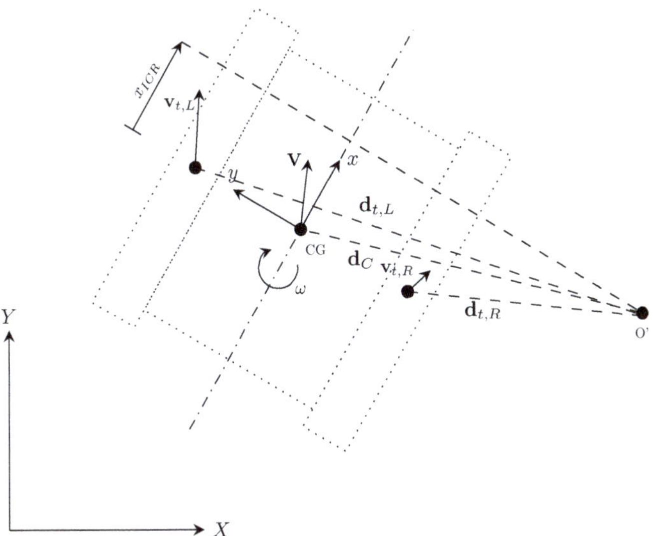

Figure 4. Instantaneous center of rotation geometry.

Coordinates of the ICR can be defined in the local frame as

$$ICR = (x_{ICR}, y_{ICR}) = \left(-d_{Cx}, -d_{Cy}\right). \tag{15}$$

Thus, Equation (14) can be rewritten in the following way:

$$\frac{V_x}{y_{ICR}} = -\frac{V_y}{x_{ICR}} = \omega. \tag{16}$$

Note that, due to the planar motion of the vehicle, $\omega = \dot{\theta}$.

The ICR coordinates can be obtained from Equation (16) as follows:

$$\begin{bmatrix} x_{ICR} \\ y_{ICR} \end{bmatrix} = \begin{bmatrix} -\dot{y}/\dot{\theta} \\ \dot{x}/\dot{\theta} \end{bmatrix}.$$

In a case of straight line motion, both the lateral velocity $\dot{y}$ and the angular velocity $\dot{\theta}$ vanish; thus, the ICR shifts to infinity along the y-axis. During turning maneuvers, the ICR moves along x-axis by an amount of $|x_{ICR}|$. Shift of the x_{ICR} beyond the vehicle geometry causes loss of motion stability.

Following Reference [11], we impose the nonholonomic constraint obtained from Equation (16)

$$V_y + x_{ICR}\dot{\theta} = 0.$$

The above expression can be presented in Pfaffian form [39]

$$\begin{bmatrix} -\sin\theta & \cos\theta & x_{ICR} \end{bmatrix} \begin{bmatrix} \dot{X} \\ \dot{Y} \\ \dot{\theta} \end{bmatrix} = \mathbf{A}(\mathbf{q})\dot{\mathbf{q}} = \mathbf{0}. \tag{17}$$

Now, we choose a full-rank matrix $\mathbf{S}(\mathbf{q}) \in \mathbb{R}^{3x2}$, in which columns are in null space of $\mathbf{A}(\mathbf{q})$, that is

$$\mathbf{S}^T(\mathbf{q})\mathbf{A}(\mathbf{q}) = \mathbf{0},$$

and $\mathbf{S}(\mathbf{q})$ can be, for example, defined as follows:

$$\mathbf{S}(\mathbf{q}) = \begin{bmatrix} c\theta & x_{ICR}s\theta \\ s\theta & -x_{ICR}c\theta \\ 0 & 1 \end{bmatrix}.$$

Then, it is possible to define the generalized velocities by means of $\mathbf{S}$ and an auxiliary vector $\boldsymbol{v}(t) \in \mathbb{R}^2$

$$\dot{\mathbf{q}} = \mathbf{S}(\mathbf{q})\boldsymbol{v}(t), \tag{18}$$

where the auxiliary vector is $\boldsymbol{v}(t) = \begin{bmatrix} V & \dot{\theta} \end{bmatrix}^T$.

4. Dynamics

In this section, we provide a model of tracked vehicle dynamics for a comprehensive description of the vehicle dynamics required for the development of the controller and identification scheme.

4.1. Forces and Moments Acting on the Tracked Vehicle

In the current analysis, we assume that the service brake is not applied, and the friction brakeforce is not generated. Additionally, the aerodynamic loads are neglected since the vehicle travels with a very low speed and the cross-sectional area of the vehicle is assumed to be small.

A moving vehicle is subjected only to track-terrain interaction forces, which are illustrated in Figure 5 and can be classified as follows:

- tractive forces $\mathbf{F}_R$ and $\mathbf{F}_L$,
- longitudinal resistance forces $\mathbf{R}_R$ and $\mathbf{R}_L$,
- lateral forces $\mathbf{F}_{y,R}$ and $\mathbf{F}_{y,L}$, and
- moment of turning resistance $\mathbf{M}_r$ induced by the resistive forces.

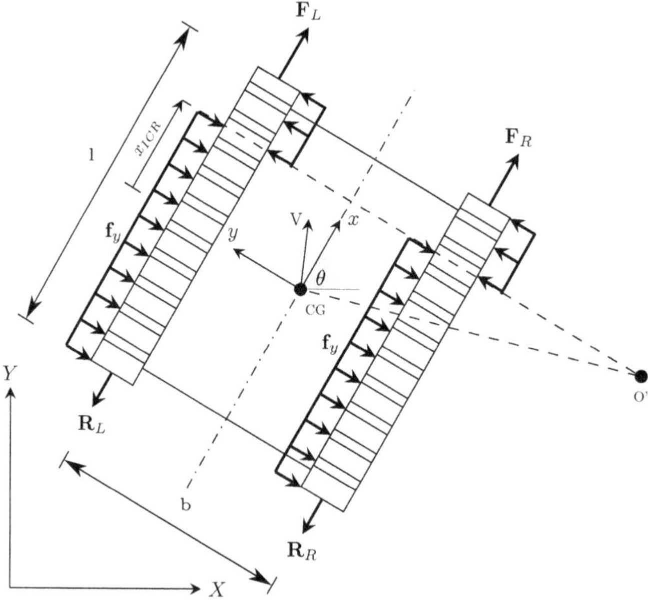

Figure 5. Free-body diagram.

4.1.1. Tractive Force

The concept of tractive force has been already discussed in Section 2.1. In this study, it is assumed that rotation of the tracks, which results in the development of traction, is caused by the torque transmitted to the sprockets from a pair of DC motors—one per each track of the vehicle. Transmission factor is assumed to be ideal.

4.1.2. Longitudinal and Lateral Resistance Forces

Friction term for longitudinal resistance force can be calculated with the following expression [5]:

$$r_l = \frac{\mu_l m g}{2},$$

where μ_l is the coefficient of longitudinal resistance, and g is the gravitational acceleration. To define the direction of the longitudinal friction forces, the following function is employed [14]:

$$G(F, f, \dot{x}) = \begin{cases} -f sign(\dot{x}), & \dot{x} \neq 0, \\ 0, & \dot{x} = 0, \ |F| \leq f, \\ -f sign(F), & \dot{x} = 0, \ |F| > f, \end{cases} \tag{19}$$

where $sign(\cdot)$ denotes the signum function.

Then, the longitudinal resistance force can be defined as follows:

$$R_{l,i} = G(F_i, r_l, v_{t,i}), \qquad i = \{R, L\}.$$

The lateral friction distribution and force can be obtained according the following equations:

$$f_y = \frac{\mu_t m g}{l}, \qquad F_y = -2 sign(\dot{y}) f_y x_{ICR},$$

where μ_t denotes the coefficient of lateral resistance.

4.1.3. Turning Moment and Moment of Turning Resistance

The turning moment is induced by the forces acting in the longitudinal direction, i.e.,

$$M = (F_L - R_{l,L})\frac{b}{2} - (F_R - R_{l,R})\frac{b}{2} = (F_L - F_R)\frac{b}{2}.$$

The resistance moment can be obtained by integrating over the track length of the distribution given by the following expression from Reference [3]:

$$m_r = 2 f_y \left(\frac{l^2}{4} - x_{ICR}^2 \right) \tag{20}$$

Direction of the moment of turning resistance can be again determined with Equation (19), namely

$$M_r = G(M, m_r, \dot{\theta}). \tag{21}$$

4.1.4. Drive Model

In this study, it is assumed that the vehicle is actuated with two DC motors that drive the sprockets through the transmission gear and are controlled with a simple PID controller. The relationship between the motor torque τ and the rotor current i_a is considered to be linear [12]:

$$\tau_m = k_m i_a,$$

where k_m is a motor torque constant. Voltage and current in the motor circuits can be approximated with the differential Equation [12]:

$$u_a = L_a \frac{d}{dt} i_a + R_a i_a + k_e \omega_m,$$

where L_a and R_a are the inductance and resistance of the motor, respectively, k_e denotes the electromotive force coefficient, and ω_m is the angular velocity at the output. Under the assumption that the transmission is ideal, we can write:

$$\tau = nk_m i_a,$$
$$\omega_m = n\omega_s,$$

where n is the transmission ratio, and ω_s is the rotational speed of sprocket.

Assuming that the voltage u_a is the motor control input, the following equations can describe the powertrain system:

$$\boldsymbol{\tau} = nk_m \mathbf{i}_a,$$
$$\mathbf{u}_a = L_a \frac{d}{dt}\mathbf{i}_a + R_a \mathbf{i} + k_e \boldsymbol{\omega}_m,$$

where $\boldsymbol{\tau} = \begin{bmatrix}\tau_L & \tau_R\end{bmatrix}^T$, $\mathbf{u}_a = \begin{bmatrix}u_L & u_R\end{bmatrix}^T$, $\mathbf{i}_a = \begin{bmatrix}i_L & i_R\end{bmatrix}^T$, and $\boldsymbol{\omega} = \begin{bmatrix}\omega_L & \omega_R\end{bmatrix}^T$.

4.2. Equations of Motion

Taking into account the nonholonomic constraint, we can obtain the equations of motion of the tracked vehicle through Lagrange-Euler formula with Lagrange's multipliers similar to approach given in Reference [40]:

$$\frac{d}{dt}\left(\frac{\partial L}{\partial \dot{\mathbf{q}}}\right) - \frac{\partial L}{\partial \mathbf{q}} + \mathbf{A}(\mathbf{q})\boldsymbol{\lambda} = \mathbf{Q}, \tag{22}$$

where $\boldsymbol{\lambda}$ is vector of the Lagrange's multipliers, $\mathbf{A}(\mathbf{q})$ is the nonholonomic constraint defined in Equation (17), $\mathbf{Q}$ is the vector of generalized forces, and $L(\dot{\mathbf{q}}, \mathbf{q}, t)$ is the Lagrangian defined as

$$L = E_T - E_V, \tag{23}$$

where E_T and E_V are the kinetic and potential energy, respectively.

First, we obtain the Lagrangian of the system. Since it is assumed that the vehicle is in planar motion, it can be assumed that $E_V = 0$ and Equation (23) takes the following form:

$$L(\dot{\mathbf{q}}, \mathbf{q}, t) = E_T(\dot{\mathbf{q}}, \mathbf{q}).$$

Assuming that the energy of rotating tracks can be neglected, the kinetic energy of the system is given by

$$E_T = \frac{1}{2}m\mathbf{v}^T\mathbf{v} + \frac{1}{2}I\omega^2, \tag{24}$$

where m is mass of the vehicle, and I denotes its moment of inertia about the COM. As $\mathbf{v}^T\mathbf{v} = \|V\|^2$, and the value of velocity magnitude is independent of the reference frame, one can rewrite Equation (24) in the following form

$$E_T = \frac{1}{2}m\left(\dot{X}^2 + \dot{Y}^2\right) + \frac{1}{2}I\dot{\theta}^2. \tag{25}$$

Hence, the derivatives of kinetic energy can be computed as

$$\frac{d}{dt}\left(\frac{\partial E_T}{\partial \dot{\mathbf{q}}}\right) = \begin{bmatrix} m\ddot{X} \\ m\ddot{Y} \\ I\ddot{\theta} \end{bmatrix} = \mathbf{M}\ddot{\mathbf{q}},$$

$$\frac{\partial E_T}{\partial \mathbf{q}} = \mathbf{0},$$

where

$$\mathbf{M} = \begin{bmatrix} m & 0 & 0 \\ 0 & m & 0 \\ 0 & 0 & I \end{bmatrix}.$$

Vector of the generalized forces can be decomposed into actuating forces generated by motors and resistive forces causing dissipation of energy, which yields

$$\mathbf{Q} = \mathbf{B}(\mathbf{q})\boldsymbol{\tau} - \mathbf{C}(\mathbf{q}, \dot{\mathbf{q}}),$$

with

$$\mathbf{C}(\mathbf{q}, \dot{\mathbf{q}}) = \begin{bmatrix} 2R_l c\theta + F_y s\theta \\ 2R_l s\theta - F_y c\theta \\ M_r \end{bmatrix}, \quad \mathbf{B}(\mathbf{q}) = \frac{1}{r}\begin{bmatrix} c\theta & c\theta \\ s\theta & s\theta \\ b/2 & -b/2 \end{bmatrix}, \quad \boldsymbol{\tau} = \begin{bmatrix} \tau_L \\ \tau_R \end{bmatrix},$$

where r is the sprocket radius, and τ_L, τ_R are the torques provided by the left and right motors, respectively. Having all terms of Equation (22) defined, the mathematical model of system dynamics is obtained [11,13]:

$$\mathbf{M}\ddot{\mathbf{q}} + \mathbf{C}(\mathbf{q}, \dot{\mathbf{q}}) + \mathbf{A}(\mathbf{q})\boldsymbol{\lambda} = \mathbf{B}(\mathbf{q})\boldsymbol{\tau}. \tag{26}$$

For control purposes, it is convenient to express the generalized velocities $\dot{\mathbf{q}}$ in terms of pseudo-velocity $v(t)$. Differentiating Equation (18), we have

$$\ddot{\mathbf{q}} = \dot{\mathbf{S}}(\mathbf{q})v(t) + \mathbf{S}(\mathbf{q})\dot{v}(t) \tag{27}$$

Next, Equations (18) and (27) are substituted into Equation (26), and both sides of the obtained equation are multiplied by $\mathbf{S}^T(\mathbf{q})$, which leads to the modified mathematical description of the system dynamics

$$\begin{aligned} \dot{\mathbf{q}} &= \mathbf{S}(\mathbf{q})v(t), \\ \dot{v}(t) &= \widetilde{\mathbf{M}}^{-1}\Big(\widetilde{\mathbf{B}}(\mathbf{q})\boldsymbol{\tau} - \widetilde{\mathbf{E}}(\mathbf{q}, \dot{\mathbf{q}})v(t) - \widetilde{\mathbf{C}}(\mathbf{q}, \dot{\mathbf{q}})\Big), \end{aligned} \tag{28}$$

where

$$\widetilde{\mathbf{M}} = \mathbf{S}^T\mathbf{M}\mathbf{S} = \begin{bmatrix} m & 0 \\ 0 & mx_{ICR}^2 + I \end{bmatrix}, \quad \widetilde{\mathbf{E}}(\mathbf{q}, \dot{\mathbf{q}}) = \mathbf{S}^T\mathbf{M}\dot{\mathbf{S}} = \begin{bmatrix} 0 & mx_{ICR}\dot{\theta} \\ -mx_{ICR}\dot{\theta} & mx_{ICR}\dot{x}_{ICR} \end{bmatrix},$$

$$\widetilde{\mathbf{C}}(\mathbf{q}, \dot{\mathbf{q}}) = \mathbf{S}^T\mathbf{C} = \begin{bmatrix} 2R_l \\ F_y x_{ICR} + M_r \end{bmatrix}, \quad \widetilde{\mathbf{B}}(\mathbf{q}) = \mathbf{S}^T\mathbf{B} = \frac{1}{r}\begin{bmatrix} 1 & 1 \\ -b/2 & b/2 \end{bmatrix}.$$

5. System Overview

The introduced kinematics and dynamics of the tracked vehicle are used to design the control system and the system identification framework. This section aims to provide a high-level overview of the overall system design. In the beginning, a block diagram of the system is presented, and a brief description of each subsystem and their interfaces is provided. Furthermore, the measurement system is described.

5.1. Block Diagram of the System

The overall system consists of five sub-modules representing different functionalities. Figure 6 shows the block diagram of the system.

The base element of the whole structure is the tracked vehicle platform represented in the diagram with the vehicle model block. The model is described in Sections 2–4.

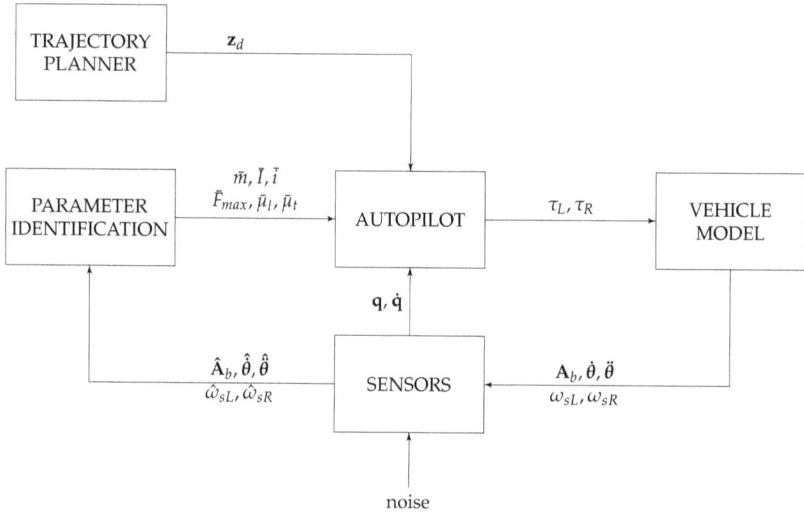

Figure 6. Block diagram of the tracked vehicle system.

The vehicle is controlled with an autopilot sending computed torque demand to the motor controller. The autopilot obtains a reference trajectory from the trajectory planer, which, in this research, was composed of a couple of the predefined trajectories. At the same time, information about the current vehicle state is fed to the autopilot from the on-board sensors. The autopilot computes the control inputs based on the difference between the desired trajectory and the actual state of the vehicle.

Concurrently, measurements obtained with sensors are provided to the parameter identification framework. The updated values of the system parameters are estimated and passed to the autopilot to improve the trajectory tracking performance of the platform.

The sensors block function is introduced to mimic the behavior of real sensors, including dynamics and noise, providing more realistic simulations. More detailed discussion on the sensors can be found in the following section.

5.2. Measurements and Sensors

The control scheme, as well as the identification process, requires knowledge about the current system state. Such information can be obtained through measurement of observable states. Today, there is a wide choice of measurement instrumentation in the market, varying in quality and price. In this research, the authors aimed to choose the most affordable solution possible.

5.2.1. Position, Velocity and Acceleration

One of the most popular devices utilized to obtain information about the vehicle motion is the inertial measurement unit (IMU), which combines multiple sensors, such as accelerometers, gyroscopes, and magnetometers. At the output, IMU provides the vehicle acceleration and rotation rate, that can be further integrated to obtain the velocities and position in space. Although IMU measurements are quite accurate and provide information at high frequency, they suffer from accumulating error in time. IMUs are prone to the influence of noise (needed signals (speed and acceleration) for pose computation are derivations of the base signal), which accelerates the error accumulation. Odometry as an inertial navigation system (INS) that uses IMU measurements, for example, accumulates the pose error with the square of the traversed distance.

To improve the accuracy of IMU measurements, an inertial navigation system (INS) combined with GPS technology can be introduced. GPS measurements are generally less accurate and sampled at lower frequencies. However, it is quite a common strategy to combine data from different sources each having its errors and sampling frequencies to get better estimations [41]. In GPS/INS, the data obtained from IMU is fused with the GPS measurements using Kalman Filter to correct the IMU bias and increase the overall system accuracy [42].

Identification of the moment of inertia requires information on the vehicle angular acceleration. This can be obtained with two accelerometers mounted at different distances from the rotation center [43]. Let us assume that the accelerometers are placed at distances r_1 and r_2 from the vehicle center of rotation, respectively (see Figure 7), and a_{y1} and a_{y2} denote the tangential accelerations measured with the accelerometers. Distance between the devices can be computed as $D = r_2 - r_1$, and then the angular acceleration is given by $\ddot{\theta} = (a_{y1} - a_{y2})/D$.

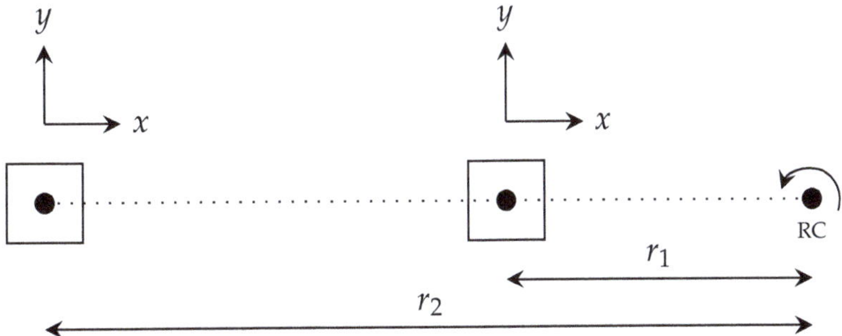

Figure 7. Positioning of the accelerometers for rotational measurements.

In the simulation, the INS/GPS sensors are modeled with the IMU and INS/GPS objects available with MATLAB® Sensor Fusion and Tracking Toolbox. INS/GPS sensor noise is modeled as a white noise process. For the simulations conducted in this research, the following INS/GPS parameters are set:

- yaw accuracy: σ = 1 deg,
- position accuracy: σ = 1 m, and
- velocity accuracy: σ = 0.1 m/s.

For the IMU sensor, accelerometer and gyroscope performance is also specified. Values of the parameters employed in the simulation were found in the datasheets of the IMU sensors available in the market

- gyro bias: 0.01 rad/s,
- accelerometer bias: 0.002 m/s^2, and
- axis cross-coupling: 0.001.

Assuming that s_{real} is the real signal value and s_{meas} is the measured value, then the measurement error is given by $s = s_{real} - s_{meas}$.

5.2.2. Rotational Velocity of the Sprockets and Motor Torque

Another quantity that needs to be measured in the system is the rotational speed of the tracks. There are two possible ways of obtaining this information. First, the shaft decoders can be utilized. The second approach is to estimate the rotation rate from the motor current measurement (applicable when the electric motors actuate the system). Torque and, consequently, rotational speed measurement can be performed with sensors available on the market.

In the simulation environment, sprocket rotational speed, as well as torque measurements, are obtained by introducing the additive white noise to the computed signal value. In Simulink, the noise is incorporated into the signal with *AWGN Channel* block that can be found in Communications Toolbox. In this study, we defined the variance of white noise added to the input signals. For both torque and sprocket rotational speed, the variance was set to $\sigma^2 = (0.1)^2$.

6. Control

In this chapter, the derivation of the control scheme for the tracked vehicle is described. Our control objective is to accurately guide the vehicle along the desired trajectory. Below, a thorough description of the employed method is presented.

6.1. Static State Feedback

The tracked vehicle is a nonlinear system, therefore, to be able to apply linear control techniques, it is vital to algebraically transform a nonlinear dynamics into a linear form.

First, nonlinear static state feedback is applied to compensate for the system dynamics. From Equation (28), we can obtain the expression for torque:

$$\boldsymbol{\tau} = \widetilde{\mathbf{B}}^{-1}(\mathbf{q})\left(\widetilde{\mathbf{M}}\dot{\boldsymbol{v}} + \widetilde{\mathbf{E}}(\mathbf{q},\dot{\mathbf{q}})\boldsymbol{v}(t) + \widetilde{\mathbf{C}}(\mathbf{q},\dot{\mathbf{q}})\right).$$

Following Reference [11], we choose the new control variable $\boldsymbol{u} = \begin{bmatrix} u_1 & u_2 \end{bmatrix}^T$ and transform the torque control signal so that input $\boldsymbol{u}$ becomes proportional to the system acceleration response $\dot{\boldsymbol{v}}$, namely

$$\boldsymbol{\tau} = \widetilde{\mathbf{B}}^{-1}\left(\widetilde{\mathbf{M}}\mathbf{u} + \widetilde{\mathbf{E}}\boldsymbol{v} + \widetilde{\mathbf{C}}\right),$$

and results in a torque control signal of the following explicit form:

$$\begin{bmatrix} \tau_1 \\ \tau_2 \end{bmatrix} = \begin{bmatrix} \frac{r}{2}(mu_1 + mv_2^2 + 2R_l) \\ +\frac{r}{b}\left((mx_{ICR}^2 + I)u_2 - mx_{ICR}v_1v_2 + mx_{ICR}\dot{x}_{ICR}v_2 + F_y + M_r\right) \\ \\ \frac{r}{2}(mu_1 + mv_2^2 + 2R_l) \\ -\frac{r}{b}\left((mx_{ICR}^2 + I)u_2 - mx_{ICR}v_1v_2 + mx_{ICR}\dot{x}_{ICR}v_2 + F_y + M_r\right) \end{bmatrix}. \quad (29)$$

Thus, a second-order kinematic model is obtained:

$$\begin{aligned} \dot{\mathbf{q}} &= \mathbf{S}\boldsymbol{v}, \\ \dot{\boldsymbol{v}} &= \mathbf{u}, \end{aligned}$$

which gives:

$$\begin{aligned} \dot{X} &= v_1\cos\theta + v_2x_{ICR}\sin\theta, \\ \dot{Y} &= v_1\sin\theta - v_2x_{ICR}\cos\theta, \\ \dot{\theta} &= v_2, \\ \dot{v}_1 &= u_1, \\ \dot{v}_2 &= u_2. \end{aligned}$$

6.2. Input-Output Linearization

In the next step, the dynamic state feedback is applied so that the system becomes input-output decoupled.

Let us consider a smooth affine nonlinear system:

$$\begin{aligned} \dot{x} &= f(x) + G(x)u, \\ z &= h(x), \end{aligned}$$

with x, u, and z being the system state, input, and output, respectively. Moreover, we assume that the number of inputs and outputs is equal. In case of linearization via static feedback, we seek for the control law of the following form [18]

$$u = a(x) + B(x)r,$$

where r is an external auxiliary input of the same size as u, and $B(x)$ is non-singular decoupling matrix. Sometimes it is not possible to solve the problem by means of static feedback. In such case, a dynamic feedback compensator might be successfully utilized: [18]

$$\begin{aligned} u &= a(x,\xi) + B(x,\xi)r, \\ \dot{\xi} &= c(x,\xi) + D(x,\xi)r, \end{aligned}$$

where ξ is the state compensator.

In the case of the tracked vehicle, a new set of linearizing outputs needs to be chosen for the decoupling matrix to be non-singular [11]. Hence, we choose to observe the position of a point D located on the x-axis at distance d from the body-frame origin. Therefore, the coordinates of the point D in the inertial frame are given by

$$\mathbf{z} = \begin{bmatrix} X + d\cos\theta \\ Y + d\sin\theta \end{bmatrix}. \tag{30}$$

Then, the dynamic extension is introduced on the input u_1

$$\begin{aligned} u_1 &= \xi, \\ \dot{\xi} &= \eta_1, \\ u_2 &= \eta_2 \end{aligned} \tag{31}$$

with η_1, η_2 being the new control inputs.

In the input-output decoupling algorithm, the output $\mathbf{z}$ is differentiated until the new input $\boldsymbol{\eta}$ appears explicitly in the equations

$$\dot{\mathbf{z}} = \begin{bmatrix} \dot{X} - ds\theta\dot{\theta} \\ \dot{Y} + dc\theta\dot{\theta} \end{bmatrix} = \begin{bmatrix} v_1 c\theta + v_2 x_{ICR} s\theta - ds\theta v_2 \\ v_1 s\theta - v_2 x_{ICR} c\theta + dc\theta v_2 \end{bmatrix} = \begin{bmatrix} v_1 c\theta \\ v_1 s\theta \end{bmatrix}, \tag{32a}$$

$$\ddot{\mathbf{z}} = \begin{bmatrix} \dot{v}_1 c\theta - v_1 s\theta\dot{\theta} \\ \dot{v}_1 s\theta + v_1 c\theta\dot{\theta} \end{bmatrix} = \begin{bmatrix} u_1 c\theta - v_1 v_2 s\theta \\ u_1 s\theta + v_1 v_2 c\theta \end{bmatrix}, \tag{32b}$$

$$\begin{aligned} \dddot{\mathbf{z}} &= \begin{bmatrix} \dot{u}_1 c\theta - u_1 v_2 s\theta - \dot{v}_1 v_2 s\theta - v_1 \dot{v}_2 s\theta - v_1 v_2 c\theta\dot{\theta} \\ \dot{u}_1 s\theta + u_1 v_2 c\theta + \dot{v}_1 v_2 c\theta + v_1 \dot{v}_2 c\theta - v_1 v_2 s\theta\dot{\theta} \end{bmatrix} \\ &= \begin{bmatrix} \eta_1 c\theta - 2\xi v_2 s\theta - v_1 \eta_2 s\theta - v_1 v_2^2 c\theta \\ \eta_1 s\theta + 2\xi v_2 c\theta + v_1 \eta_2 c\theta - v_1 v_2^2 s\theta \end{bmatrix}. \end{aligned} \tag{32c}$$

Subsequently, Equation (32c) can be rearranged to a more convenient form:

$$\begin{aligned} \dddot{\mathbf{z}} &= \begin{bmatrix} c\theta & -v_1 s\theta \\ s\theta & v_1 c\theta \end{bmatrix} \boldsymbol{\eta} + \begin{bmatrix} -2\xi v_2 s\theta - v_1 v_2^2 c\theta \\ 2\xi v_2 c\theta - v_1 v_2^2 s\theta \end{bmatrix} \\ &= \alpha(\mathbf{q}, \boldsymbol{v})\boldsymbol{\eta} + \beta(\mathbf{q}, \boldsymbol{v}, \xi). \end{aligned}$$

Further, we rearrange the above equation to obtain the control law

$$\boldsymbol{\eta} = \alpha^{-1}(\mathbf{q}, \boldsymbol{v})[\mathbf{r} - \beta(\mathbf{q}, \boldsymbol{v}, \xi)], \tag{33}$$

where we introduce the trajectory jerk reference $\mathbf{r} = \dddot{\mathbf{z}}$. From the determinant of α

$$\det[\alpha(\mathbf{q}, \boldsymbol{v})] = v_1,$$

we can conclude that the decoupling matrix is non-singular apart from the situation when the linear speed of the vehicle is zero, i.e., $v_1 = 0$, which does not negatively influence the tracking performance for the persistent trajectories.

Combining Equation (31) with Equation (33) yields the final form of fully linearizing, input-output decoupling controller:

$$\begin{aligned} \dot{\xi} &= r_1 \cos\theta + r_2 \sin\theta + v_1 v_2^2, \\ u_1 &= \xi, \\ u_2 &= \frac{1}{v_1}(-r_1 \sin\theta + r_2 \cos\theta) - 2\xi\frac{v_2}{v_1}. \end{aligned} \tag{34}$$

6.3. Exponentially Stabilizing Feedback for Tracking

Once full-state linearization is obtained, the control design can be completed with a globally stabilizing feedback for the desired smooth trajectory $z_d(t)$ [18]:

$$r_i = \dddot{z}_{di} + k_{ai}(\ddot{z}_{di} - \ddot{z}_i) + k_{vi}(\dot{z}_{di} - \dot{z}_i) + k_{pi}(z_{di} - z_i), \qquad i = \{1,2\}.$$

The feedback gains are chosen so that the polynomials,

$$\lambda^3 + k_{ai}\lambda^2 + k_{vi}\lambda + k_{pi}, \tag{35}$$

are Hurwitz polynomials, and $\mathbf{z}$, $\dot{\mathbf{z}}$, $\ddot{\mathbf{z}}$ can be computed with Equations (30) and (32a–c). Thus, we obtained the control scheme that results in the following open-loop transfer function of the system [11]

$$F(s) = C(s)P(s) = \frac{k_a s^2 + k_v s + k_p}{s^3}.$$

The block diagram of the proposed controller is illustrated in Figure 8.

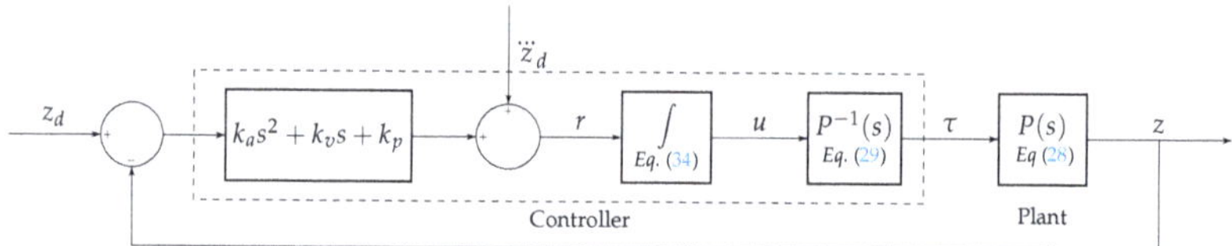

Figure 8. Block diagram of the trajectory tracking control system.

It should be noted that Hurwitz polynomials guarantee the stability of the developed controller. In the case of perfect knowledge of the vehicle dynamics model, the proposed control law perfectly performs trajectory tracking. In case of uncertain model dynamics parameters, the vehicle closed-loop dynamics will be nonlinear and coupled. The effect of the estimation error is a torque disturbance that could be rejected by the controller augmented with adaptive capabilities and by avoiding unfeasible commands via proper path planning.

7. Parameter Identification

The vehicle mass, inertia, and terrain behavior can vary due to mission operations or moving through unknown terrain. Imprecise knowledge of the vehicle model can lead to poor control performance. However, model parameters can be estimated on-line and provided to the controller and/or the trajectory planner to improve the system performance.

Selection of the model structure and parametrization are crucial steps of the system identification. Within this research, the models of ground vehicle dynamics and vehicle-terrain interaction described above are used for parametrization. In particular, values of the vehicle mass, inertia, and also parameters of the vehicle-terrain interaction, such as slip, resistance coefficients, cohesion F_{max}, and shear deformation modulus K, are estimated on the fly. Analysis of the model structure enabled selection of the proper estimation technique for each of the mentioned parameters.

7.1. Estimations of Mass, Inertia and Motion Resistance

The EF RLS is a common approach for estimation of mass, inertia, and road grade based on the model of longitudinal and rotational dynamics of the ground vehicle [23,24], and it was applied to estimate mass, inertia, and motion resistance coefficients in the current research.

Let us first give a brief overview of the system identification methods used in the current research.

7.1.1. Recursive Least Squares with Exponential Forgetting

Recursive Least Squares with Exponential Forgetting (RLS) algorithm is the most common approach since it is simple and can be applied to a large variety of on-line estimation problems.

Assume that a process is generated by the following model:

$$y(t) = \boldsymbol{\phi}^T(t)\boldsymbol{\vartheta} + \varepsilon(t), \tag{36}$$

where y is the observed variable, $\boldsymbol{\phi}$ is the vector of known functions, ε is the measurement noise, and $\boldsymbol{\vartheta}$ is the vector of unknown parameters.

The estimation $\hat{\boldsymbol{\vartheta}}(k)$ of unknown parameter vector $\boldsymbol{\vartheta}$ at a step k can be obtained recursively using the following equations [23,44]:

$$\begin{aligned}
\hat{\boldsymbol{\vartheta}}(k) &= \hat{\boldsymbol{\vartheta}}(k) + \boldsymbol{K}(k)\varepsilon(k), \\
\varepsilon(k) &= y(k) - \boldsymbol{\phi}^T(k)\hat{\boldsymbol{\vartheta}}(k-1), \\
\boldsymbol{K}(k) &= \frac{P(k-1)\boldsymbol{\phi}(k)}{\lambda + \boldsymbol{\phi}^T(k)P(k-1)\boldsymbol{\phi}(k)}, \\
P(k) &= \frac{1}{\lambda}\left[P(k-1) - \frac{P(k-1)\boldsymbol{\phi}(k)\boldsymbol{\phi}^T(k)P(k-1)}{\lambda + \boldsymbol{\phi}^T(k)P(k-1)\boldsymbol{\phi}(k)}\right],
\end{aligned}$$

where $\boldsymbol{K}(k)$ is the adaptation gain vector, and $P(k)$ is the covariance matrix.

7.1.2. Identification Parametrization

Following the common way, here, we estimate the mass and longitudinal resistance coefficient based on the longitudinal dynamics. If the torque converter and the driveline are fully engaged, it can be assumed that all the torque from the engine is fully passed to the track. Another assumption made here is that the service brake is not applied during identification procedures, and the friction brake force is not generated. Further, we assume that there is no slip, which is an acceptable assumption, for the most part. It is also assumed that the aerodynamic loads can be neglected since the vehicle travels with a very low speed. Under these assumptions, the longitudinal dynamics can be presented in the following simplified form:

$$m\ddot{X} = \frac{\tau_L + \tau_R}{r} - 2R_l.$$

We want to rearrange the above equation and represent it in the form of Equation (36). Note that R_l is also dependent on m, as well as μ_l. We obtain

$$\ddot{X} = \frac{\tau_L + \tau_R}{r}\frac{1}{m} - g\mu_l,$$

where $y = \ddot{X}$, $\boldsymbol{\phi} = \left[(\tau_L + \tau_R)/r \quad -g\right]^T$ and $\hat{\boldsymbol{\vartheta}} = \left[1/m \quad \mu_l\right]^T$.

The inertia and lateral resistance indices can be evaluated from the rotational dynamics model. To simplify parametrization of the identification, let us write the rotational dynamics of the tracked vehicle in the body fixed reference frame in the following way [3]:

$$(I)\ddot{\theta} = \frac{b(\tau_L - \tau_R)}{2r} - M_r. \tag{37}$$

Recalling the expression for M_r from Equations (20) and (21) and rearranging the Equation (37) into the form of Equation (36), we can obtain

$$\frac{b(\tau_L - \tau_R)}{2r} = \ddot{\theta}I - \frac{2sign(\dot{\theta})mg}{l}\left(\frac{l^2}{4} - x_{ICR}^2\right)\mu_t,$$

where

$$\begin{aligned} y &= \frac{b(\tau_L - \tau_R)}{2r}, \\ \phi &= \left[\ddot{\theta} \quad \frac{-2sign(\dot{\theta})mg}{l}\left(\frac{l^2}{4} - x_{ICR}^2\right)\right]^T, \\ \vartheta &= \left[I \quad \mu_t\right]^T. \end{aligned}$$

7.2. Slip Estimation

As it is discussed earlier, the Unscented Kalman Filter (UKF) demonstrates quite accurate estimation of parameters and can be effectively used for longitudinal slip identification. A brief description of the UKF algorithm is provided in the next subsection, followed by the derivation of state and measurement equations for the tracked vehicle.

7.2.1. Unscented Kalman Filter

UKF is an effective technique for estimating the state of a nonlinear dynamic system. Let us assume that a system has the following nonlinear dynamics:

$$\begin{aligned} \mathbf{x}_{k+1} &= f(\mathbf{x}_k, \mathbf{u}_k) + \mathbf{v}_k, \\ \mathbf{z}_k &= h(\mathbf{x}_k) + \mathbf{w}_k, \end{aligned}$$

where $\mathbf{x}_k$ is a process state vector at time t_k, $\mathbf{u}_k$ is a vector of the control inputs, $\mathbf{z}_k$ is a measurement vector, and $f(\mathbf{x}_k, \mathbf{u}_k)$ and $h(\mathbf{x}_k)$ are state transition function and measurement function, respectively. Finally, $\mathbf{v}_k$ and $\mathbf{w}_k$ denote the additive white noise of the covariance determined by $\mathbf{Q}_k$ and $\mathbf{R}_k$, respectively.

Approximations associated with the linearization process, e.g., in Extended Kalman Filter (EKF), may lead to noticeable errors in the covariance and posterior mean of the transformed random variable, which, subsequently, can cause a sub-optimal performance or even divergence of the filter [45].

The approximation drawbacks described above are eliminated with the UKF, where the prior and posterior mean and covariance are represented by minimal set of weighted samples, so called sigma points. The UKF utilizes an unscented transformation method for calculating the statistics of a random variable which undergoes a nonlinear transformation. In the unscented transform, prior sigma points projection through nonlinear function gives results very close to the real transformed distribution [46]. The applied method for the sigma points derivation is described below.

Let us assume that mean and covariance of the random variable $\mathbf{x}$ are denoted with $\hat{\mathbf{x}}$ and $\mathbf{P}_\mathbf{x}$, respectively. The variable is propagated through the nonlinear functions $\mathbf{x}_{k+1} = f(\mathbf{x}_k)$ and $\mathbf{z}_k = h(\mathbf{x}_k)$. For this purpose, a matrix $\mathcal{X}$ of sigma points (Equation (38)) is formed, and the corresponding weights W_i (Equation (39)) are computed. The number of sigma points is defined with the expression: $2L + 1$, where L is dimension of the $\mathbf{x}$ vector [46]:

$$\begin{aligned} \mathcal{X}_0 &= \hat{\mathbf{x}}, \\ \mathcal{X}_i &= \hat{\mathbf{x}} + \left(\sqrt{(L+\lambda)\mathbf{P_x}}\right)_i \quad i = 1, \cdots, L, \\ \mathcal{X}_i &= \hat{\mathbf{x}} - \left(\sqrt{(L+\lambda)\mathbf{P_x}}\right)_i \quad i = L+1, \cdots, 2L, \end{aligned} \tag{38}$$

$$\begin{aligned} W_0^{(m)} &= \lambda/(L+\lambda), \\ W_0^{(c)} &= \lambda/(L+\lambda) + (1 - \alpha_{ukf}^2 + \beta_{ukf}), \\ W_i^{(m)} &= W_i^{(c)} = 1/[2(L+\lambda)] \quad i = 1, \cdots, 2L, \end{aligned} \tag{39}$$

where $\lambda = \alpha_{ukf}^2(L + \kappa_{ukf})$ and κ_{ukf} are the scaling parameters, α_{ukf} determines how sigma points are spread around mean, and $\beta_{ukf} = 2$ for Gaussian distributions and should be chosen basing on the prior knowledge about the random variable distribution.

Further, points are propagated through the nonlinear functions

$$\mathcal{X}_{k+1|k}^i = f(\mathcal{X}_k^i, \mathbf{u}_k), \qquad i = 0, \cdots, 2L, \tag{40a}$$

$$\mathcal{Z}_{k+1|k}^i = h(\mathcal{X}_{k+1|k}^i). \tag{40b}$$

Subsequently, the mean and the covariance of $\mathbf{x}$ and $\mathbf{z}$ are computed:

$$\hat{\mathbf{x}}_{k+1}^- = \sum_{i=0}^{2L} W_i^{(m)} \mathcal{X}_{k+1|k}^i, \tag{41a}$$

$$\hat{\mathbf{z}}_{k+1}^- = \sum_{i=0}^{2L} W_i^{(m)} \mathcal{Z}_{k+1|k}^i, \tag{41b}$$

$$\mathbf{P}_{k+1|k} = \sum_{i=0}^{2L} W_i^{(c)} (\mathcal{X}_{k+1|k}^i - \hat{\mathbf{x}}_{k+1}^-)(\mathcal{X}_{k+1|k}^i - \hat{\mathbf{x}}_{k+1}^-)^T + \mathbf{Q}_k, \tag{42a}$$

$$\mathbf{P}_{\tilde{\mathbf{z}}\tilde{\mathbf{z}}} = \sum_{i=0}^{2L} W_i^{(c)} (\mathcal{Z}_{k+1|k}^i - \hat{\mathbf{z}}_{k+1}^-)(\mathcal{Z}_{k+1|k}^i - \hat{\mathbf{z}}_{k+1}^-)^T + \mathbf{R}_k, \tag{42b}$$

$$\mathbf{P}_{\tilde{\mathbf{x}}\tilde{\mathbf{z}}} = \sum_{i=0}^{2L} W_i^{(c)} (\mathcal{X}_{k+1|k}^i - \hat{\mathbf{x}}_{k+1}^-)(\mathcal{Z}_{k+1|k}^i - \hat{\mathbf{z}}_{k+1}^-)^T. \tag{42c}$$

In the last step, the Kalman gain is computed:

$$\mathbf{K}_{k+1} = \mathbf{P}_{\tilde{\mathbf{x}}\tilde{\mathbf{z}}} \mathbf{P}_{\tilde{\mathbf{z}}\tilde{\mathbf{z}}}^{-1},$$

and the new state and covariance estimate are obtained:

$$\hat{\mathbf{x}}_{k+1} = \hat{\mathbf{x}}_{k+1}^- + \mathbf{K}_{k+1}\left(\mathbf{z}_{k+1} - \hat{\mathbf{z}}_{k+1}^-\right),$$

$$\mathbf{P}_{k+1} = \mathbf{P}_{k+1|k} + \mathbf{K}_{k+1} \mathbf{P}_{\tilde{\mathbf{z}}\tilde{\mathbf{z}}} \mathbf{K}_{k+1}^T.$$

7.2.2. Identification Parametrization

The UKF is designed to recover the slip parameters i_L and i_R from the vehicle states. Therefore, the augmented state vector is formed:

$$\mathbf{x} = \begin{bmatrix} X & Y & \theta & i_L & i_R & \alpha \end{bmatrix}^T,$$

and the vector of estimates $\begin{bmatrix}\hat{X} & \hat{Y} & \hat{\theta} & \hat{i}_L & \hat{i}_R & \hat{\alpha}\end{bmatrix}^T$ is obtained from the available measurements:

$$\mathbf{z} = \begin{bmatrix}X_m & Y_m & \theta_m & \omega_{tm,L} & \omega_{tm,R}\end{bmatrix}^T,$$

where $\omega_{tm,L}$ and $\omega_{tm,R}$ are the control inputs. The following process model is adopted from the kinematic model of the tracked vehicle from Equation (3):

$$\mathbf{x}_{k+1} = \begin{bmatrix} X_k + \Delta T \cdot 0.5r\left[(1-i_{L,k})\omega_{t,L}^{k+1} + (1-i_{R,k})\omega_{t,R}^{k+1}\right](\cos\theta_k - \sin\theta_k \tan\alpha_k) \\ Y_k + \Delta T \cdot 0.5r\left[(1-i_{L,k})\omega_{t,L}^{k+1} + (1-i_{R,k})\omega_{t,R}^{k+1}\right](\sin\theta_k + \cos\theta_k \tan\alpha_k) \\ \theta_k + \Delta T\frac{r}{b}\left[-(1-i_{L,k})\omega_{t,L}^{k+1} + (1-i_{R,k})\omega_{t,R}^{k+1}\right] \\ i_{L,k} \\ i_{R,k} \\ \alpha_k \end{bmatrix}.$$

7.3. Soil Parameters Estimation

It was shown previously that Generalized Newton–Raphson (GNR) method identifies unknown soil parameter with a high accuracy and rapid convergence [47]. The GNR is employed in this study for the soil parameters estimation that impact the tractive effort, i.e., cohesion F_{max} and shear deformation modulus K. Below, we provide a brief overview of this method and the parametrization of the identification of soil parameters.

7.3.1. Generalized Newton–Raphson

The system equation can then be expressed as a function of the parameter vector $\mathbf{p}$ and the measurement vector $\mathbf{x}_i$ [31]:

$$\begin{aligned} f_1(p_1, p_2, \cdots, p_n, x_1(t_1), x_2(t_1), \cdots, x_m(t_1)) &= 0, \\ f_2(p_1, p_2, \cdots, p_n, x_1(t_2), x_2(t_2), \cdots, x_m(t_2)) &= 0, \\ &\cdots \\ f_q(p_1, p_2, \cdots, p_n, x_1(t_q), x_2(t_q), \cdots, x_m(t_q)) &= 0. \end{aligned}$$

In the GNR method, q independent equations are required to find the parameter vector $\mathbf{p}$, where $q > n$

$$\begin{bmatrix} p_1 \\ p_2 \\ \vdots \\ p_n \end{bmatrix}_{i+1} \approx \begin{bmatrix} p_1 \\ p_2 \\ \vdots \\ p_n \end{bmatrix}_i - \begin{bmatrix} \frac{\partial f_1}{\partial p_1} & \cdots & \frac{\partial f_1}{\partial p_n} \\ \frac{\partial f_2}{\partial p_1} & \cdots & \frac{\partial f_2}{\partial p_n} \\ \vdots & \ddots & \vdots \\ \frac{\partial f_q}{\partial p_1} & \cdots & \frac{\partial f_q}{\partial p_n} \end{bmatrix}_i^{-1} \begin{bmatrix} f_1 \\ f_2 \\ \vdots \\ f_q \end{bmatrix}_i.$$

The above system of equation is solved recursively until the convergence condition is met.

The main advantage of GNR algorithm is robustness and fast convergence. However, an initial guess of the parameter can influence the convergence rate. Especially, if the initial derivative of a function is close to zero, the convergence speed is low [31]. The advantage of the GNR method over classic Newton–Raphson method is that the former is considered more robust to the measurement noise due to the higher number of samples included in the equation [48].

7.3.2. Identification Parametrization

To obtain the estimation model, we need to consider a function in the form

$$f(\mathbf{p}, \mathbf{x}) = 0.$$

For the tractive force expressed as in Equation (7), we have:

$$f(F_{max}, K, \tau, i) = 0, \tag{43}$$

where: $\mathbf{p} = \begin{bmatrix} F_{max} & K \end{bmatrix}^T$ is the vector of parameters to be identified, and $\mathbf{x} = \begin{bmatrix} \tau & i \end{bmatrix}^T$ is the measurement vector, where τ is the total torque produced by the motors, and i is the slip. Explicitly, Equation (43) yields:

$$F_{max}\left(1 - \frac{K}{il}\left(1 - e^{-il/K}\right)\right) - \frac{\tau}{r} = 0,$$

and Jacobian elements can be obtained with the following expressions:

$$\begin{aligned} \frac{\partial f}{\partial F_{max}} &= 1 - \frac{K}{il}\left(1 - e^{-il/K}\right), \\ \frac{\partial f}{\partial K} &= \frac{F_{max}}{Kil}\left(K\left(1 - e^{il/K}\right) + il\right)e^{-il/K}. \end{aligned}$$

During the identification process, the algorithm considers q time samples of the measurements to find the solution for Equation (43) using the following equation:

$$\begin{bmatrix} F_{max} \\ K \end{bmatrix}_{i+1} \approx \begin{bmatrix} F_{max} \\ K \end{bmatrix}_i - \begin{bmatrix} \frac{\partial f}{\partial F_{max}} & \frac{\partial f}{\partial K} \end{bmatrix}_i^{-1} f_i.$$

8. Results

This section presents the simulation outcomes and discussion on the obtained results. The system model was created in MATLAB®/Simulink environment. In the first part of this section, validation of the control system performance is provided. Next, system response to parameters uncertainties is investigated. In the closing part of this section, identification algorithms implemented in the system are evaluated.

Table 1 provides information on the vehicle parameters adopted in the simulation. Default soil parameters are included in Table 2 and are used throughout all the simulations unless stated otherwise.

Table 1. Vehicle parameters.

Parameter	Value	Unit
Mass, m	1450	kg
Inertia, I	1180	$\text{kg} \cdot \text{m}^2$
Tread, b	1.7	m
Track width, w	0.3	m
Track contact length, l	2	m
Sprocket radius, r	0.3	m
Gear ratio, n	1/380	

Table 2. Soil parameters for heavy clay [5].

Parameter	Value	Unit
Cohesion, c	70	kPa
Angle of internal shearing resistance, Φ	38.4	deg
Shear deformation parameter, K	0.02	m
Coefficient of longitudinal resistance, μ_l	0.6	
Coefficient of lateral resistance, μ_t	0.8	

Results presented in this chapter were obtained with a fixed-step solver with the time step $dt = 0.01$ s. Differential equations were solved with ODE4 (Runge-Kutta) algorithm.

8.1. Validation of the Control System

At the beginning, the control system behavior was investigated. Controller gains were chosen experimentally with regard to the stability condition Equation (35). Final values of the feedback gains can be found in Table 3.

Table 3. Final values of the controller gains.

Gain	Final Value
k_p	50
k_v	70
k_a	15

In the first test, all model parameters are assumed to be known, and the input signals represent the ideal values of the system states.

Figure 9 shows the system response to the circular and straight line trajectory demand.

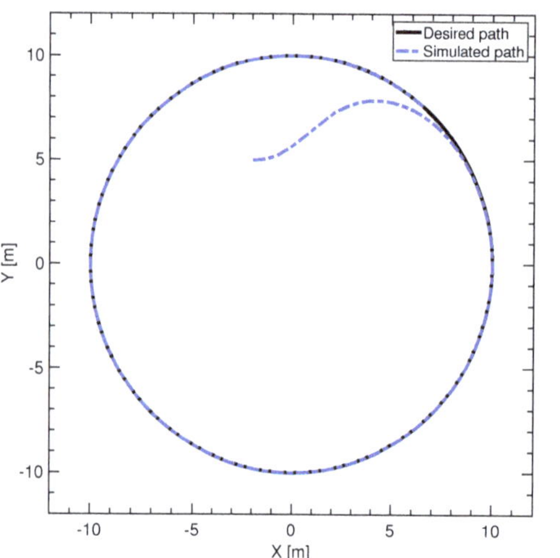

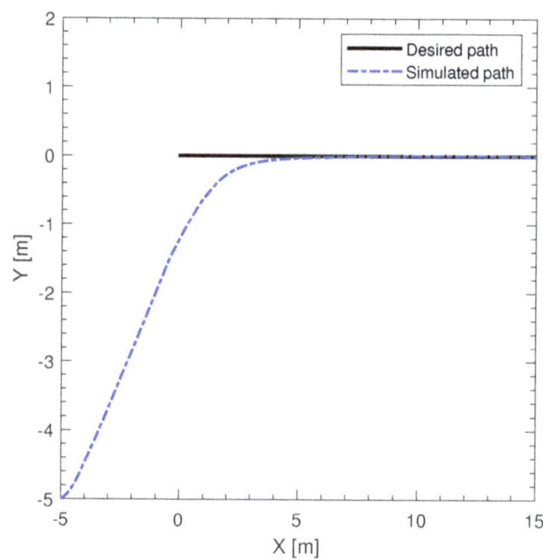

Figure 9. Trajectory following tests: circular (**left**) and straight (**right**) path.

From Figure 10, one can see that once on path, the vehicle travels with constant linear and angular velocities that demonstrated on the left and right subplots, respectively.

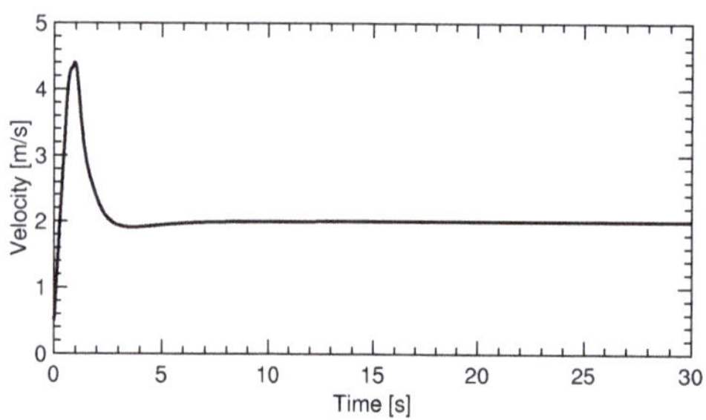

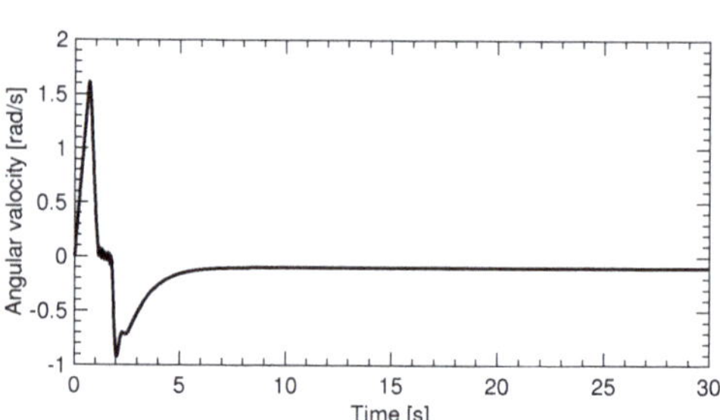

Figure 10. Trajectory following test: forward (**left**) and rotational (**right**) velocity.

According to the results obtained for the circular trajectory (shown as an example), it can be observed that the controller successfully guides the vehicle to the desired path and then continues following the path with no position, velocity, and acceleration error, which is shown in Figure 11.

The behavior of the system parameters, such as torque, sprocket rotational velocity, and slip of the track, is also of interest. These parameters are shown in Figure 12. One can observe that, as soon the vehicle reaches the constant forward and rotational velocities, all the parameters remain of constant magnitudes. Constant turning rate is achieved due to the fixed difference between the left and the right sprocket angular velocities—approximately 1 rad/s. Additionally, one can observe in Figure 12 that the slip of the tracks is directly related to the sprocket torque, as slip characteristics have a similar shape to those obtained for the torque.

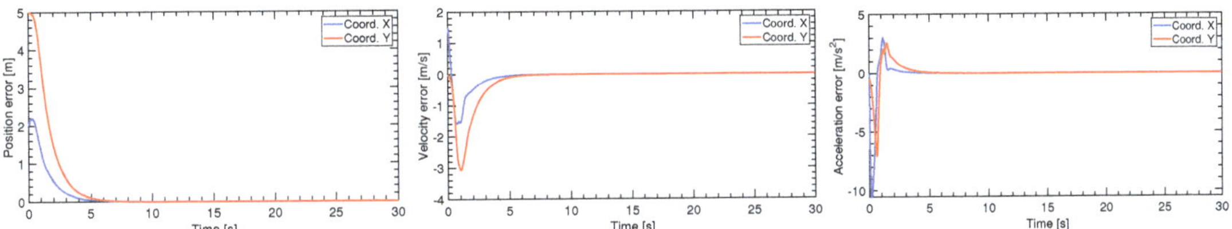

Figure 11. Trajectory following test: position (**left**), velocity (**middle**), acceleration (**right**) errors.

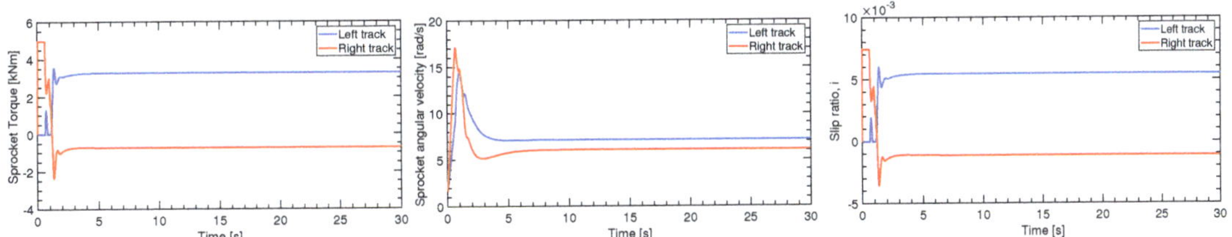

Figure 12. Trajectory following test: torque behavior (**left**), sprocket rotational velocity (**middle**), slip of the track (**right**).

8.2. System Behavior under Parameter Uncertainties

In this section, the closed-loop behavior is analyzed under unknown uncertainties introduced in the tracked vehicle dynamics. Different scenarios are considered: in the first scenario, the uncertainties brought in the mass and inertia; in the second scenario, the uncertainties are in the longitudinal and lateral resistance coefficients; and, in the last considered scenario, the uncertainties are introduced in the soil parameters.

8.2.1. Incorrect Mass and Inertia of the Vehicle

In the first scenario, mass and inertia of the vehicle used in the controller for torque command calculation differ from the actual values. Figure 13 shows the tracking performance of the controller for the circular and the straight path tests.

The mass provided to the controller is 55% of the true vehicle mass m, and moment I is 110% of the actual value. For true values, take a look at Table 1. The introduced uncertainty is quite big, which is purposely caused to obtain a clear view of uncertainty impact produced on the system behavior.

Figure 14 demonstrates the velocity tracking behavior of the controller.

A more detailed look into the position, velocity, and acceleration errors is given in Figure 15. According to these results, one can conclude that the inaccurate knowledge of mass and inertia causes the inaccurate trajectory tracking, but the controller performance remains at the acceptable level.

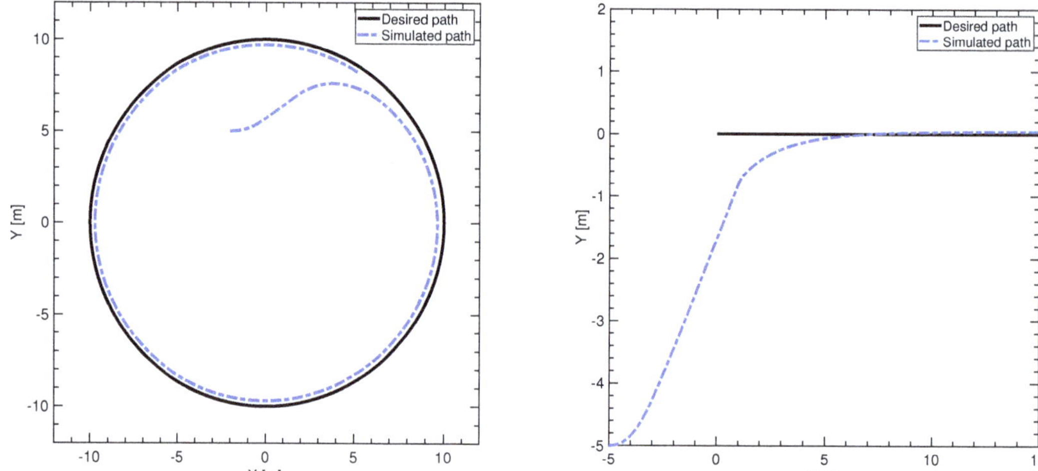

Figure 13. Trajectory following test in case of incorrect knowledge of incorrect mass and inertia: circular (**left**) and straight paths (**right**).

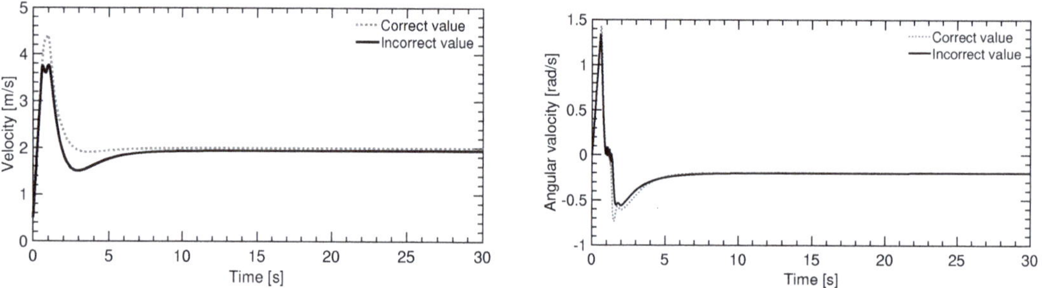

Figure 14. Trajectory following test in case of incorrect knowledge of incorrect mass and inertia: instantaneous forward velocity (**left**) and rotational velocity(**right**).

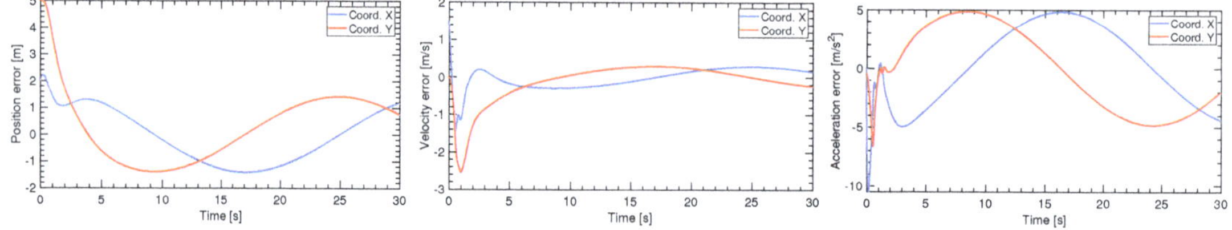

Figure 15. Trajectory following test in case of incorrect knowledge of incorrect mass and inertia: position (**left**), velocity (**middle**), acceleration (**right**) errors.

The torque, sprocket rotational velocity, and the slip of the track produced by the controller designed for the uncertain system is demonstrated in Figure 16, together with the results for a controller designed for the system without uncertainties. The behavior is quite similar, with the only difference in the sprocket rotational velocity during the initial stage (first 10 s).

To separate the effects produced by the presence of uncertainty in mass and inertia, the root mean square error (RMSE) of the vehicle trajectory for various levels of uncertainty in the mass and the inertia values as compared to the trajectory obtained for the correct model is calculated and plotted in Figure 17. These results manifest that inaccuracies in mass estimation have a greater impact on the trajectory tracking capabilities of the vehicle than the moment of inertia.

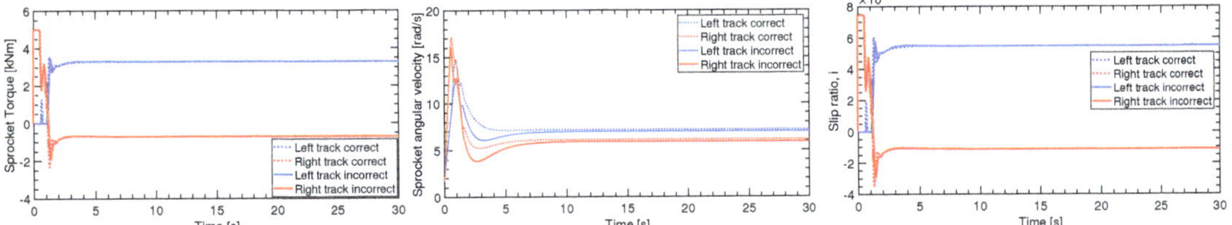

Figure 16. Trajectory following test in case of incorrect knowledge of incorrect mass and inertia: torque behavior (**left**), sprocket rotational velocity (**middle**), slip of the track (**right**).

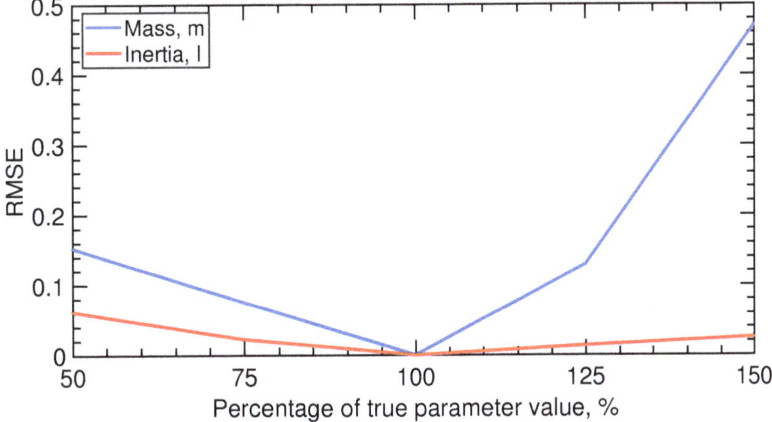

Figure 17. Root mean square error (RMSE) of the vehicle trajectory.

8.2.2. Incorrect Resistance Coefficients

In the next considered scenario, it is assumed that the longitudinal and lateral resistance coefficients vary depending on the X-Y plane location. Such scenarios are relevant for the cases when vehicle is moving through unknown rugged terrain where different types of vehicle-terrain interactions can be met.

Maps of the coefficient distributions over the surface are synthesized based on the experimental data provided in Reference [5] and shown in Figure 18.

The dependencies of the resistance coefficients for the longitudinal and lateral coefficients were generated with the following functions, respectively:

$$\mu_l = 0.7 + 0.02X\cos(0.3Y), \tag{44}$$

and

$$\mu_t = 0.6 + 0.03X\sin(0.2X)e^{-0.01Y}. \tag{45}$$

The functions (44) and (45) were selected quite arbitrarily, just to test the ability of the controller to cope with uncertainties.

Behavior of the coefficients along the vehicle trajectory is given in Figure 19 with solid lines. Reference values adopted in the controller are marked with the dotted lines.

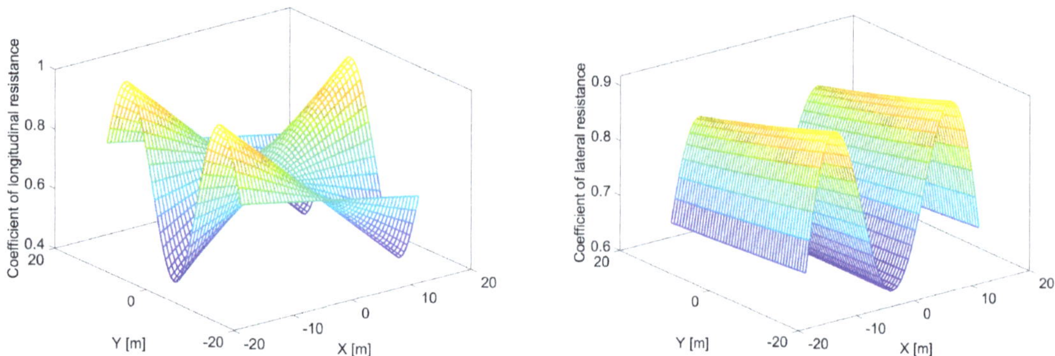

Figure 18. Maps of resistance coefficients: longitudinal (**left**) and lateral (**right**).

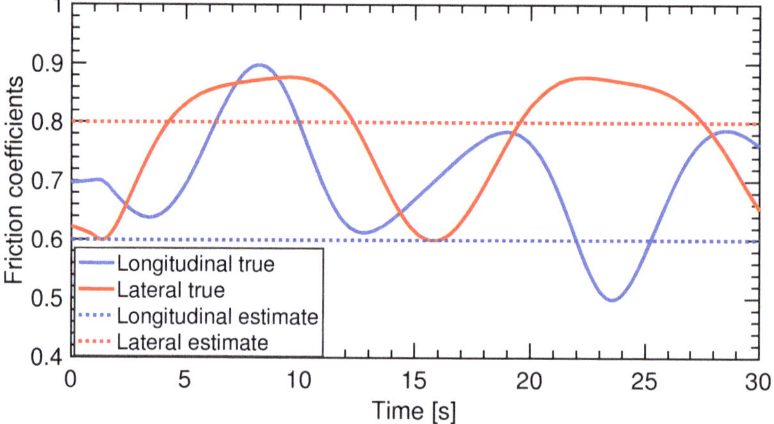

Figure 19. Friction coefficients values on the vehicle path (circular path).

The trajectory tracking tests under the discussed above variations of the resistance coefficients are provided in Figure 20.

From presented results, one can observe that the performance of the controller is more sensitive to the uncertainties in resistance coefficients for the rotational motion. Namely, in the circular test, vehicle trajectory diverges from the reference one for the sections of the path, where the real values of the coefficients differ significantly from the values assumed in the controller. The controller performance for the straight-line motion is still good enough even for the inaccurate coefficient approximations since the lateral resistance coefficient does not affect significantly on the trajectory tracking precision in the longitudinal direction and the longitudinal resistance coefficient uncertainty cause inaccuracies rather for the velocity.

The circular test is selected for a further analysis since the uncertainty in resistance coefficients can produce more effect on the system dynamics in this case.

Instantaneous forward and rotational velocities are given in Figure 21, while the controller errors are provided in Figure 22.

From the figures, one can conclude that the tracking errors are observed where the assumed friction model has mismatches with the true values (Figure 19).

The system parameter states, which are the torque, the sprocket rotational velocity, and the slip of the track, are demonstrated in Figure 23 and compared with the parameters when the correct coefficients are available. It can be noted that the controller cannot reach a steady-state, and the oscillatory response of the system reflects the changes in the resistant properties of the terrain.

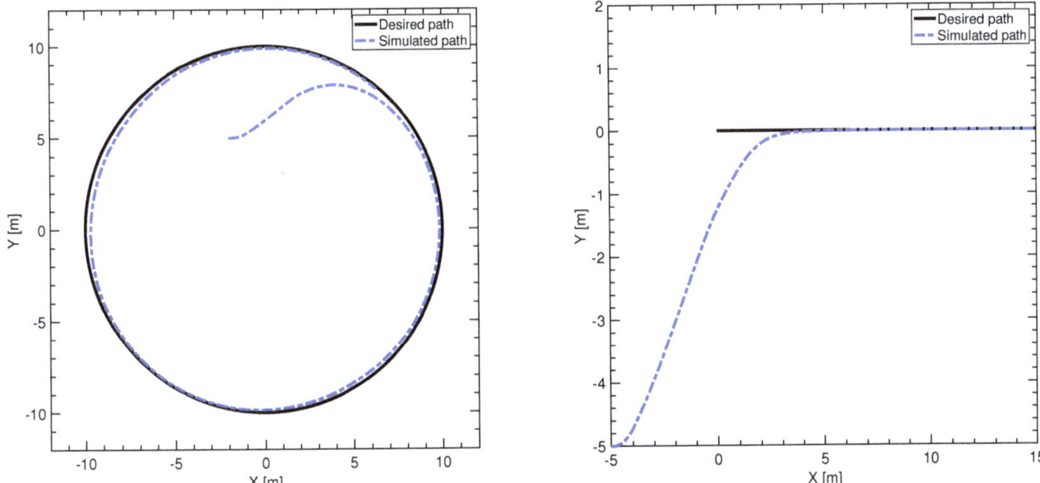

Figure 20. Trajectory following test in case of incorrect knowledge of resistance coefficients: circular (**left**) and straight paths (**right**).

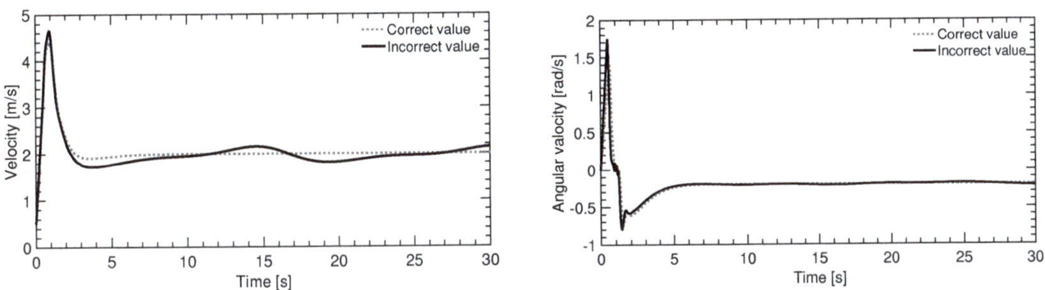

Figure 21. Trajectory following test in case of incorrect knowledge of resistance coefficients: instantaneous forward velocity (**left**) and rotational velocity (**right**).

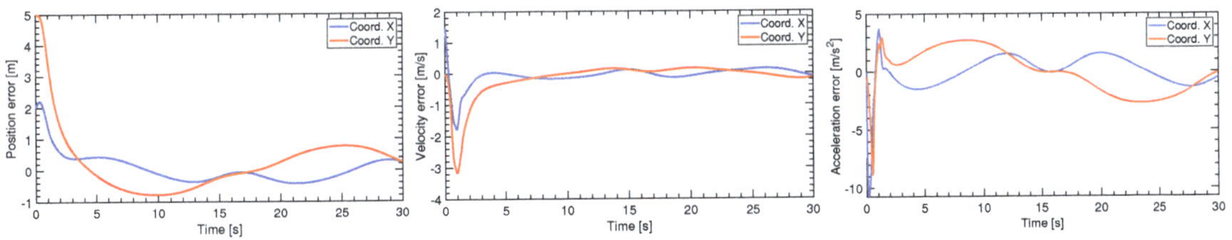

Figure 22. Trajectory following test (circular path) in case of incorrect knowledge of resistance coefficients: position error (**left**), velocity error (**middle**), and acceleration error (**right**).

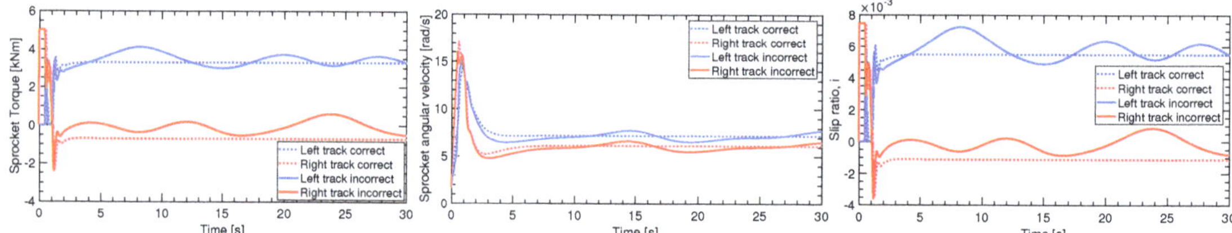

Figure 23. Trajectory following test (circular path) in case of incorrect knowledge of resistance coefficients: torque behavior (**left**), sprocket rotational velocity (**middle**), and slip of the track (**right**).

8.2.3. Incorrect Estimation of Maximum Tractive Effort

In this section, the last scenario is considered, namely the maximum tractive effort that the vehicle can develop on the certain type of soil is incorrectly predicted, which results in the trajectory demand that the vehicle is unable to follow. The expected maximum tractive effort was computed for the heavy clay (parameters provided in Table 2), while the actual parameters in this simulation corresponded to sandy loam with $c = 9.65$ kPa and $\phi = 35^o$.

In this scenario, we performed the same simulation tests, namely the circular and straight path following, which are shown in Figures 24 and 25 .

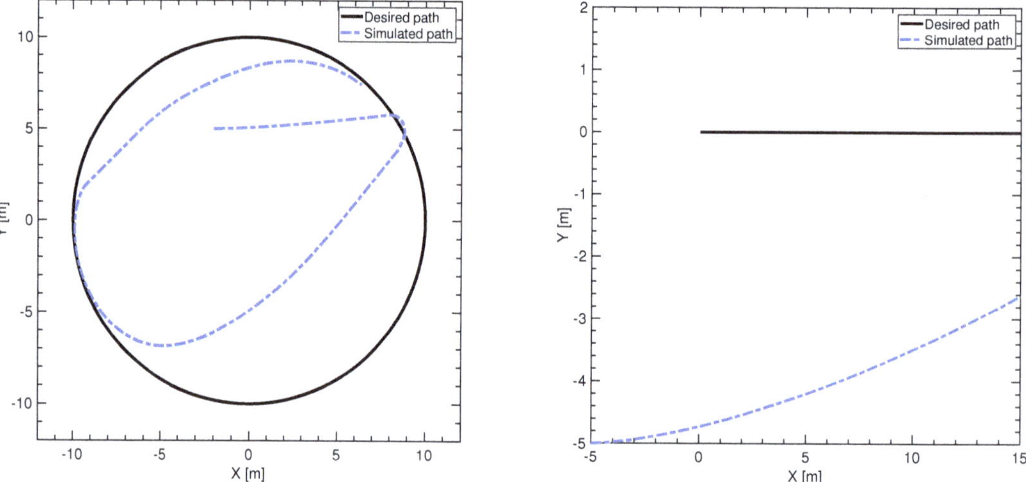

Figure 24. Trajectory following test in case of incorrect tractive force estimation: circular path (**left**) and straight path (**right**).

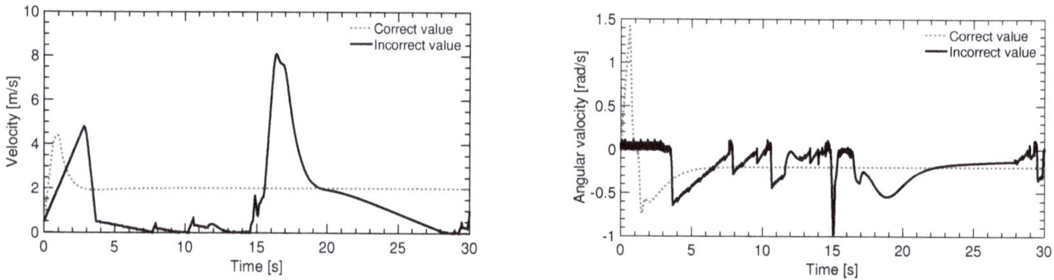

Figure 25. Trajectory following test in case of incorrect tractive force estimation: velocity V (**left**) and rotational velocity ω (**right**).

As it is clearly seen in the figures, the vehicle cannot follow the desired trajectory and causes large errors in positioning and following the desired velocities (Figure 26).

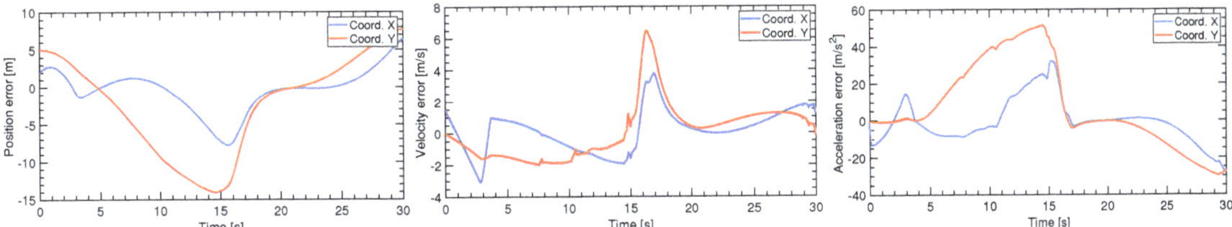

Figure 26. Trajectory following test (circular path) in case of incorrect tractive force estimation: position error (**left**), velocity error (**middle**), and acceleration error (**right**).

The demanded torque is too big for the system capabilities (Figure 27) and causes control saturation. The vehicle cannot provide proper acceleration to follow the circular path. The controller even is not capable of maintaining a steady velocity. From Figure 25, one can conclude that, in the 7th second of the simulation, the vehicle practically stops when it reaches the desired trajectory for the first time. Then, it rapidly accelerates after a couple of seconds to mitigate a huge position error, shown in Figure 26, that has been accumulated in the process.

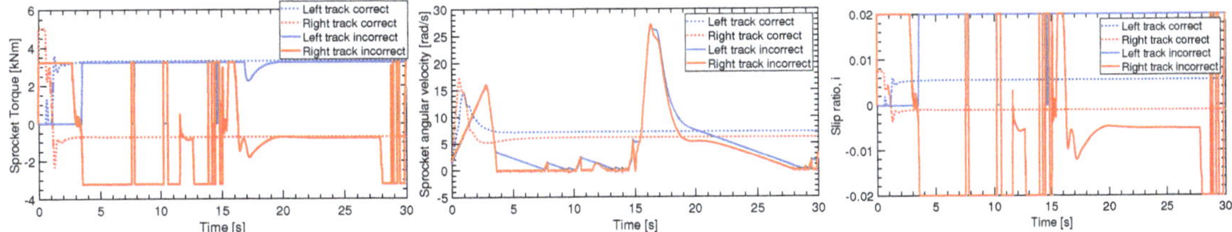

Figure 27. Trajectory following test (circular path) in case of incorrect tractive force estimation: torque behavior (**left**), sprocket rotational velocity (**middle**), and slip of the track (**right**).

To summarize this section, it can be concluded that incorrect estimation of the vehicle or soil-track interaction parameters can reduce the control system performance. As a result, the vehicle cannot follow the desired trajectory. In particular, inaccurate predictions of mass, inertia or friction coefficients introduce errors between the obtained trajectory of the vehicle and the reference trajectory. This, in turn, can cause extensive energy consumption as the motors have to generate bigger moments to guide the vehicle back on the path. Furthermore, a wrong assumption of the soil parameters can even lead to total inability to follow the desired trajectory.

To address the issues discussed above, the identification framework is implemented and tested. The results are presented in the next section.

8.3. Validation of Parameters Identification

This section presents results obtained with parameter identification algorithms, which were previously introduced in Section 7.

It should be indicated here that some of the identified parameters, such as mass, moment of inertia, and longitudinal and lateral resistance coefficients, can be used to adjust controller on the fly, thus providing an adaptive augmentation to the baseline controller. However, some of the parameters, which are track slip ratios and soil parameters can be used only for the trajectory modifications due to new environmental conditions, thus giving a contribution to an adaptive path planning framework. However, in the current study, we consider the trajectory to be predefined and unchangeable. So, the identified mass, moment of inertia, and longitudinal and lateral resistance coefficients are used to improve controller performance, and track slip ratios, maximum tractive force, and shear deformation are only identified and are not used to improve the motion performance of the vehicle.

8.3.1. Identification of Mass, Moment of Inertia and Resistance Coefficients

As discussed in Section 7.1.1, EF RLS is selected for identification of mass and moment of inertia of the vehicle. Forgetting factor $\lambda = 0.998$ is utilized in the current simulations.

Identification of mass and moment of inertia is demonstrated in Figures 28 and 29. Identification of the longitudinal and lateral resistance coefficients is provided in Figures 30 and 31. The identification maneuver can be found in Figure 32.

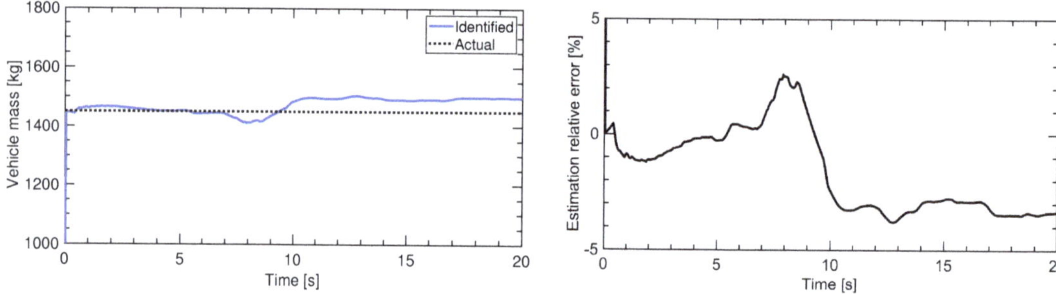

Figure 28. Identification of mass: identified value (**left**) and relative error (**right**).

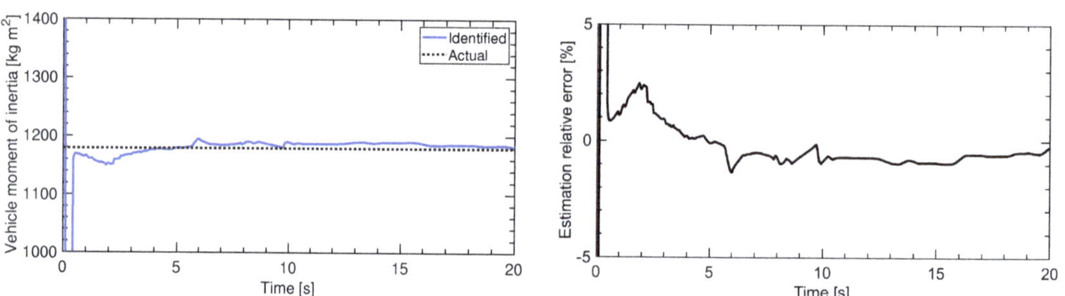

Figure 29. Identification of moment of inertia: identified value (**left**) and relative error (**right**).

The identified values were used to improve the controller performance under uncertainties provided earlier in Section 8.2. Data obtained with the identification scheme was then fed into the controller to form the adaptation capabilities of the control scheme. The trajectory tracking of the controller with adaptive augmentation is compared with the baseline controller in Figure 32. One can notice that the identification of the parameters decreases position error as compared to the baseline controller.

It can be observed that both mass and inertia of the vehicle were estimated quite accurately. However, predictions of moment of inertia is more accurate.

The precision of mass identification is better during the first 7 s (the error is less than 1%), while the vehicle being driven along the straight path. The identification error increases after the vehicle starts the spiral motion. This is caused by the fact that the mass identification scheme developed in Section 7 assumes the longitudinal motion. Even under this assumption, the proposed scheme helps to to improve the trajectory tracking (Figure 32); in particular, the error is kept at the satisfactory level, namely below 5%. The identification precision might be further improved either by changing the assumptions of longitudinal motion or by synchronization of the updating system parameters with the identification maneuvers by using some top level governing algorithm similar to that proposed in Reference [49].

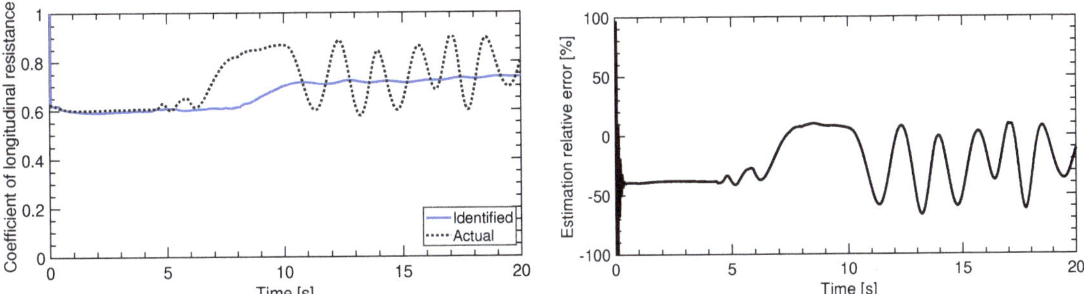

Figure 30. Longitudinal resistance coefficient: identified value (**left**) and relative error (**right**).

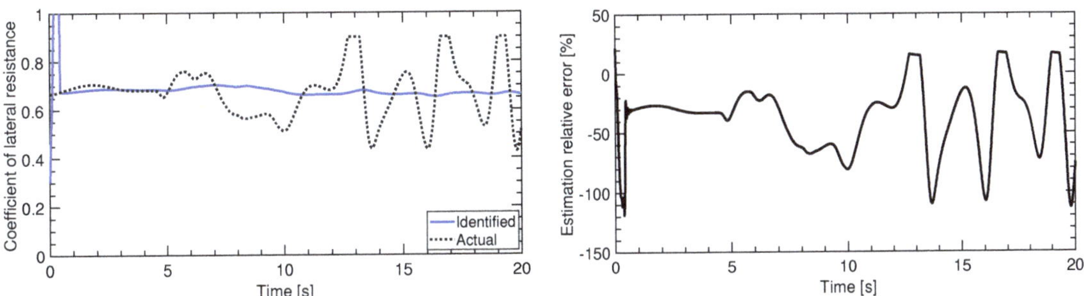

Figure 31. Lateral resistance coefficient: identified value (**left**) and relative error (**right**).

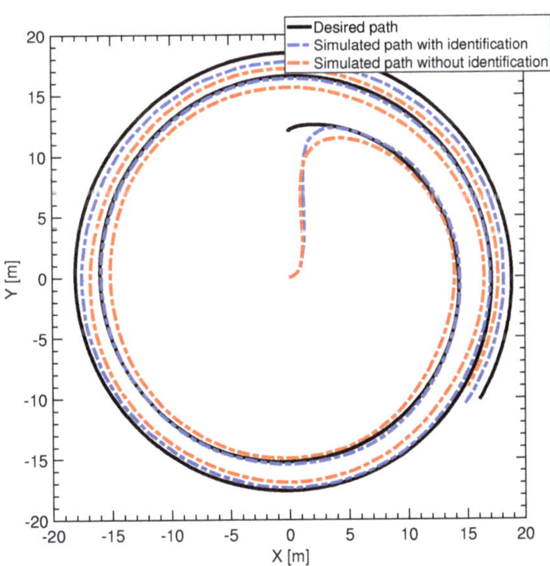

Figure 32. Trajectory of the vehicle: with identified parameters and incorrect parameters.

Estimations of the coefficients of longitudinal and lateral resistance are at a quite reasonable level. The proposed scheme identifies the resistance coefficients quite good when the coefficients vary slowly. Starting from the $t = 6$ s, the resistance coefficients start to change very fast and the EF RLS cannot follow such an abrupt variations. This is quite well known limitation of the RLS method. Nevertheless, the EF RLS provides good time average estimates of μ_L and μ_t. The variation of the resistant coefficients, simulated in the current experiment (see Equations (44) and (45)), is not quite realistic, namely, in the real motion, the the resistance coefficients are not expected to vary so fast; thus, the proposed scheme should be applicable for real-time applications.

8.3.2. Identification of Slip Parameter

Estimation of the slip parameter is performed with the UKF-based approach, which utilizes kinematics of the vehicle for state equations and is described in Section 7.2. Identification of the track slip ratio together with the actual value is coplotted in Figure 33.

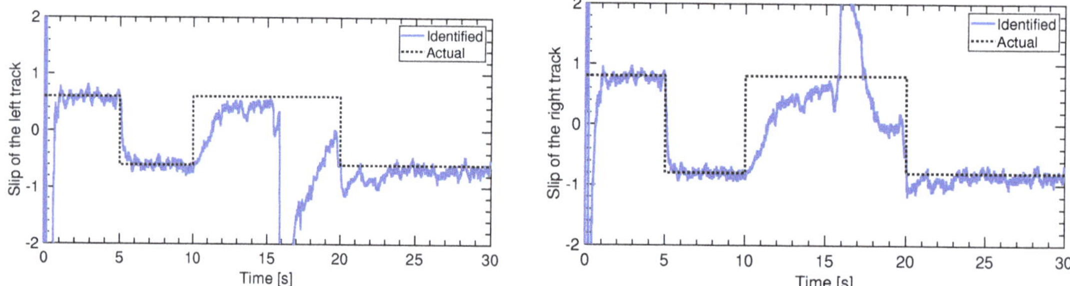

Figure 33. Track slip ratio identification: Left track (**left**) and Right track (**right**).

In the simulation, UKF and controller are fed with noisy measurement data. It can be noticed that the filter is capable of capturing the changing values of the slip ratio for both the left and right track. It provides an unbiased estimation of the states. The temporary divergence from the actual value between 15th and 20th second of the simulation was caused by the rapid speed change of the vehicle. Thus, for a relatively slow motion of the vehicle, the proposed approach manifests quite a good result.

8.3.3. Identification of Soil Parameters

Soil parameters, i.e., maximum tractive force and shear deformation parameter were identified using the GNR method presented in Section 7.3. Results are shown in Figures 34 and 35 . It can be concluded that the algorithm is able to obtain very accurate estimations of the parameters and adapts to the changes almost instantly.

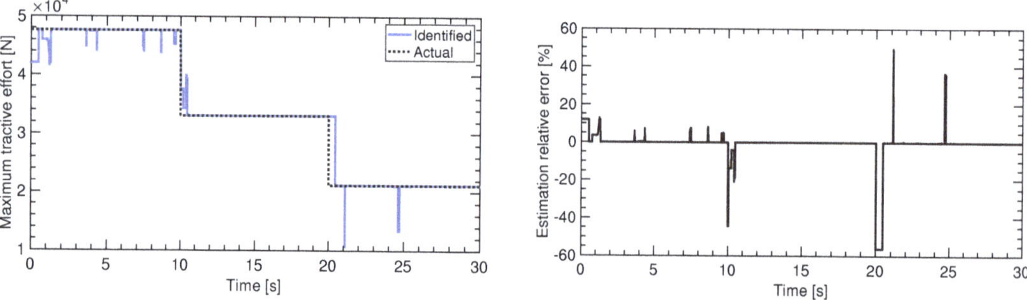

Figure 34. Identification of max. tractive force: identified value (**left**) and relative error (**right**).

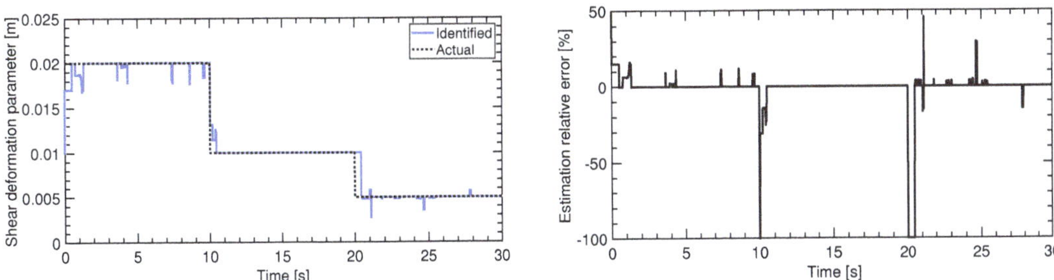

Figure 35. Identification of shear deformation parameter: identified value (**left**) and relative error (**right**).

Different settings of the window size for the number of measurement samples were tested. It was concluded that fifty samples provided robust behavior and were not computationally demanding at the same time. Additionally, the maximum number of iterations was set to twenty. In case when the algorithm was unable to converge, the last converged value was provided at the output. The number of iterations for each time step is presented in Figure 36.

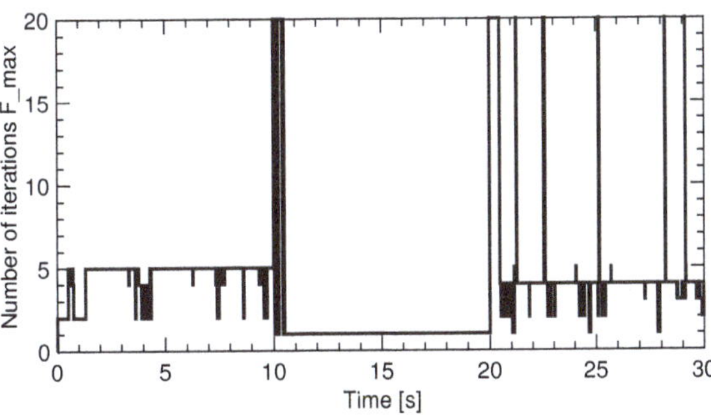

Figure 36. Generalized Newton–Raphson (GNR)—number of algorithm iterations.

Comparing the prediction error (Figures 34 and 35) with the number of iterations in Figure 36 one can conclude that high error is mainly obtained when the algorithm was unable to converge. Thus, fine-tuning of the number of iterations could further improve the algorithm performance. However, it can be done only by sacrificing the computational power of the on-board computer and should be decided for a particular real-world application based on its characteristics.

It should be noted that conversely to the inertial parameters, soil parameters identification yields better results for the experiment conducted when the vehicle is moving in a steady-state and not performing rapid maneuvers. This illustrates once again the need for different identification methods (including different maneuvers) for different state parameters and the presence of a governing algorithm to supervise learning [49].

9. Conclusions

The increasing level of autonomy of unmanned ground vehicle dictates a higher demand for operation performance, including trajectory tracking precision. Presence of model uncertainties can significantly reduce ground vehicle performance when then vehicle is traversing an unknown terrain or the vehicle inertial parameters are changed due to a mission schedule or external disturbances. Current research addressed the problem of trajectory tracking capabilities of a tracked vehicle under uncertainties in inertial parameters and in the vehicle-terrain interaction model.

At the beginning, a deep insight into the dynamics of a tracked vehicle was provided, which facilitates better understanding of the system vulnerabilities and possible control problems.

Next, the controller aiming to obtain trajectory tracking capabilities was developed. The control system design tackled the challenges brought by the nonlinear nature of the tracked vehicle dynamics. In case of perfect knowledge of motion parameters, the proposed control law perfectly follows the desired trajectory. However, when the vehicle model parameters are unknown or incorrectly estimated, the vehicle closed-loop dynamics will still be nonlinear. The influence of an uncertainties in the system dynamics is a torque disturbance that should be compensated by a proper control design.

Sensitivity of the controller to the model uncertainties was analysed and provided the valuable intuition into the consequences, such as controller divergence, offset from the desired trajectory, etc. It was shown that a precise approximation of the parameters employed in the control scheme improved tracking capabilities.

In addition, three identification schemes were proposed to address different issues of parameters estimation. First, the EF RLS was utilized to obtain improved estimates of the mass and moment of inertia of the vehicle, as well as friction coefficients. The algorithm yields very good estimation of the vehicle mass and inertia. Regarding the friction coefficients, the RLS method adapted at a slower pace to the dynamically changing terrain characteristics still providing an acceptable average from the process.

Identified values were further used to form an adaptive augmentation loop and to improve the tracking performance of the controller. It was demonstrated that providing estimations to the controller helps to follow the desired path more precisely.

Next, the Unscented Kalman Filter was designed to estimate the slip ratio of the tracks. Only kinematics of the vehicle were included in the state equation. The filter was able to track the actual slip values accurately.

Finally, the generalized Newton–Raphson algorithm was employed for accurate estimations of the soil parameters.

The estimated values of vehicle-terrain interaction did not effect directly on the controller; thus, knowledge of these parameters was not used to improve the control performance. Such parameters can be used for improved path and trajectory planning in the future; however, for the purposes of this work, the trajectory was assumed to be defined a priori.

The proposed approach helps to improve vehicle tracking capabilities regardless of a control approach via identifying uncertainties of the tracked vehicle dynamics. The developed framework can tackle the trajectory tracking problem not only as a direct problem but also as a reciprocal problem. Indeed, the identification module provides estimations of uncertain parameters to the controller for appropriate trajectory tracking. At the same time, estimated vehicle-terrain interaction parameters can be used for generating feasible trajectories. Such bilateral tackling of trajectory tracking is essential for future autonomous vehicle missions.

Future work will look into the algorithms proposed in the paper towards development of an adaptive trajectory planner and validation on a testbed.

Author Contributions: The authors contributed to the article as follows: conceptualization, N.S., D.I.I. and A.T.; methodology, N.S. and D.I.I.; software, N.S.; validation, N.S.; investigation, N.S.; writing—original draft preparation, N.S. and D.I.I.; writing—review and editing, N.S., D.I.I., A.C.Z. and A.T.; supervision, D.I.I. and A.T.; project administration, A.T. All authors have read and agreed to the published version of the manuscript.

Funding: This research received no external funding.

Acknowledgments: The authors would like to thank Jeremy Baxter from QinetiQ for fruitful discussions helped to crystallize the problem statement and for providing thoughtful comments on the research.

Conflicts of Interest: The authors declare no conflict of interest.

References

1. Astrom, K.J.; Wittenmark, B. *Adaptive Control*, 2nd ed.; Addison-Wesley Longman Publishing Co., Inc.: Petaluma, CA, USA, 1994.
2. Wellhausen, L.; Dosovitskiy, A.; Ranftl, R.; Walas, K.; Cadena, C.; Hutter, M. Where Should I Walk? Predicting Terrain Properties From Images Via Self-Supervised Learning. *IEEE Robot. Autom. Lett.* **2019**, *4*, 1509–1516. [CrossRef]
3. Shiller, Z.; Serate, W.; Hua, M. Trajectory planning of tracked vehicles. In Proceedings of the IEEE International Conference on Robotics and Automation, Atlanta, GA, USA, 2–6 May 1993; Volume 3, pp. 796–801.
4. Ma, Y.; Li, Y.; Liang, H. Design of sliding mode controller on steering control of skid steering 6 × 6 unmanned vehicle. In Proceedings of the 2017 IEEE International Conference on Unmanned Systems (ICUS), Beijing, China, 27–29 October 2017; pp. 272–276.
5. Wong, J.Y. *Theory of Ground Vehicles*, 4th ed.; Wiley: Hoboken, NJ, USA, 2008.
6. Elshazly, O.; Abo-Ismail, A.; Abbas, H.S.; Zyada, Z. Skid steering mobile robot modeling and control. In Proceedings of the 2014 UKACC International Conference on Control (CONTROL), Loughborough, UK, 9–11 July 2014; pp. 62–67.
7. Wu, Y.; Wang, T.; Liang, J.; Chen, J.; Zhao, Q.; Yang, X.; Han, C. Experimental kinematics modeling estimation for wheeled skid-steering mobile robots. In Proceedings of the 2013 IEEE International Conference on Robotics and Biomimetics (ROBIO), Shenzhen, China, 12–14 December 2013; pp. 268–273.
8. Sutoh, M.; Iijima, Y.; Sakakieda, Y.; Wakabayashi, S. Motion Modeling and Localization of Skid-Steering Wheeled Rover on Loose Terrain. *IEEE Robot. Autom. Lett.* **2018**, *3*, 4031–4037. [CrossRef]
9. Pacejka, H.B. *Tire and Vehicle Dynamics*; Butterworth-Heinemann Distributed in conjunction with SAE International: Oxford, UK; Waltham, MA, USA; Warrendale, PA, USA, 2012.
10. Meng, H.; Xiong, L.; Gao, L.; Yu, Z.; Zhang, R. Tire-Model-Free Control for Steering of Skid Steering Vehicle. In Proceedings of the 2018 IEEE Intelligent Vehicles Symposium (IV), Changshu, China, 26–30 June 2018; pp. 1590–1595.
11. Caracciolo, L.; De Luca, A.; Iannitti, S. Trajectory tracking control of a four-wheel differentially driven mobile robot. In Proceedings of the 1999 IEEE International Conference on Robotics and Automation (Cat. No.99CH36288C), Detroit, MI, USA, 10–15 May 1999; Volume 4, pp. 2632–2638.
12. Kozlowski, K.; Pazderski, D. Modeling and control of a 4-wheel skid-steering mobile robot. *Int. J. Appl. Math. Comput. Sci.* **2004**, *14*, 477–496.
13. Zou, T.; Angeles, J.; Hassani, F. Dynamic modeling and trajectory tracking control of unmanned tracked vehicles. *Robot. Auton. Syst.* **2018**, *110*, 102–111. [CrossRef]
14. Ahmadi, M.; Polotski, V.; Hurteau, R. Path tracking control of tracked vehicles. In Proceedings of the 2000 ICRA. Millennium Conference. IEEE International Conference on Robotics and Automation. Symposia Proceedings (Cat. No.00CH37065), San Francisco, CA, USA, 24–28 April 2000; Volume 3, pp. 2938–2943.
15. Tang, S.; Yuan, S.; Hu, J.; Li, X.; Zhou, J.; Guo, J. Modeling of steady-state performance of skid-steering for high-speed tracked vehicles. *J. Terramech.* **2017**, *73*, 25–35. [CrossRef]
16. Economou, J.T.; Colyer, R.E. Modelling of skid steering and fuzzy logic vehicle ground interaction. In Proceedings of the 2000 American Control Conference, ACC (IEEE Cat. No.00CH36334), Chicago, IL, USA, 28–30 June 2000; Volume 1, pp. 100–104.
17. Garimella, G.; Funke, J.; Wang, C.; Kobilarov, M. Neural network modeling for steering control of an autonomous vehicle. In Proceedings of the 2017 IEEE/RSJ International Conference on Intelligent Robots and Systems (IROS), Vancouver, BC, Canada, 24–28 September 2017; pp. 2609–2615.
18. De Luca, A.; Oriolo, G.; Samson, C. Feedback control of a nonholonomic car-like robot. In *Robot Motion Planning and Control*; Laumond, J.P., Ed.; Springer: Berlin/Heidelberg, Germany, 1998; pp. 171–253.
19. Inoue, R.S.; Cerri, J.P.; Terra, M.H.; Siqueira, A.A.G. Robust recursive control of a skid-steering mobile robot. In Proceedings of the 2013 16th International Conference on Advanced Robotics (ICAR), Montevideo, Uruguay, 25–29 November 2013; pp. 1–6.
20. Liu, J.; Han, W.; Liu, C.; Peng, H. A New Method for the Optimal Control Problem of Path Planning for Unmanned Ground Systems. *IEEE Access* **2018**, *6*, 33251–33260. [CrossRef]
21. Liu, J.; Han, W.; Zhang, Y.; Chen, Z.; Peng, H. Design of an Online Nonlinear Optimal Tracking Control Method for Unmanned Ground Systems. *IEEE Access* **2018**, *6*, 65429–65438. [CrossRef]
22. Jun, J.; Hua, M.; Benamar, F. A trajectory tracking control design for a skid-steering mobile robot by adapting its desired instantaneous center of rotation. In Proceedings of the 53rd IEEE Conference on Decision and Control, Los Angeles, CA, USA, 15–17 December 2014; pp. 4554–4559.
23. Lin, N.; Zong, C.; Shi, S. The Method of Mass Estimation Considering System Error in Vehicle Longitudinal Dynamics. *Energies* **2018**, *12*, 52. [CrossRef]
24. Vahidi, A.; Stefanopoulou, A.; Peng, H. Recursive least squares with forgetting for online estimation of vehicle mass and road grade: Theory and experiments. *Veh. Syst. Dyn.* **2005**, *43*, 31–55. [CrossRef]
25. Hsu, L.Y.; Chen, T.L. Vehicle Dynamic Prediction Systems with On-Line Identification of Vehicle Parameters and Road Conditions. *Sensors* **2012**, *12*, 15778–15800. [CrossRef] [PubMed]
26. Rhode, S.; Gauterin, F. Vehicle mass estimation using a total least-squares approach. In Proceedings of the 2012 15th International IEEE Conference on Intelligent Transportation Systems, Anchorage, AK, USA, 16–19 September 2012; pp. 1584–1589.
27. Dar, T.M.; Longoria, R.G. Slip estimation for small-scale robotic tracked vehicles. In Proceedings of the 2010 American Control Conference, Baltimore, MD, USA, 30 June–2 July 2010; pp. 6816–6821.

28. Zhou, B.; Peng, Y.; Han, J. UKF based estimation and tracking control of nonholonomic mobile robots with slipping. In Proceedings of the 2007 IEEE International Conference on Robotics and Biomimetics (ROBIO), Sanya, China, 15–18 December 2007; pp. 2058–2063.
29. Cui, M.; Liu, W.; Liu, H.; Lü, X. Unscented Kalman Filter-based adaptive tracking control for wheeled mobile robots in the presence of wheel slipping. In Proceedings of the 2016 12th World Congress on Intelligent Control and Automation (WCICA), Guilin, China, 12–15 June 2016; pp. 3335–3340.
30. Song, Z.; Zweiri, Y.H.; Seneviratne, L.D.; Althoefer, K. Non-linear observer for slip estimation of tracked vehicles. *Proc. Inst. Mech. Eng. Part D J. Automob. Eng.* **2008**, *222*, 515–533. [CrossRef]
31. Song, Z.; Hutangkabodee, S.; Zweiri, Y.H.; Seneviratne, L.D.; Althoefer, K. Identification of soil parameters for unmanned ground vehicles track-terrain interaction dynamics. In Proceedings of the SICE 2004 Annual Conference, Sapporo, Japan, 4–6 August 2004; Volume 3, pp. 2255–2260.
32. Hutangkabodee, S.; Zweiri, Y.H.; Seneviratne, L.D.; Altho, K. Multi-solution Problem for Track-Terrain Interaction Dynamics and Lumped Soil Parameter Identification. In *Field and Service Robotics*; Corke, P., Sukkariah, S., Eds.; Springer: Berlin/Heidelberg, Germany, 2006; pp. 517–528.
33. Hutangkabodee, S.; Zweiri, Y.H.; Seneviratne, L.D.; Althoefer, K. Validation of Soil Parameter Identification for Track-Terrain Interaction Dynamics. In Proceedings of the 2007 IEEE/RSJ International Conference on Intelligent Robots and Systems, San Diego, CA, USA, 29 October–2 November 2007; pp. 3174–3179.
34. Angelova, A.; Matthies, L.; Helmick, D.; Perona, P. Learning and prediction of slip from visual information. *J. Field Robot.* **2007**, *24*, 205–231. [CrossRef]
35. Helmick, D.; Angelova, A.; Matthies, L. Terrain Adaptive Navigation for planetary rovers. *J. Field Robot.* **2009**, *26*, 391–410. [CrossRef]
36. Bekker, M. *Introduction to Terrain-Vehicle Systems*; University of Michigan Press: Ann Arbor, MI, USA, 1969.
37. Janosi, Z.; Hanamoto, B. Istituto elettrotecnico nazionale Galileo Ferraris. The Analytical Determination of Drawbar Pull as a Function of Slip for Tracked Vehicles in Deformable Soils. In Proceedings of the 1st International Conference of Terrain-Vehicle Systems, Turin, Italy, 1 June 1961; pp. 707–726.
38. Landau, L.; Lifshitz, E. Chapter VI—Motion of a Rigid Body. In *Mechanics*, 3rd ed.; Landau, L., Lifshitz, E., Eds.; Butterworth-Heinemann: Oxford, UK, 1976; pp. 96–130. [CrossRef]
39. Choset, H.; Burgard, W.; Hutchinson, S.; Kantor, G.; Kavraki, L.E.; Lynch, K.M.; Thrun, S. *Principles of Robot Motion: Theory, Algorithms, and Implementation*; MIT Press: Cambridge, MA, USA, 2005.
40. Baruh, H. *Analytical Dynamics*; WCB/McGraw-Hill: Boston, MA, USA, 1999.
41. Ignatyev, D.I.; Shin, H.S.; Tsourdos, A. Bayesian calibration for multiple source regression model. *Neurocomputing* **2018**, *318*, 55–64. [CrossRef]
42. Grewal, M. *Global Positioning Systems, Inertial Navigation, and Integration*; Wiley-Interscience: Hoboken, NJ, USA, 2007.
43. Kionix. Using Two Tri-Axis Accelerometers for Rotational Measurements. 2015. Available online: http://kionixfs.kionix.com/en/document/AN019%20Using%20Two%20Tri-Axis%20Accelerometers%20for%20Rotational%20Measurements.pdf (accessed on 11 January 2021).
44. Soderstrom, T.; Stoica, P. *System Identification (Prentice Hall International Series in Systems and Control Engineering)*; Prentice Hall: Englewood Cliffs, NJ, USA, 1989.
45. Van Der Merwe, R.; Doucet, A.; De Freitas, N.; Wan, E. The Unscented Particle Filter. In *NIPS'00: Proceedings of the 13th International Conference on Neural Information Processing Systems*; MIT Press: Cambridge, MA, USA, 2000; pp. 563–569.
46. Julier, S.J.; Uhlmann, J.K. New extension of the Kalman filter to nonlinear systems. In Proceedings of the Signal Processing, Sensor Fusion, and Target Recognition VI, Orlando, FL, USA, 21–24 April 1997; Volume 3068.
47. Zweiri, Y.H.; Seneviratne, L.D.; Althoefer, K. Parameter estimation for excavator arm using generalized Newton method. *IEEE Trans. Robot.* **2004**, *20*, 762–767. [CrossRef]
48. Zweiri, Y.H. Identification schemes for unmanned excavator arm parameters. *Int. J. Autom. Comput.* **2008**, *5*, 185–192. [CrossRef]
49. Ignatyev, D.I.; Shin, H.S.; Tsourdos, A. Two-layer adaptive augmentation for incremental backstepping flight control of transport aircraft in uncertain conditions. *Aerosp. Sci. Technol.* **2020**, *105*, 106051. [CrossRef]

 *electronics* 

Article

High Velocity Lane Keeping Control Method Based on the Non-Smooth Finite-Time Control for Electric Vehicle Driven by Four Wheels Independently

Qinghua Meng [1,*], Xin Zhao [1], Chuan Hu [2] and Zong-Yao Sun [3]

1 School of Mechanical Engineering, Hangzhou Dianzi University, Hangzhou 310018, China; 40341@hdu.edu.cn
2 Department of Mechanical Engineering, University of Alaska Fairbanks, Fairbanks, AK 99775, USA; chuan.hu.2013@gmail.com
3 College of Engineering, Qufu Normal University, Qufu 276826, China; sunzongyao@sohu.com
* Correspondence: mengqinghua@hdu.edu.cn

Abstract: In order to improve the output response and robustness of the lane keeping controller for the electric vehicle driven by four wheels independently (EV-DFWI), the article proposes a lane keeping controller based on the non-smooth finite-time (NoS-FT) control method. Firstly, a lane keeping control (LKC) model was built for the EV-DFWI. Secondly, a tracking method and error weight superposition method to track error computing for the lane keeping control based on the LKC model are proposed according to the lane line information. Thirdly, a NoS-FT controller was constructed for lane keeping. It is proved that the NoS-FT controller can stabilize the system by the direct Lyapunov method. Finally, the simulations were carried out to verify that the NoS-FT controller can keep the vehicle running in the desired lane with the straight road, constant curvature road, varied curvature road, and S-bend road. The simulation results show that the NoS-FT controller has better effectiveness than the PID controller. The contributions of this article are that two kinds of tracking error computing methods of lane keeping control are proposed to deal with different conditions, and a Non-FT lane keeping controller is designed to keep the EV-DFWI running in the desired lane suffering external disturbances.

Keywords: lane keeping control (LKC); non-smooth finite-time control; previewed tracking; error weight superposition; electric vehicle (EV)

Citation: Meng, Q.; Zhao, X.; Hu, C.; Sun, Z.-Y. High Velocity Lane Keeping Control Method Based on the Non-Smooth Finite-Time Control for Electric Vehicle Driven by Four Wheels Independently. *Electronics* **2021**, *10*, 760. http://doi.org/10.3390/electronics10060760

Academic Editor: Mahmut Reyhanoglu

Received: 3 March 2021
Accepted: 21 March 2021
Published: 23 March 2021

1. Introduction

The electric vehicle (EV), especially the electric vehicle driven by four wheels independently (EV-DFWI), has the potential capacity to reduce energy consumption, enhance traffic safety, and preserve environmental pollution [1,2]. Therefore, the EV has been produced all over the world. Researchers have also studied many technologies for the electric vehicle driven by in-wheel motors including cooling system methods [3], energy regeneration approaches [4], handling stability improvement methods [5], traction control systems [6], etc. Because the EV has higher electrification than the traditional fuel vehicle, new control technologies and equipment are easily applied [7]; therefore, the EV becomes more and more intelligent. Lane keeping control (LKC) technology is one of the intelligent technologies that has been used in EVs; however, the technology needs to be studied further to improve the response, robustness, etc. The basic principle of LKC is to ensure that an EV accurately follows the desired lane via different kinds of sensors and controllers. With the help of various sensors, an EV can perceive and identify the driving environment around the vehicle and the driving state of the vehicle itself in real time. Then, the LKC system makes correct decisions and planning of vehicle motion control based on all sensors' information to guide the vehicle actuators to make unified and coordinated movement for trajectory tracking. Now more and more researchers have paid attention to the field throughout the world. Some researchers constructed lane keeping control models such as model predictive control [8], the fuzzy Takagi–Sugeno model [9], the linear parameter varying model [10], etc. Many lane keeping control methods have been proposed. The researchers designed a kind of fault-tolerant lane-keeping controllers for the automated vehicles [11]. A lane keeping assistant system was proposed to track the desired path with minimized trajectory overshoot, and an optimal controller was designed to minimize the cost function in [12]. The paper [13] proposed the concept of driver steering override for lane keeping assistant systems. An approach for a robust multi-rate lane-keeping control with predictive virtual lanes was proposed in [14]. In [15], a fuzzy-logic-based switching control law was constructed for the lane keeping assistance system. The paper [16] proposed a simple adaptive lane keeping controller based on an improved vehicle dynamic expression. The paper [17] used a multi-rate Kalman filter to deal with the asynchronous and irregular sampling time, and constructed a lane keeping system based on a kinematic model. The paper [18] studied active disturbance rejection control for the lane keeping system to achieve satisfactory performance. The paper [19] presented a lane keeping system for an autonomous vehicle, which used an image sensor to obtain the lane information.

The aforementioned literature mainly implemented the LKC by controlling the steering system, which could influence the vehicle running stability. With the development of control theory, the sliding mode control methods were improved in theory and application greatly for the robustness and fast response [20,21]. Levant et al. studied the *k*-order filter and time delay in sliding mode [22]. Fridman et al. studied different structures of sliding mode [23]. Zhang et al. investigated an alternative non-recursive finite-time trajectory tracking control methodology for a class of nonlinear systems via higher-order sliding modes [24]. These researchers developed sliding mode control methods in theory. Many researchers also improved the sliding mode control methods for application in vehicles. A state-saturated-like second-order sliding-mode algorithm was proposed by using the saturation technique and the back stepping-like method in [25], and the sliding mode control method was applied to in-wheel electric vehicles in [26]. In paper [27], a sliding mode controller for friction compensation of a three-wheeled omni-directional mobile robot was designed based on a reduced-order extended state observer. In [28], a new fast non-singular terminal sliding mode surface without any constraint was proposed and applied to trajectory tracking control for the wheeled mobile robots. Location information and angular speed were used in the sliding mode controller to solve the lane keeping problem in [29]. The paper reconstructed the unmeasured auxiliary states and disturbances synchronously online by constructing a higher-order extended state observer; an output feedback sliding mode control method was proposed based on this approach for motion control in [30]. In [31], the authors designed a sliding mode controller for the lane keeping control and applied it to four-wheel independently actuated autonomous vehicles. In [32], the authors designed a sliding mode controller for autonomous vehicles that considered input saturation. An adaptive sliding mode control based on a higher-order nonlinear disturbance observer was proposed for underactuated mechanical systems in [33]. The paper [34] designed a fault-tolerant control method based on sliding mode control and control allocation algorithm.

However, the sliding mode control method may generate chatter that could influence the service life of the actuator. On the other hand, besides the sliding mode control method, some other finite-time control methods are also developed. The paper [35] proposed a novel control strategy to unify the construction of Lyapunov functions for finite-time stability theorem. A finite-time controller for four-wheel steering of an electric vehicle was designed to improve the vehicle stability [36]. A finite-time controller was designed to stabilize the electric vehicle if a tire blowouts [37]. The paper [38] investigated the finite-time boundedness of a class of neutral type switched systems with time-varying delays.

A non-smooth control method has grabbed researchers' attention in recent years. The non-smooth control is a kind of nonlinear control method between smooth control and non-continuous control method. The method has fast convergence and strong anti-disturbance features that are useful in practice. Many researchers have achieved some results. In [39], two non-smooth control laws, high-gain finite-time guidance law, and composite guidance law were designed to improve the disturbance rejection for the missile-target interception problem. The first one assumes that the system uncertainty is bounded by a constant. The second one includes a disturbance observer and finite-time state feedback. The disturbance observer was used to estimate the system uncertainty, and the finite-time state feedback was used to stabilize the system. A non-smooth control method combining the active front-wheel steering control method with the direct yaw moment control method was proposed to ensure the stability of the electric vehicle driven by four wheels independently in [40]. A non-smooth composite control approach that could stabilize the system in finite time was proposed to improve the anti-disturbance performance of permanent magnet synchronous motor in [41].

The development of non-smooth control theory and application provides the possibility for the lane keeping control system. Therefore, we propose a novel lane keeping controller based on the non-smooth finite-time (NoS-FT) control method for the EV-DFWI under high vehicle velocity in this article. The main contributions of this article lie in the following aspects.

- Two kinds of tracking error computing methods of the lane keeping, previewed tracking and error weight superposition, are proposed to deal with different conditions for EV-DFWI.
- An NoS-FT lane keeping controller was designed, which can stabilize the vehicle to run in the desired lane when suffers external disturbance. The controller is proved by the Lyapunov method.

The article is organized as follows. An LKC model of EV-DFWI and two kinds of tracking error computing methods of lane keeping control are presented in Section 2. A lane keeping controller is designed based on the NoS-FT method in Section 3. Section 4 details the simulation of the designed NoS-FT controller compared with the PID controller, which is followed by the conclusions in Section 5.

2. Modeling of the LKC for the EV-DFWI

In order to express the relationship between the LKC model and the following designed controller, the EV-DFWI is simplified into a two-DOF model, which just includes lateral motion and yaw motion with lateral force—shown in Figure 1. The model can be expressed as follows:

$$\begin{aligned}\dot{\beta} &= -\frac{(C_f+C_r)}{mv}\beta - (\frac{(l_fC_f-l_rC_r)}{mv^2}+1)\gamma + \frac{C_f}{mv}\delta_f + \frac{F_w}{mv} \\ \dot{\gamma} &= -\frac{(l_fC_f-l_rC_r)}{I_z}\beta - \frac{(l_f^2C_f+l_r^2C_r)}{v}\gamma + \frac{l_fC_f}{I_z}\delta_f + \frac{M_z}{I_z} + \frac{F_wl_w}{I_z},\end{aligned} \tag{1}$$

where β is the sideslip angle of vehicle, γ is the yaw rate, m is the mass of vehicle, v is the longitudinal velocity of the centroid, I_z is the rotary inertia around the Z axis, C_f and C_r are the front tire cornering stiffness and rear tire cornering stiffness, l_f and l_r are the distances from the centroid to the front axle and rear axle, respectively, δ_f is the wheel angle of the front wheel, F_w is the lateral force, l_w is the distance from vehicle centroid to the lateral wind force center, and M_z is the additional yaw moment generated by different torques of four in-wheel motors.

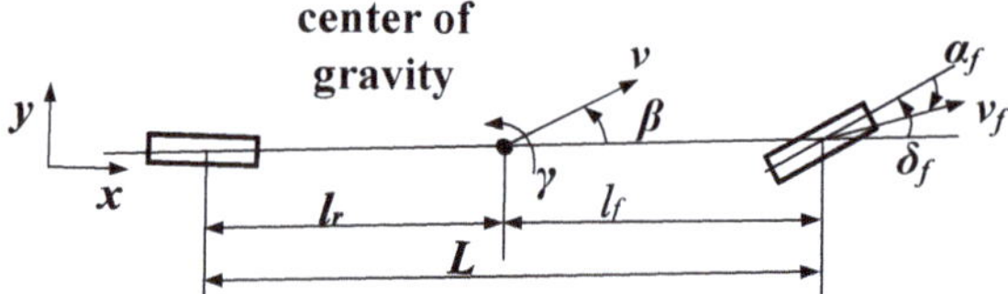

Figure 1. The bike model of the electric vehicle driven by four wheels independently (EV-DFWI).

If an EV-DFWI is controlled to run along the desired lane, the controller must output yaw moment M_z according to the tracking error of the lane. Therefore, how to obtain the tracking error of the lane is very important. Generally speaking, the tracking error is obtained in two ways. When the cameras collect enough information about the lane lines, the reference path polynomial is fitted according to the direction of the lane line and the current location of the EV-DFWI. Then the displacement offset between the next time previewed point and the reference target path point is regarded as the controller input. The tracking error can be obtained accurately in this way; however, if the lane lines are collected incompletely, an error weight superposition method is used to determine the total error as the controller input. The structure of the LKC is shown in Figure 2. Next, the core modules of the LKC are discussed in detail.

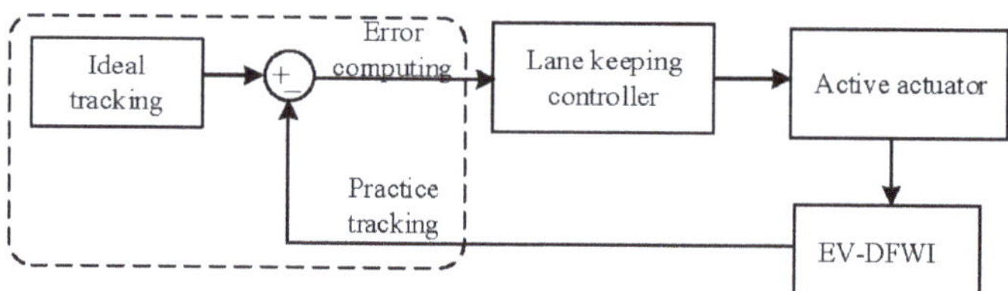

Figure 2. The structure of the lane keeping control.

Tracking Error Computing of the LKC

As previously mentioned, the tracking error is obtained in two ways according to the lane lines collection. The two tracking error computing methods are shown in Figure 3.

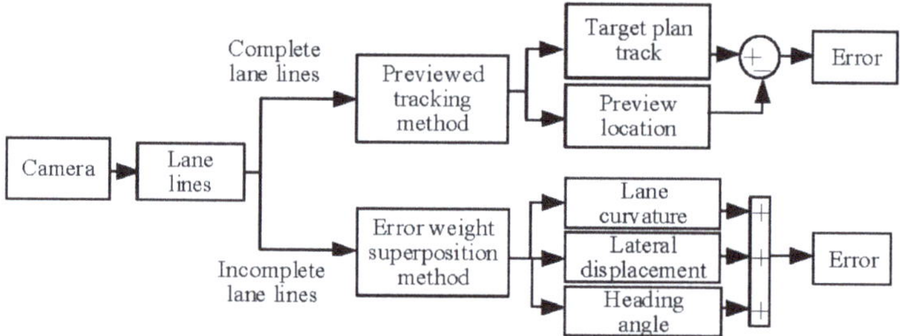

Figure 3. The computing methods for tracking error of the lane keeping control.

(1) Previewed tracking method for tracking error computing

In order to acquire an ideal path, the lane line information and the current location of the vehicle should be acquired completely. The ideal path is not always the center line of the lane for a running EV. That is to say, the correction function of the LKC method will keep the vehicle running in the lane according to its current location, not along the center line of the lane. The ideal path is shown in Figure 4.

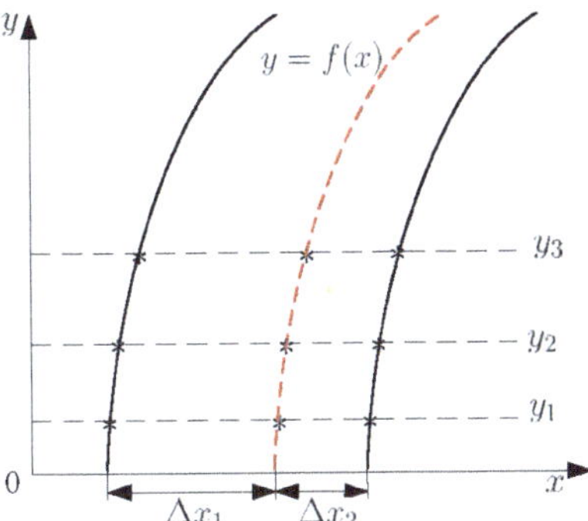

Figure 4. The ideal path of a HEV-DFWI.

As shown in Figure 4, if the vehicle's distances from the left and right lane lines are Δx_1 and Δx_2 respectively, the ideal path is the red dotted line. The ideal path can be obtained as follows.

a. The locations of the sampled points on the left lane line are (x_{11}, y_1), (x_{12}, y_2), (x_{13}, y_3), $\cdots$. The initial offset distance Δx_1 is added to every sampled points to obtain the locations of the first group points $(x_{11} + \Delta x_1, y_1)$, $(x_{12} + \Delta x_1, y_2)$, $(x_{13} + \Delta x_1, y_3)$, $\cdots$.

b. The locations of the sampled points on the right lane line are (x_{21}, y_1), (x_{22}, y_2), (x_{23}, y_3), $\cdots$. The initial offset distance Δx_2 is subtracted from every sampled points to obtain the locations of the second group points $(x_{21} - \Delta x_2, y_1)$, $(x_{22} - \Delta x_2, y_2)$, $(x_{23} - \Delta x_2, y_3)$, $\cdots$.

c. For the same Y-axle values, the locations of the third group points are obtained as $(\frac{1}{2}((x_{11} + \Delta x_1) + (x_{21} - \Delta x_2)), y_1)$, $(\frac{1}{2}((x_{12} + \Delta x_1) + (x_{22} - \Delta x_2)), y_2)$, $(\frac{1}{2}((x_{13} + \Delta x_1) + (x_{23} - \Delta x_2)), y_3)$, $\cdots$.

d. A curve is matched according to the locations of the third group points, which is the ideal path under the current condition.

An absolute coordinate, oxy, and vehicle coordinate, OXY, shown in Figure 5, were built to construct the driver preview tracking model. As shown in Figure 5, the vehicle's location is (x_t, y_t) in the absolute coordinate at t instant. The target trajectory function is $y_{tra} = f(x)$. The previewed time is $t_0 = \frac{d}{v}$, where d is the previewed distance, v is the vehicle velocity. Then after time t_0, the vehicle's abscissa is

$$x_p(t + t_0) = x(t) + t_0 v \cos(\beta + \eta), \tag{2}$$

where x_p is the vehicle abscissa of the next instant, η is the yaw angle. According to the target trajectory function, after time t_0, the target trajectory ordinate with this abscissa is

$$y_{tra}(x_p) = f(x_p(t + t_0)) = f(x(t) + t_0 v \cos(\beta + \eta)). \tag{3}$$

According to the displacement computing function, after previewed time t_0, the ordinate is

$$y_p(t + t_0) = y(t) + t_0 \dot{y}(t) + \frac{1}{2}\ddot{y}(t){t_0}^2. \tag{4}$$

Then, the offset λ is the error between the previewed ordinate and the computed ordinate according to the target trajectory.

$$\lambda = f(x(t) + t_0 v \cos(\beta + \eta)) - [y(t) + t_0 \dot{y}(t) + \frac{1}{2}\ddot{y}(t){t_0}^2]. \tag{5}$$

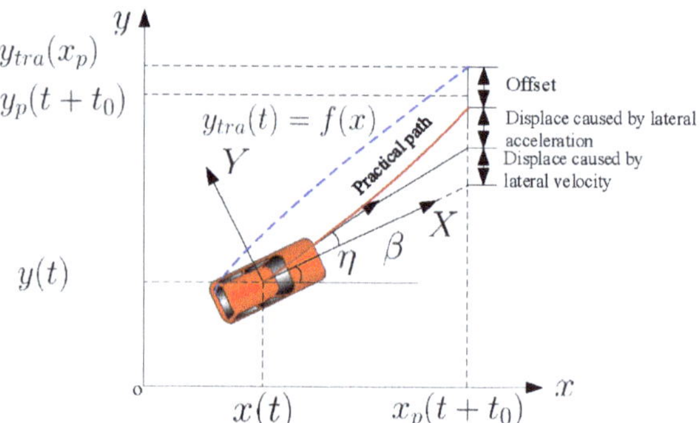

Figure 5. The vehicle locations in the absolute coordinate and vehicle coordinate (β is the sideslip angle of vehicle, η is the yaw angle).

(2) Error weight superposition method for tracking error computing

The lane line information collected by the cameras may be not enough because of muddy road, insufficient light, etc. Then the polynomial of the lane line can not be fitted to obtain the target trajectory. In this case, an error weight superposition method is proposed to obtain the total error as the controller input. The error weight superposition method considers the influence to lane departure from the road curvature, lateral position relative to the lane center, and vehicle yaw angle. The controller input is composed of every factor with respective weights.

If a vehicle is controlled to run within a lane, the road curvature and the departure from the center of the road should be provided. In this case, we can understand this issue from the relative curvature. For example, a vehicle generates a yaw angle when it runs along a straight road. The lane lines collected by cameras may still curve. Then the actuator regards this case as the vehicle running on a curve road unless the heading angle is corrected to obtain the straight lane lines by the cameras. Therefore, the offset between the vehicle and road is composed of three parts, the road curvature, offset between vehicle's location and road center line, and vehicle yaw angle. The lane center line can be obtained by the left and right lane lines. Then the curvature of the lane center line determines the lane curvature, which can be calculated by

$$k_{curve} = \frac{\delta}{h}, \tag{6}$$

where h is the horizontal offset, δ is the vertical offset, as shown in Figure 6. If the vehicle's vertical center line deviates from the road center line as shown in the figure, and runs into the curve lane along the tangent, the vehicle must be adjusted to return to the road center line and runs along the curve line. Therefore, the direction angle which needs to be adjusted can be described as

$$\theta = p_1\frac{\delta}{h} + p_2\lambda, \tag{7}$$

where p_1 and p_2 are the coefficients.

However, the vehicle may not run parallel to the lane lines, or run into the curved lane along the tangent, i.e., the vehicle deviates the road center line with a yaw angle as shown in Figure 7. In this case, the vehicle must be adjusted to return to the road center line for running along the curve line. The yaw angle should also be adjusted. Therefore, the total error e can be calculated as

$$e = p_1\frac{\delta}{h} + p_2\lambda + p_3\eta, \tag{8}$$

where p_3 is the coefficient. Then, we can calculate the necessary adjusted angle for keeping the car running along the lane by Equation (8).

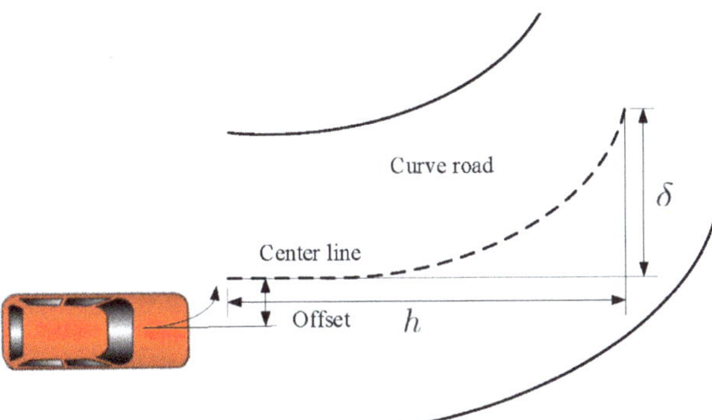

Figure 6. The road curvature computing (h is the horizontal offset, δ is the vertical offset).

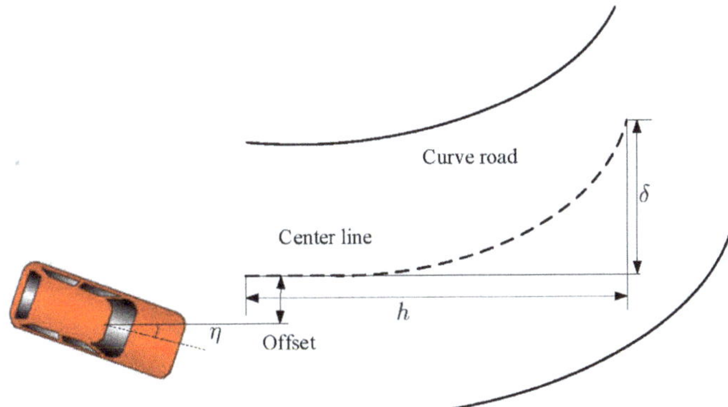

Figure 7. The vehicle runs into a curve lane with a yaw angle (h is the horizontal offset, δ is the vertical offset, η is the yaw angle).

3. Design of the Lane Keeping Controller Based on NoS-FT Control Method

The classic lane control method usually outputs an additional steering wheel angle by the PID algorithm according to the relative relation between the vehicle location information and lane center line. This control method is mature and reliable, but not robust. When the system suffers a sudden change of external force, it will generate an obvious oscillation, and the oscillation is difficult to decrease. Some scholars proposed some improved control algorithms in the references. These algorithms have better robust, but need larger memory space, and become complex which may lead to poor stability.

The purpose of this article is to design a NoS-FT controller to keep the EV-DFWI running along the desired route. The designed controller generates a additional direct yaw moment M_z for this purpose. In fact, The additional yaw moment is the torque difference among the four driving wheels. This method is faster and more direct.

The objective of the controller is to ensure the total error e being zero. Therefore, the necessary adjusted angle of the body is the state, and the additional yaw moment is the controller input. The relation between the total error and additional yaw moment is simplified to a first-order plant.

$$\dot{x}_e = u_{dym} = M_z. \tag{9}$$

Then, the designed NoS-FT lane keeping controller is

$$u_{dym} = -k_u \text{sign}(x_e)|x_e|^{\alpha_u}, 0 < \alpha_u < 1, k_u > 0. \tag{10}$$

According to Equations (9) and (10), we obtain

$$\dot{x}_e = -k_u \text{sign}(x_e)|x_e|^{\alpha_u}. \tag{11}$$

The solution of Equation (11) is

$$x_e(t) = \begin{cases} \text{sign}(x_e(0))(|x_e(0)|^{1-\alpha_u} - k_u(1-\alpha_u)t)^{\frac{1}{1-\alpha_u}}, 0 < t \leq \frac{|x_e(0)|^{1-\alpha_u}}{(1-\alpha_u)k_u} \\ 0, \quad t > \frac{|x_e(0)|^{1-\alpha_u}}{(1-\alpha_u)k_u} \end{cases}. \tag{12}$$

Next, we prove the system can be stabilized by the designed NoS-FT controller via the direct Lyapunov method. The selected Lyapunov function is

$$V(x_e) = \frac{1}{2}x_e^2. \tag{13}$$

The derivation of Equation (13) is

$$\dot{V}(x_e) = x_e \dot{x}_e. \tag{14}$$

Substituting Equation (11) into Equation (14), one obtains

$$\dot{V}(x_e) = -k_u x_e \text{sign}(x_e)|x_e|^{\alpha_u} = -k_u|x_e|^{1+\alpha_u} < 0, \tag{15}$$

which means that the designed NoS-FT controller can stabilize the vehicle's driving deviation to zero, i.e., the controller can keep the vehicle running in the desired lane.

4. Simulation and Analysis

In this section, the simulation of the control system (11) was conducted under a vehicle velocity of 90 km/h. The designed NoS-FT controller was compared with the PID controller in the simulation to verify its efficiency. We classify the vehicle running conditions into four types: straight road, constant curvature road, varied curvature road, and S-bend road. Next, we present the simulations of the four conditions separately. The vehicle parameters used in the simulation are in Table 1.

Table 1. Vehicle parameters.

Parameters	Value
Mass m/kg	1800
Rotary inertia $I_z/\text{kg} \cdot \text{m}^2$	3000
Length between front axle and centroid l_f/m	1.2
Length between rear axle and centroid l_r/m	1.8
Lateral stiffness of front axle $C_f/\text{N} \cdot \text{rad}^{-1}$	–1500
Lateral stiffness of rear axle $C_r/\text{N} \cdot \text{rad}^{-1}$	–1200

We first simulated the straight running condition. In the simulation, the car was located on the right of the white line that separates the road into two lanes, as shown in Figure 8. The simulation results are shown in Figure 9. From Figure 9, we can see that the lateral displacement under the NoS-FT controller is smaller than the PID controller. The displacement under the NoS-FT controller tends to zero after 75 m of longitudinal displacement, but the displacement under the PID controller tends to zero after 150 m of longitudinal displacement.

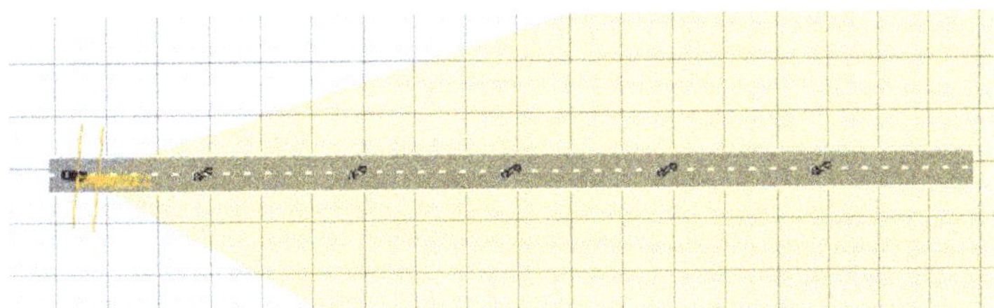

Figure 8. The straight road used in the simulation (the X axle is the longitudinal displacement, the Y axle is the lateral displacement).

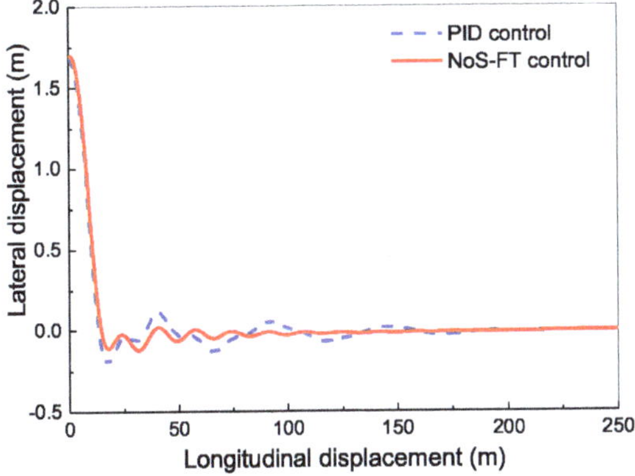

Figure 9. The displacements of the car on a straight road controlled by different controllers.

The second simulation was carried out using the constant curvature road, which is shown in Figure 10. In this simulation, the car was located in one lane, and the input steering wheel angle was zero. The simulation results are shown in Figures 11 and 12. The NoS-FT controller and PID controller both keep the car running along the constant curvature as shown in Figure 11, and the NoS-FT controller has a smaller offset than the PID controller as shown in Figure 12. The offset is limited within −0.1 to 0.15 m under the NoS-FT controller, but within −0.5 to 0.28 m under the PID controller.

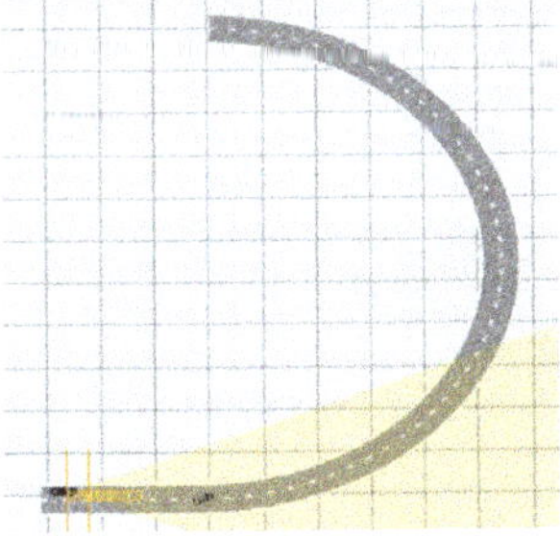

Figure 10. The car runs on a constant curvature road (the X axle is the longitudinal displacement, the Y axle is the lateral displacement).

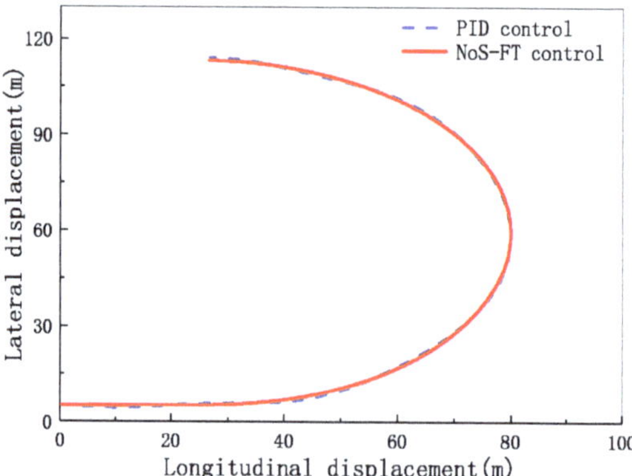

Figure 11. The traveling track of the car runs on a constant curvature road under different controllers.

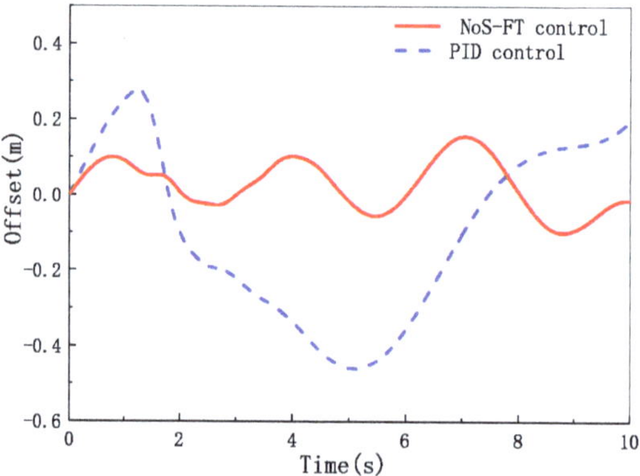

Figure 12. The offset of the car runs on a constant curvature road under different controllers.

Another common condition is that the car runs on a varied curvature road. In the simulation, the input steering wheel angle is zero. The road model is shown in Figure 13. The car runs on the straight part, then runs into the varied curvature part. Figure 14 is the car's displacements under different controllers, which shows that the NoS-FT and PID controllers are both effective for the varied curvature road. Figure 15 shows that the offset is between −0.12 and 0.1 m under the NoS-FT controller, which is much smaller than the PID controller that is between −0.35 and 0.4 m.

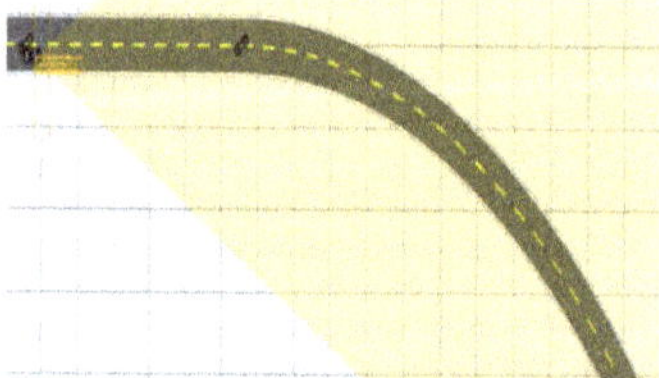

Figure 13. The car runs on a varied curvature road (the X axle is the longitudinal displacement, the Y axle is the lateral displacement).

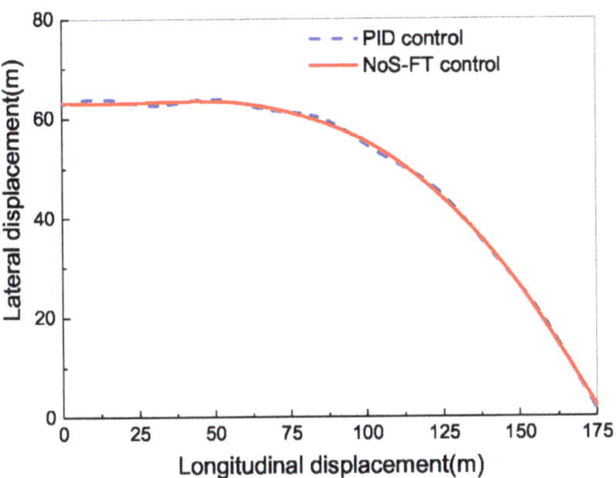

Figure 14. The traveling track of the car runs on a varied curvature road under different controllers.

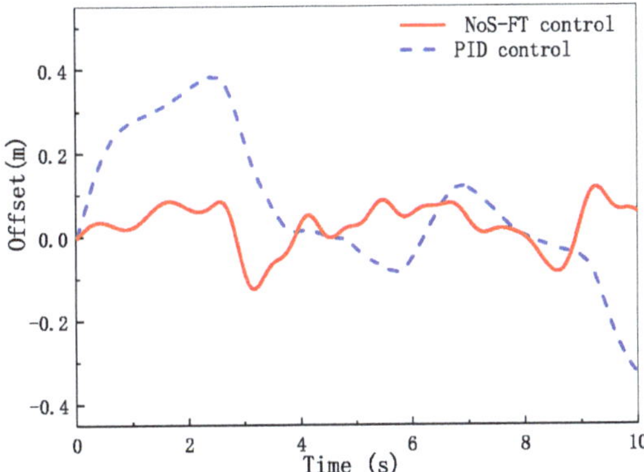

Figure 15. The offset of the car runs on a varied curvature road under different controllers.

The last condition is the S-bend curvature road, which is shown in Figure 16. The car runs in one lane, and the car's steering wheel angle is zero. The car's displacements under different controllers are shown in Figure 17. From which we can see that the two controllers can keep the car running along the desired lane. But the offset is within −0.12 to 0.15 m under the NoS-FT controller, which is much smaller than −0.35 to 0.55 m under the PID controller as shown in Figure 18.

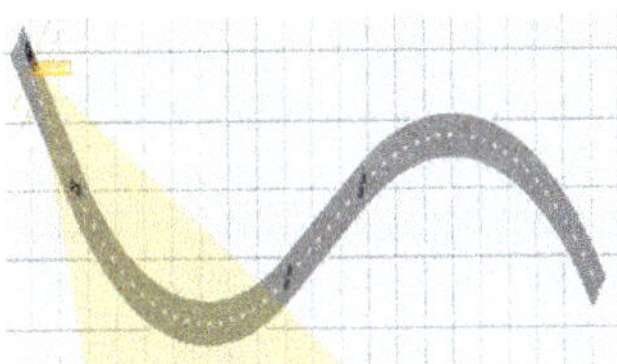

Figure 16. The car runs on a S-bend curvature road (the X axle is the longitudinal displacement, the Y axle is the lateral displacement).

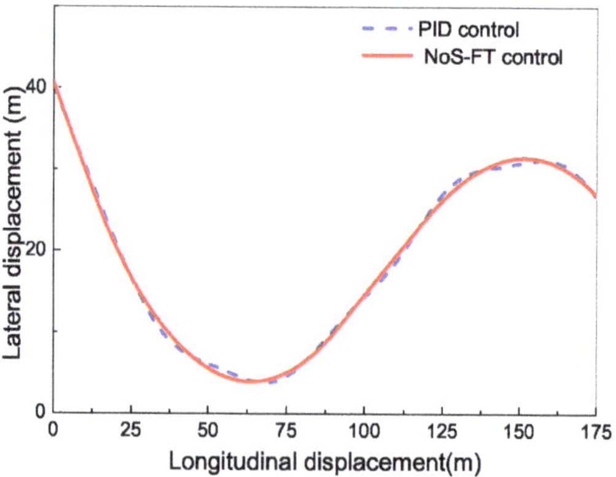

Figure 17. The traveling track of the car runs on a S-bend curvature road under different controllers.

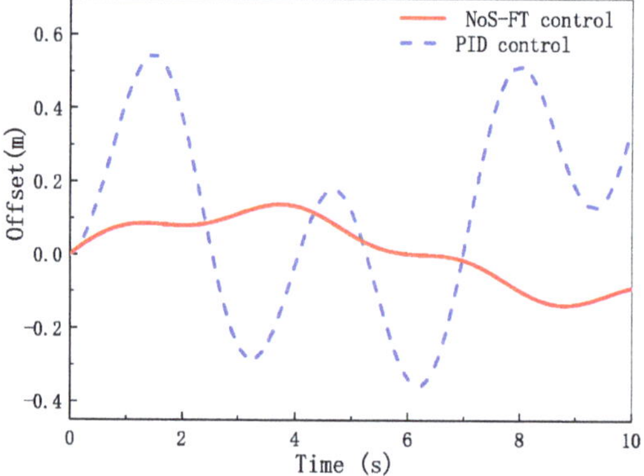

Figure 18. The total track error of the car runs on a S-bend curvature road under different controllers.

5. Conclusions

In this article, an LKC model of an EV-DFWI is built for to design a controller to improve the output response and robustness of the LKC. Then, based on the model, both a previewed tracking method and error weight superposition method are proposed to compute the tracking errors for the following designed controller to enable the vehicle to run along the desired lane. Based on the NoS-FT control method, the lane keeping controller is designed to control four driving wheels to reduce the total tracking error. The designed controller is proved by the direct Lyapunov method that it can stabilize the vehicle in theory. Through the simulation of the designed NoS-FT controller and PID controller under the common four conditions, this paper achieves the following conclusions:

- The designed NoS-FT controller can stabilize the lateral displacement of the EV-DFWI to zero faster than the PID controller running along a straight road.
- The designed NoS-FT controller has a smaller offset and better effectiveness than the PID controller under other different conditions.

Author Contributions: Conceptualization, Q.M.; methodology, Q.M.; software, X.Z.; formal analysis, X.Z.; writing—original draft preparation, Q.M.; writing—review and editing, C.H. and Z.-Y.S. All authors have read and agreed to the published version of the manuscript.

Funding: This research was supported by Zhejiang Provincial Natural Science Foundation of China under Grant No. LZ21E050002, National Natural Science Foundation of China under Grant No. 61773237, Postgraduate Education and Teaching Reform Project of Hangzhou Dianzi University under Grant No. JXGG2020YB007.

Conflicts of Interest: The authors declare no conflict of interest.

Abbreviations

The following abbreviations are used in this manuscript:

EV-DFWI	Electric Vehicle Driven by Four Wheels Independently
EV	Electric Vehicle
NoS-FT	Non-Smooth Finite-Time
LKC	Lane Keeping Control

References

1. Zhang, J.; Sun, W.; Du, H. Integrated Motion Control Scheme for Four-Wheel-Independent Vehicles Considering Critical Conditions. *IEEE Trans. Veh. Technol.* **2019**, *68*, 7488–7497. [CrossRef]
2. Merabet, A. Advanced Control for Electric Drives: Current Challenges and Future Perspectives. *Electronics* **2020**, *9*, 1762. [CrossRef]
3. Lim, D.H.; Lee, M.Y.; Lee, H.S.; Kim, S.C. Performance Evaluation of an In-Wheel Motor Cooling System in an Electric Vehicle/Hybrid Electric Vehicle. *Energies* **2014**, *7*, 961–971. [CrossRef]
4. Xu, Q.; Zhou, C.; Huang, H.; Zhang, X. Research on the Coordinated Control of Regenerative Braking System and ABS in Hybrid Electric Vehicle Based on Composite Structure Motor. *Electronics* **2021**, *10*, 223. [CrossRef]
5. Yim, S. Vehicle Stability Control with Four-Wheel Independent Braking, Drive and Steering on In-Wheel Motor-Driven Electric Vehicles. *Electronics* **2020**, *9*, 1934.
6. Han, K.; Choi, M.; Lee, B.; Choi, S.B. Development of a Traction Control System Using a Special Type of Sliding Mode Controller for Hybrid 4WD Vehicles. *IEEE Trans. Veh. Technol.* **2018**, *67*, 264–274. [CrossRef]
7. Meng, Q.; Sun, Z.Y.; Shu, Y.; Liu, T. Lateral motion stability control of electric vehicle via sampled-data state feedback by almost disturbance decoupling. *Int. J. Control* **2019**, *92*, 734–744. [CrossRef]
8. Li, Z.; Cui, G.; Li, S.; Zhang, N.; Tian, Y.; Shang, X. Lane Keeping Control Based on Model Predictive Control Under Region of Interest Prediction Considering Vehicle Motion States. *Int. J. Automot. Technol.* **2020**, *21*, 1001–1011. [CrossRef]
9. Chen, W.; Zhao, L.; Wang, H.; Huang, Y. Parallel Distributed Compensation/H(infinity)Control of Lane-keeping System Based on the Takagi-Sugeno Fuzzy Model. *Chin. J. Mech. Eng.* **2020**, *33*, 1–13. [CrossRef]
10. Salt Ducajú, J.M.; Salt Llobregat, J.J.; Cuenca, Á.; Tomizuka, M. Autonomous Ground Vehicle Lane-Keeping LPV Model-Based Control: Dual-Rate State Estimation and Comparison of Different Real-Time Control Strategies. *Sensors* **2021**, *21*, 1531. [CrossRef] [PubMed]
11. Suryanarayanan, S.; Tomizuka, M. Appropriate Sensor Placement for Fault-Tolerant Lane-Keeping Control of Automated Vehicles. *IEEE/ASME Trans. Mechatron.* **2007**, *12*, 465–471. [CrossRef]
12. Junyeon, H.; Kunsoo, H.; Hyukmin, N.; Hogi, J.; Hyungjin, K.; Paljoo, Y. Evaluation of lane keeping assistance controllers in HIL simulations. *IFAC Proc.* **2008**, *41*, 9491–9496.
13. Katzourakis, D.I.; Lazic, N.; Olsson, C.; Lidberg, M.R. Driver Steering Override for Lane-Keeping Aid Using Computer-Aided Engineering. *IEEE-ASME Trans. Mechatron.* **2015**, *20*, 1543–1552. [CrossRef]
14. Son, Y.S.; Kim, W.; Lee, S.H.; Chung, C.C. Robust Multirate Control Scheme With Predictive Virtual Lanes for Lane-Keeping System of Autonomous Highway Driving. *IEEE Trans. Veh. Technol.* **2015**, *64*, 3378–3391. [CrossRef]
15. Merah, A.; Hartani, K.; Draou, A. A new shared control for lane keeping and road departure prevention. *Veh. Syst. Dyn.* **2016**, *54*, 86–101. [CrossRef]
16. Shu, P.; Sagara, S.; Wang, Q.; Oya, M. Improved adaptive lane-keeping control for four-wheel steering vehicles without lateral velocity measurements. *Int. J. Robust Nonlinear Control* **2017**, *27*, 4154–4168. [CrossRef]
17. Kang, C.M.; Lee, S.H.; Chung, C.C. Multirate Lane-Keeping System With Kinematic Vehicle Model. *IEEE Trans. Veh. Technol.* **2018**, *67*, 9211–9222. [CrossRef]
18. Chu, Z.; Sun, Y.; Wu, C.; Sepehri, N. Active disturbance rejection control applied to automated steering for lane keeping in autonomous vehicles. *Control Eng. Pract.* **2018**, *74*, 13–21. [CrossRef]
19. Kuo, C.Y.; Lu, Y.R.; Yang, S.M. On the Image Sensor Processing for Lane Detection and Control in Vehicle Lane Keeping Systems. *Sensors* **2019**, *19*, 1665. [CrossRef]
20. Zhang, X.; Lin, H. Backstepping Fuzzy Sliding Mode Control for the Antiskid Braking System of Unmanned Aerial Vehicles. *Electronics* **2020**, *9*, 1731. [CrossRef]
21. Ahn, T.; Lee, Y.; Park, K. Design of Integrated Autonomous Driving Control System that Incorporates Chassis Controllers for Improving Path Tracking Performance and Vehicle Stability. *Electronics* **2021**, *10*, 144. [CrossRef]

22. Levant, A.; Yu, X. Sliding Mode Based Differentiation and Filtering. *IEEE Trans. Autom. Control* **2018**, *63*, 3061–3067. [CrossRef]
23. Fridman, L. Technical Committee on Variable Structure and Sliding Mode Control. *IEEE Control Syst. Mag.* **2018**, *38*, 17–18.
24. Zhang, C.; Yang, J.; Yan, Y.; Fridman, L.; Li, S. Semiglobal Finite-Time Trajectory Tracking Realization for Disturbed Nonlinear Systems via Higher-Order Sliding Modes. *IEEE Trans. Autom. Control* **2019**, *65*, 2185–2191. [CrossRef]
25. Ding, S.; Mei, K.; Li, S. A new second-order sliding mode and its application to nonlinear constrained systems. *IEEE Trans. Autom. Control* **2018**. [CrossRef]
26. Ding, S.; Lu, L.; Wei, X.Z. Sliding Mode Direct Yaw-Moment Control Design for In-Wheel Electric Vehicles. *IEEE Trans. Ind. Electron.* **2017**, *64*, 6752–6762. [CrossRef]
27. Ren, C.; Li, X.; Yang, X.; Ma, S. Extended State Observer-Based Sliding Mode Control of an Omnidirectional Mobile Robot With Friction Compensation. *IEEE Trans. Ind. Electron.* **2019**, *66*, 9480–9489. [CrossRef]
28. Zhai, J.Y.; Song, Z.B. Adaptive sliding mode trajectory tracking control for wheeled mobile robots. *Int. J. Control* **2019**, *92*, 2255–2262. [CrossRef]
29. Lei, J. Research on a kind of sliding mode lane keeping control for automated vehicles based on hybrid information of position and angular velocity. *Optik* **2016**, *127*, 9344–9360. [CrossRef]
30. Mao, J.; Yang, J.; Li, S.; Yan, Y.; Li, Q. Output feedback-based sliding mode control for disturbed motion control systems via a higher-order ESO approach. *IET Control Theory Appl.* **2018**, *12*, 2118–2126. [CrossRef]
31. Hu, C.; Qin, Y.; Cao, H.; Song, X.; Jiang, K.; Rath, J.J.; Wei, C. Lane keeping of autonomous vehicles based on differential steering with adaptive multivariable super-twisting control. *Mech. Syst. Signal Process.* **2019**, *125*, 330–346. [CrossRef]
32. Hu, C.; Wang, Z.; Taghavifar, H.; Na, J.; Qin, Y.; Guo, J.; Wei, C. MME-EKF-based path-tracking control of autonomous vehicles considering input saturation. *IEEE Trans. Veh. Technol.* **2019**, *68*, 5246–5259. [CrossRef]
33. Majumder, K.; Patre, B.M. Adaptive sliding mode control for asymptotic stabilization of underactuated mechanical systems via higher-order nonlinear disturbance observer. *J. Vib. Control* **2019**, *25*, 2340–2350. [CrossRef]
34. Argha, A.; Su, S.W.; Zheng, W.X.; Celler, B.G. Sliding-mode fault-tolerant control using the control allocation scheme. *Int. J. Robust Nonlinear Control* **2019**, *29*, 6256–6273. [CrossRef]
35. Sun, Z.Y.; Yun, M.M.; Li, T. A new approach to fast global finite-time stabilization of high-order nonlinear system. *Automatica* **2017**, *81*, 455–463. [CrossRef]
36. Meng, Q.; Sun, Z.Y.; Li, Y. Finite-time Controller Design for Four-wheel-steering of Electric Vehicle Driven by Four In-wheel Motors. *Int. J. Control Autom. Syst.* **2018**, *16*, 1814–1823. [CrossRef]
37. Meng, Q.; Qian, C.; Sun, Z.Y. Finite-time stability control of an electric vehicle under tyre blowout. *Trans. Inst. Meas. Control* **2019**, *41*, 1395–1404. [CrossRef]
38. Lin, X.; Yang, Z.; Li, S. Finite-time boundedness and finite-time weighted L-2-gain analysis for a class of neutral type switched systems with time-varying delays. *Int. J. Syst. Sci.* **2019**, *50*, 1703–1717. [CrossRef]
39. Ding, S.; Zhang, Z.; Chen, X. Guidance law design based on non-smooth control. *Trans. Inst. Meas. Control* **2013**, *35*, 1116–1128. [CrossRef]
40. Meng, Q.; Zhao, T.; Qian, C.; Sun, Z.; Ge, P. Integrated stability control of AFS and DYC for electric vehicle based on non-smooth control. *Int. J. Syst. Sci.* **2018**, *49*, 1518–1528. [CrossRef]
41. Xia, C.; Li, S.; Shi, Y.; Zhang, X.; Sun, Z.; Yin, W. A Non-Smooth Composite Control Approach for Direct Torque Control of Permanent Magnet Synchronous Machines. *IEEE Access* **2019**, *7*, 45313–45321. [CrossRef]

MDPI

Article

A Generic Interface Enabling Combinations of State-of-the-Art Path Planning and Tracking Algorithms

Johannes Rumetshofer [1,2,*], Michael Stolz [1,2] and Daniel Watzenig [1,2]

1 Institute of Automation and Control, Graz University of Technology, 8010 Graz, Austria; michael.stolz@tugraz.at (M.S.); daniel.watzenig@tugraz.at (D.W.)
2 Virtual Vehicle Research GmbH, 8010 Graz, Austria
* Correspondence: j.rumetshofer@tugraz.at

Abstract: In the development of Level 4 automated driving functions, very specific, but diverse, requirements with respect to the operational design domain have to be considered. In order to accelerate this development, it is advantageous to combine dedicated state-of-the-art software components, as building blocks in modular automated driving function architectures, instead of developing special solutions from scratch. However, e.g., in local motion planning and control, the combination of components is still limited in practice, due to necessary interface alignments, which might yield sub-optimal solutions and additional development overhead. The application of generic interfaces, which manage the data transfer between the software components, has the potential to avoid these drawbacks and hence, to further boost this development approach. This publication contributes such a generic interface concept between the local path planning and path tracking systems. The crucial point is a generalization of the lateral tracking error computation, based on an introduced error classification. It substantiates the integration of an internal reference path representation into the interface, to resolve the component interdependencies. The resulting, proposed interface enables arbitrary combinations of components from a comprehensive set of state-of-the-art path planning and tracking algorithms. Two interface implementations are finally applied in an exemplary automated driving function assembly task.

Keywords: ODD-based AD function design; path tracking; path planning; software architecture; interface design

Citation: Rumetshofer, J.; Stolz, M.; Watzenig, D. A Generic Interface Enabling Combinations of State-of-the-Art Path Planning and Tracking Algorithms. *Electronics* **2021**, *10*, 788. https://doi.org/10.3390/electronics10070788

Academic Editor: Soon Ki Jung

Received: 24 February 2021
Accepted: 23 March 2021
Published: 26 March 2021

1. Introduction

Automated driving (AD) has been a huge challenge over the last decades in research as well as in industry. However, estimations and expectations for final breakthroughs on the market have been continuously shifted, contrary to many announcements. An analytic look at the available products on the market reveals that SAE Level 2 [1] is still state-of-the-art in passenger cars, outdone in few applications to Level 3 for very restrictive applications (e.g., [2]). Level 4 automation, in particular for urban road networks, is still an open challenge (see [2]) and therefore in focus of current research. From a general perspective, the step from Level 3 to 4 is a shift from automated to autonomous driving. This step involves full environmental perception and decision making. Furthermore, the human driver can not be applied as safety fallback anymore. Consequently, it is a game changing development step. From implementation point of view a paradigm shift in system architecture is required since sensors must be shared between components and new concepts for multi-layer control software are required.

In contrast to the "under all conditions" requirement of Level 5 automation, which is quite problematic from a technical perspective, Level 4 automation solves the task of fully autonomous driving under clearly defined conditions, i.e., in a specified operational design domain (ODD). Different ODDs can be very diverse, e.g., valet parking and highway driving, and, hence, require a carefully matched AD function. From this point of view, a modular system architecture design (e.g., [3]) is beneficial ([2]). It enables the development of modular and, hence, reusable software components with respect to dedicated tasks of the fundamental sense-plan-act principal from robotics [4], and ODD-based assembly and tuning of these components. An attempt to overview the scientific state-of-the-art on all stages of the sense-plan-act principal, reveals a tremendous amount of highly sophisticated scientific solutions (see, for example, [5,6], (pp. 71–140) and [7]), with promising simulation and also real-world testing results in specific use cases. Although equivalent components solve the same task on a qualitative level, they diverge in their specific input/output data requirements. This fact complicates ODD-based AD function assembly, yielding potential performance interdependencies between the components. This has to be avoided since the components form a safety critical overall system. Hence, a clear strategy on the definition and implementation of component interfaces is beneficial. A common approach is to define very basic, static interfaces (cf. Autoware [8]), focusing on a minimized version of the shared information. The drawback of this approach is the need for repetitive wrapper code within the single software components adapting the interface signals to match the components' requirements. Although this wrapper code is independent from the actual component algorithm, it eventually impacts the performance of the component and of the overall combined system. A modification that affects this code involves all components. In order to overcome this drawback, these wrapper-tasks may be assigned to a dedicated interface component following, if possible, a generic design approach. This enables a clear focus of the connected components on their dedicated tasks omitting component independent input/output data processing. Simultaneously, the application of dedicated interface components supports a middelware independent component development and reduces later integration risk in combination with different middleware concepts.

The design of such an interface has to handle manifold requirements of state-of-the-art components. This obviously, on the one hand, complicates the interface design to a challenging task. On the other hand, it reveals significant benefits especially in the ODD-based AD-function assembly, since it enables straight-forward combination of arbitrary state-of-the-art components.

This publication is dedicated to the design of such an interface, between the local motion planning system and the path tracking system, which is an essential part of every AD function. The interface, hence, shall be able to connect a comprehensive set of state-of-the-art path planning and path tracking algorithms (see Figure 1), resolving performance interdependencies to the greatest possible extend. This is of special interest for example in collision avoidance, see for example [9,10], since it requires an accurate coordination of path planning and tracking.

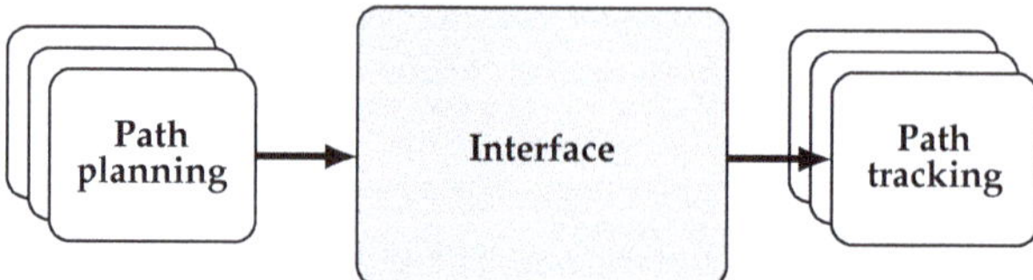

Figure 1. A generic interface shall enable arbitrary combinations of state-of-the-art path planning and tracking components, without any need for repetitive wrapper code for input/output data processing within the components.

The interface design needs an overview on state-of-the-art components and algorithms. Therefore, the publication simultaneously features a survey contribution, in particular with respect to path tracking algorithms (from a specific point of view) and specific interpolation methods. The presented classification of lateral tracking error definitions based on a comprehensive set of state-of-the-art tracking controllers is essential for the proposed interface design, but also contributes to a better understanding of the impact of tracking error definition on the performance of a tracking controller. The proposed interface design, furthermore, may serve as conceptual model for the design of other interfaces in modular system architecture for AD functions.

In detail, the publication structures as follows: Section 2 outlines the general path tracking problem and identifies the lateral tracking error computation as a major challenge in the generic interface design. In order to tackle this challenge, Section 3 introduces a lateral tracking error classification approach. This proposed classification enables a generalized approach in tracking error computation (see Section 4), based on a concise set of three elementary path operations. This set is sufficient to cover a comprehensive set ot state-of-the-art path tracking controllers. Section 5 summarizes the contribution of the preliminary findings for the aimed interface design. Section 6 extends the interface requirements from the path planning side, yielding the final interface design concept in Section 7. Finally, in Section 8 a simple, exemplary ODD-based AD function assembly tasks is considered in order to demonstrate the proposed interface design and to substantiate its benefits. Appendixs A and B provide continuative theoretical basics on path parametrization and interpolation.

2. The Path Tracking Problem

Path tracking is an important sub-problem of motion control in automated driving tasks. In contrast to the more general trajectory tracking (see for example [11,12], (p. 172)) the trajectory is split into a time-independent, spatial information—the reference path—and a time-dependent function, which defines the reference position at a specific time along this path. A lateral controller applies steering commands to approach the reference path—the so-called path tracking. Concurrently a longitudinal controller applies vehicle speed adaptions in order to track the time-dependent reference position along the path. The basic assumption behind a separated lateral and longitudinal control is almost decoupled lateral and longitudinal vehicle dynamics. In fact, they are actually coupled due to the limited traction forces of the vehicle tires and the time-dependent actuation limitations (steering rate). Whereas this assumption is valid in low- and moderate-speed driving at dry roads, respectively in high-speed driving with low steering dynamics, it is not valid in highly dynamic maneuvers, like emergency evasions or when driving on an icy road.

There exists several surveys on the classification of state-of-the-art tracking controllers like [5,13–15] and [6] (pp. 71–140). In general, the path tracking problem can be stated as follows: Given an planar reference path, which might be represented in a parametric way (cf. Appendix A) with respect to a curve parameter τ,

$$\gamma(\tau) = \begin{bmatrix} x(\tau) & y(\tau) \end{bmatrix}^T, \tag{1}$$

the path tracking controller has to compute a steering actuation, e.g., a steering wheel angle, in order to make a the position of the vehicle to follow the reference path. In order to use classical control system design approaches an appropriated error vector has to be defined,

$$\mathbf{e}(t) = \mathbf{f}_{\mathrm{e}}(\gamma(\tau), \mathbf{p}(t)) = \begin{bmatrix} e_{\mathrm{lat}}(t) & e_{\psi}(t) & \cdots \end{bmatrix}^T, \tag{2}$$

based on the reference path and the vehicle pose. The vehicle pose consists of the vehicle position $x(t), y(t)$ and heading $\psi(t)$ and possibly more vehicle states,

$$\mathbf{p}(t) = \begin{bmatrix} x(t), y(t), \psi(t), \ldots \end{bmatrix}^T. \tag{3}$$

The error vector consists of a mandatory spatial tracking error (some lateral offset error) and optionally of additional error measures like an orientation error or curvature error. Some special control approaches, e.g., Model Predictive Control (MPC), furthermore, require several lateral tracking errors corresponding to vehicle pose predictions.

With respect to a specific control error, the path tracking controller defines a steering command $\delta(t)$, which shall control the tracking error to zero, and consequently make the vehicle to follow the reference path:

$$\delta(t) = f_\delta(\mathbf{e}(t)) : \quad \mathbf{e}(t) \to \mathbf{0} \tag{4}$$

There are various reasonable possibilities to the define a lateral tracking error (see Figure 2) based on different motivations. In general, there exists no unique mappings between different lateral errors. It is an important but little-noticed fact that both the error definition and the control law design have to be considered as degrees of freedom in control system design and impact the final tracking performance.

The computation of the tracking error depends on the reference path, which is provided by the motion planning system. Consequently, its constitution and quality are defined by another component, which yields an undesired dependency on the planning system. Therefore, it is reasonable to think about outsourcing the tracking error computation to a generic interface component. To do so, it is necessary to analyze state-of-the-art path tracking algorithms with respect to the applied lateral tracking error definition. A unique classification of the used tracking error definitions and a generalization of the tracking error computation based on this classification is a major step towards the intend generic interface, which is addressed in the next sections.

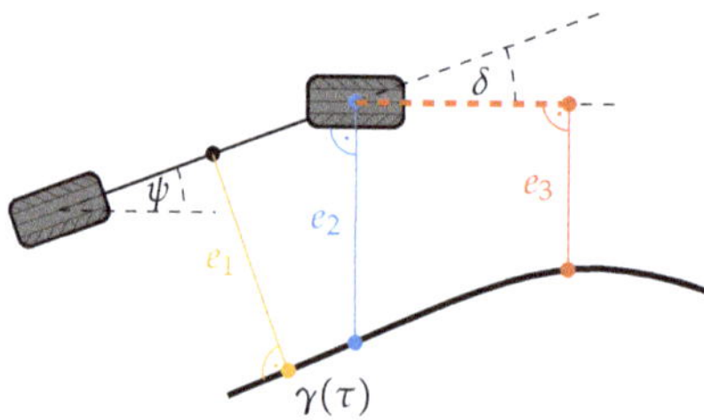

Figure 2. Different tracking error definitions of state-of-the-art tracking controllers.

3. Tracking Error Classification

Based on the analysis of a comprehensive set of state-of-the-art path tracking controllers (cf. Table 2), the introduction of three general and comprehensible classifier sets are introduced (S_{vr}: vehicle reference, S_{lh}: look-ahead (direction and distance) and S_{eo}: error orientation), which are applicable to classify these controllers and to analyze their differences.

$$S_{\text{SotA}} \in S_{\text{vr}} \times S_{\text{lh}} \times S_{\text{eo}} \tag{5}$$

These classifiers shall now be discussed in more detail, in order to summarize the corresponding motivation and the effects on the control design and control performance.

3.1. Vehicle Reference Point

The vehicle reference might be defined at an arbitrary point on the vehicle chassis. There is a set of common reference points, based on a single-track abstraction of a front-axle-steered vehicle (see Figure 3) with different motivations:

- Rear axle: A vehicle, in general, features non-holonomic dynamics. The rear axle is the point of the vehicle with the "most constrained" motion. Assuming zero lateral slip, the motion of the rear axle is aligned with the vehicle heading. Therefore, a constant zero tracking implies that also the vehicle heading is aligned to the reference path, which is a favorable tracking property. From control system theory the center of the rear axle is of interest, as it is a flat output of the system restricting on slip-free vehicle kinematics (see for example [11,16]). The turning radius of the rear axle in cornering is smaller than the turning radius of the front axle (see Figure 4). Therefore, the choice of the rear axle as a vehicle reference point, in general, implies potential undesired overshooting of the vehicle's front.
- Front axle: If the vehicle reference point is set to the center of the front axle, the non-holonomic vehicle kinematics in principle do not have to be considered in the

control design, as stopping and adjustment of the steering angle enables tracking of arbitrary reference paths within the limited turning radius. This enables a simplified control design, especially for low dynamic driving tasks as parking. A drawback of this reference point is the smaller turning radius of the rear axle in cornering (cf. rear axle reference point), which implies potentially undesired curve cutting.

- Center of gravity: The choice of the center of gravity as a vehicle reference, simplifies the setup of the vehicle's equations of motion. Therefore, it is used in many control system design approaches. From tracking perspective its position, somewhere in the middle of the car is of interest, in order to minimize the total distance of all points with respect to the reference path.
- Center of oscillations/percussion: In the center of oscillation or percussion, the translation and rotation impact of a lateral tire slip at the rear axle are in balance. Consequently, this point is of special interest in order to design control laws, which are robust with respect to lateral rear axle tires slip. The choice of this reference point is popular in tracking controllers designed for limit-handling, as racing applications (see for example [17,18]). For front-wheel-steered vehicles the position of the center of percussion with respect to the rear axle l_{CP} is:

$$l_{CP} = l_{CG} + \frac{I_{zz}}{m \times l_{CG}}, \tag{6}$$

where l_{CG} is the distance of rear axle and center of gravity, m the vehicle mass and I_{zz} the vehicle's inertia in the center of gravity with respect to the vertical vehicle axis.

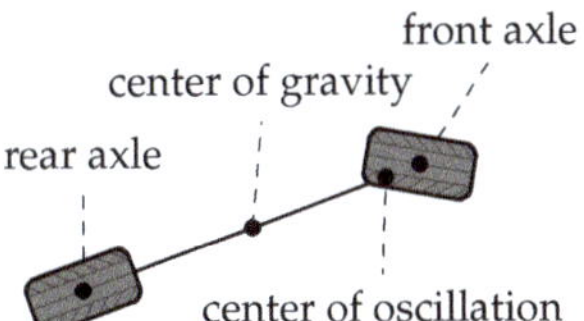

Figure 3. Common vehicle reference points for control error definition based on a single-track vehicle model.

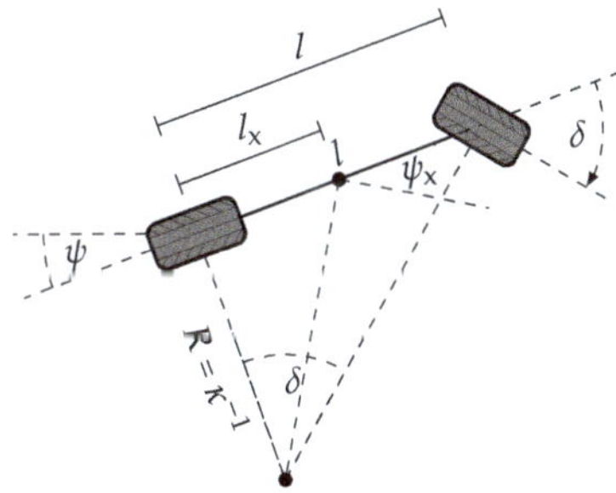

Figure 4. Direction of motion ψ_x of different reference points.

The set of vehicle reference classifiers S_{vr}, hence, can be summarized as:

$$S_{vr} = \{\text{rear}, \quad \text{front}, \quad \text{CG}, \quad \text{CP}\}. \tag{7}$$

3.2. Look-Ahead

Due to the non-holonomic motion of a vehicle, reference points on the path parallel to the vehicle are out of reach without reverse driving or spacious maneuvers. Therefore, many tracking error definitions do not directly use the vehicle reference point to compute a lateral tracking error, but some point ahead of the vehicle reference point. The application of such a so-called look-ahead, or preview (see, for example, [19]), is a very natural behavior of human drivers. The look-ahead refers to the vehicle reference point and is defined by a look-ahead direction and a look-ahead distance. The introduction of a look-ahead enables preventive reaction to sudden direction changes of the reference path, which stabilizes the vehicle motion. As this becomes more crucial for higher vehicle speeds several tracking controllers (see, for example, [19–22]) use an adaptive, velocity dependent look-ahead distance, instead of a fixed distance. This is advantageous also if the tracking controller is applied in combination with simple motion planning systems, which do not provide a smooth path (cf. Section 8), since the look-ahead damps the impact of path discontinuities. In [23], the impact of this damping characteristic is analyzed with respect to control stability in frequency domain base on a linearized vehicle model. This damping characteristic at the same time reveals the main drawback of a look-ahead application. A smooth vehicle motion is achieved at the cost of worse tracking performance in the vicinity of the actual vehicle position (e.g., curve cutting). Even in the case of perfect reference tracking $e_{\mathrm{lat}}(t) \equiv 0$ the vehicle itself actually does not follow the planned reference path. This implies that planned reference path properties, like specific safety distances to obstacles, are withdrawn. If a tracking controller is applied in combination with a comprehensive motion planning system, consequently, one should carefully think about the application of a look-ahead in the tracking error definition. Three reasonable choices for the direction of a look-ahead appear (see Figure 5):

- look-ahead towards the vehicle heading,
- look-ahead towards the direction of motion in vehicle reference point,
- and look-ahead towards the reference path in a certain distance.

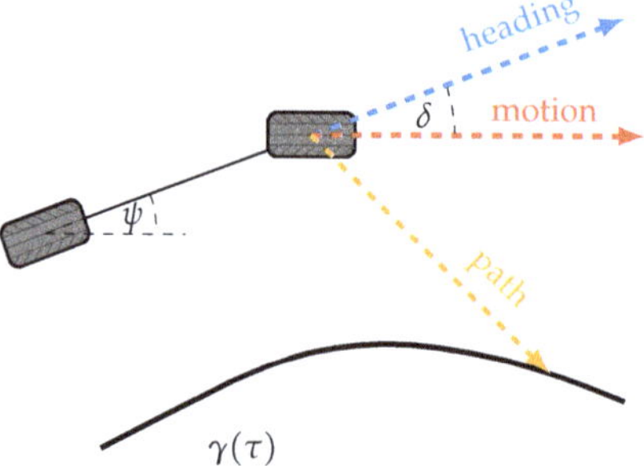

Figure 5. Different look-ahead directions based on the vehicle reference point in the center of the front axle.

Considering vehicle kinematics (see Figure 4), the absolute direction of motion ψ_{x} in a vehicle reference point at distance l_{x} in front of the rear axle (towards the vehicle heading ψ), with respect to a steering angle δ is:

$$\psi_{\mathrm{x}} = \psi + \arctan\left(\frac{l_{\mathrm{x}}}{l} \tan\delta\right). \tag{8}$$

This direction is obviously equal to the vehicle heading at the rear axle (for $l_{\mathrm{x}} = 0$) and equal to the steering direction at the front axle ($l_{\mathrm{x}} = l$). The application of a look-ahead extends the possibilities in achieving specific constitution of the final model-based tracking problem, as it is proposed for example in [12] (pp. 199–201) (front axle vehicle reference and a look-ahead in direction of motion): In this case, decoupling and input–output linearization can be achieved with a static feedback. The above considerations yield the set of look-ahead classifiers S_{lh}:

$$S_{\mathrm{lh}} = \{\text{heading}, \quad \text{motion}, \quad \text{path}\}. \tag{9}$$

3.3. Error Orientation

Based on a vehicle reference point and the optional look-ahead (direction and distance) the orientation of the tracking error needs to be defined in order to compute a lateral tracking error. Similar to the definition of look-ahead directions, reasonable orientations are:

- perpendicular to the vehicle heading,
- perpendicular to the direction of vehicle motion in the vehicle reference point,
- perpendicular to the path.

For almost straight reference paths, different error orientations approximately coincide if the vehicle drives along the reference path. On the other hand, if the vehicle first has to approach the reference path, or drives along curvy path sections, differences reveal (see Figure 6). In summary the classification set for the error orientation is:

$$S_{\text{eo}} = \{\text{heading}, \quad \text{motion}, \quad \text{path}\}. \tag{10}$$

The above definitions can be used to give a comprehensive definition of a lateral tracking error based on the classifiers *vehicle reference*, *look-ahead* (direction and distance) and *error orientation* in the combined set,

$$S = S_{\text{vr}} \times S_{\text{lh}} \times S_{\text{eo}}, \tag{11}$$

according to (7), (9) and (10). Figure 6 exemplarily illustrates all possible error definitions for the vehicle reference at the front axle, i.e., for the set $\{\text{heading}\} \times S_{\text{lh}} \times S_{\text{eo}}$, as listed in Table 1. In the case of a look-ahead towards the path and an error orientation perpendicular to the path, the lateral error has to be defined either with respect to the vehicle heading or motion (see errors $e_{\text{p,p(h)}}$ and $e_{\text{p,p(m)}}$).

Based on given coordinates of the vehicle reference point, a specified look-ahead distance and direction and error orientation define a reference point on the path. In addition to the lateral tracking error, this reference point can be used to compute further error measures like an orientation or curvature error. In [24], the choice of an appropriate heading error is discussed and its impact on the tracking performance is analyzed.

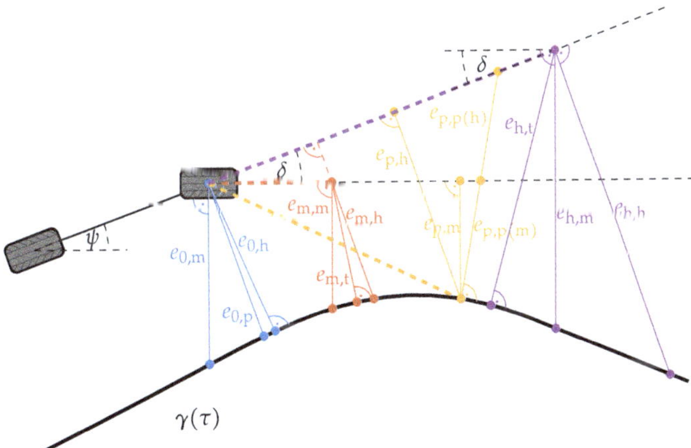

Figure 6. Possible tracking error definitions for vehicle reference at the front axle (see Table 1), i.e., the classification set $\{\text{front} \times S_{\text{lh}} \times S_{\text{eo}}\}$.

Table 1. Corresponding tracking error definition classifiers to Figure 6.

	Look-Ahead	Error Orientation
$e_{0,\mathrm{h}}$	no	heading
$e_{0,\mathrm{m}}$	no	motion
$e_{0,\mathrm{p}}$	no	path
$e_{\mathrm{h,h}}$	heading	heading
$e_{\mathrm{h,m}}$	heading	motion
$e_{\mathrm{h,p}}$	heading	path
$e_{\mathrm{m,h}}$	motion	heading
$e_{\mathrm{m,m}}$	motion	motion
$e_{\mathrm{m,p}}$	motion	path
$e_{\mathrm{p,h}}$	path	heading
$e_{\mathrm{p,m}}$	path	motion
$e_{\mathrm{p,p(h)}}$	path	path (heading)
$e_{\mathrm{p,p(m)}}$	path	path (motion)

3.4. Application to State-of-the-Art Tracking Controller

To prove the applicability of the proposed tracking error definition classification, it is applied to a comprehensive set of state-the-art tracking controllers. Table 2 lists the classification according to the specified classifiers.

Table 2. Tracking error classification of state-of-the-art path tracking controllers.

Controller	Vehicle Ref.	Look-Ahead	Error Orient.
Hoffmann (*Stanley*) [25], Kolb [26]	front	no	path
Sun [27,28], Kritayakirane [17], Chen [29], Hu [30] Bruschetta [31]	CG	no	path
Chatzikomis [14], Xu [19], Zhang [32]	CG	no	heading
Hiraoka [33]	CP	no	path
Tieber [34], Samson [35], Dominguez [36], Solea [37]	rear	no	path
Nestlinger [20], Ackermann [38], ARGO [21], Guldner [23], Yuan [22]	CG	heading	heading
Elkaim [39]	CG	heading	path
Solea [37]	rear	motion	path
Sentouh [40]	CG	motion	motion
Coulter (*Pure-pursuit*) [41]	rear	path	heading

While Table 2 shows the application of various vehicle reference points, and look-ahead directions, most tracking controllers are based on an error orientation perpendicular to the reference path or to the vehicle heading. Only one of the considered controllers [40] is based on an error orientation perpendicular to the vehicle motion in the reference point. A possible explanation for this is the fact that direct measurement of the actual vehicle motion direction is challenging due to the side slip of the tires. As an alternative, an estimation based on the yaw rate has to be used. In fact, in [40], the authors do not propose a steering controller for application in an autonomous vehicle but propose a human driver model for simulation purpose. In [42], a hybrid concept consisting of an Pure-pursuit and Stanley controller is proposed in order to combine a tracking error definition with look-ahead and without look-ahead. In [17], the control law is designed with a vehicle reference point in the center of percussion. The used lateral control error e_{CP}, however, is a projection of the distance of the center of gravity to the path e_{CG}:

$$e_{\mathrm{CP}} = e_{\mathrm{CG}} + d_{\mathrm{CG,CP}} \times \sin e_{\psi}, \tag{12}$$

according to the distance between center of gravity and center of percussion $d_{CG,CP}$ and the heading error e_ψ. Therefore, the vehicle reference point is assigned to the center of gravity. In [27,28,31], an MPC is applied for path tracking. In every MPC step, actuation signals are optimized with respect to a specific cost function on a prediction horizon. In general, the cost function includes a deviation from the reference path, i.e., a lateral tracking error. Instead of an evaluation of the tracking error at a single vehicle pose (cf. Figure 8), hence additionally an evaluation at several predicted vehicle poses is required.

The choice of the single introduced classifiers can be considered as part of the control parametrization. As Figure 6 indicates, the actual control error computation is defined by specific geometrical operations with respect to the current vehicle pose, the control parametrization and the provided reference path. A generalization of the tracking error computation is presented in the next section.

4. Tracking Error Computation

Based on a generalized reference point, which already considers an optional vehicle-oriented look-ahead (except for a look-ahead towards to reference path), it is noticeable that despite the various possibilities in the tracking error definition the actual error computation can be handled with only five geometric operations (see Tables 3 and 4 and Figure 7).

The operations **a** (intersection of two lines) and **c** (projection of a point on a line) do not depend on the reference path and have simple analytic solutions. Contrary, the operations **A** (intersection of the reference path with a line), **B** (intersection of the reference path and a circle) and **C** (point projection onto the reference path) depend on the path representation (cf. Appendix A). Hence, the reference path representation also determines the resulting computational complexity. Some general statements can be made on this path operations.

Table 3. Path operations.

	Operation
A	intersection line/path
B	intersection circle/path
C	point projection on path
a	intersection line/line
c	point projection on line

Table 4. Operations for error computation.

		Error Orientation		
		Path	**Heading**	**Motion**
look-ahead	no	C	A	A
	path	B+a	B+c	B+c
	heading	C	A	A
	motion	C	A	A

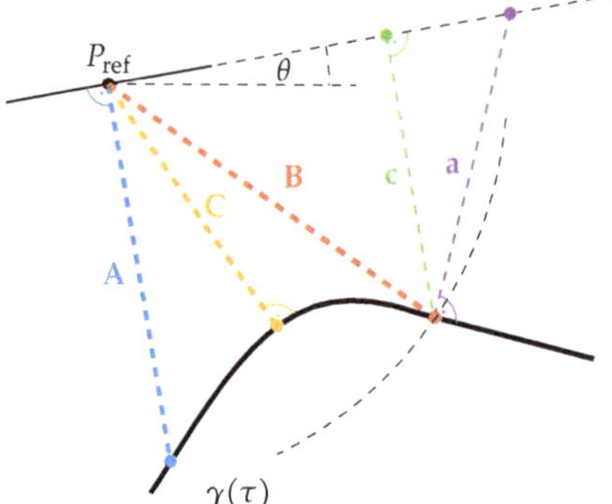

Figure 7. Visualization of path operations in the lateral tracking error computation.

4.1. Intersection of Reference Path and a Straight Line

After performing a translation and rotation of the path, this operation reveals as common root finding problem, which can, in general, not be solved analytically, for example in case of a higher order polynomial reference path. However, many well-established numeric algorithms like Newton-method, bisection-method or Brent-method, can be used to compute arbitrarily exact approximations to the solution of this problem (see, for example, [43–45]). The solution of this path operation is neither unique nor it exists for sure. Hence, tracking controllers relying on this path operation inevitably have to define a fallback solution for the control law. The fallback solution, e.g., a constant curvature turn, has to ensure the convergence to a region where the path operation yields a solution. With respect to the uniqueness, it is reasonable to use the closest solution.

4.2. Intersection of Reference Path and a Circle

From an analytic perspective, this operation can be considered equivalent to operation **A** after performing a transformation of the corresponding reference path section into polar coordinates, which is challenging for an arbitrarily parameterized path. From numerical perspective, it might also be considered as a generalization of the shortest-distance-problem (cf. operation **C**), searching for a fixed given distance to the reference path instead of the shortest distance. The solution of this path operation in general yields at least two solutions, which can be prioritized following again the most-progress-on-path principle. The solution of this path operation obviously does not exist if the shortest distance between the center point of the circle and the path exceeds the specified circle radius.While this offers a straight-forward method to test for the existence of a solution, there is, in contrast to operation **A**, no simple fallback, which can be applied for arbitrary controllers, which rely on this path operation. Therefore, a reliable fallback solution, which ensures that the vehicle is approaching the path, has to be design including the specific control law. For many controllers including the famous pure-pursuit controller [41], the solution of the shortest-distance-problem (see operation **C**) is applicable as a fallback solution.

4.3. Point Projection Onto the Reference Path

Point projection is a well investigated topic, but still an ongoing research field in geometry (see, for example, [46–49]). Although for some path representations (e.g., if the path consists of circular arcs) an analytic solution exists, most implementations apply iterative numeric methods, similar to path operation **A**, to compute approximations of the solution. The existence of a solution is not ensured. A widely used fallback for tracking controllers is to transition to the shortest-distance-problem. A solution to this fallback problem always exists and if a solution to the point projection problem exists, the solutions are equivalent. The ambiguity of the solution can be handled based on the path direction (choosing the solution which gains the most progress along the reference path).

5. Summary of Interface Requirements from Tracking Control Perspective

The comprehensive investigations of the last sections provide the theoretical backbone in order to state the general requirements for a generic planning and control interface from control side. The central idea for the interface implementation is to detach the tracking error computation from the control component. On the one hand, this avoids redundant computations in different control components. On the other hand, it enables careful coordination with respect to the representation of the provided reference path. This supports the performance of the overall system, which is especially of interest in safety critical applications like collision avoidance. According to Section 3, the definition of three classifiers (vehicle reference point, look-ahead and error orientation) is sufficient to define the lateral tracking error definition applied in state-of-the-art path tracking controllers. Furthermore, according to Section 4, three elementary path operations are sufficient to compute all possible tracking errors, which can be defined based on the introduced tracking error classifiers. Hence, these elementary path operations can be implemented with respect to the reference path representation outside of the actual path tracking component, in order to supply various different tracking controllers with the required tracking error, as illustrated in Figure 8.

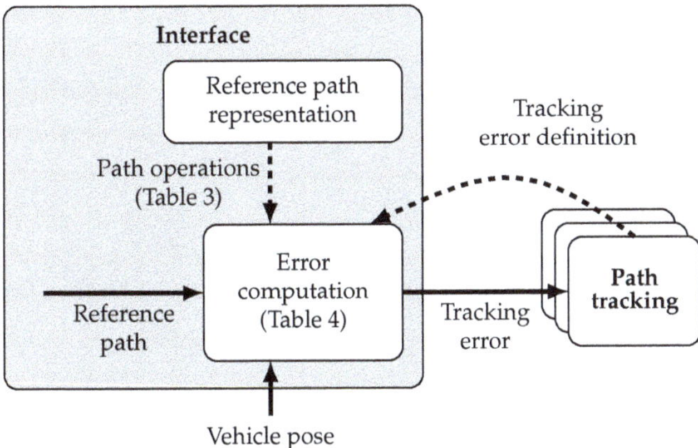

Figure 8. Interface concept to meet the output requirements defined by the state-of-the-art in path tracking controller design.

In order to complete the interface concept, the following section consider the path planning problem in general and the interface requirements resulting from the state-of-the-art in path planning.

6. Interface Requirements from Path Planning Perspective

Similar to path tracking path planning, is a sub-problem of the more general trajectory planning problem. It focuses exclusively on the spatial planning, also known as lateral planning, neglecting the temporal aspect of the motion planning. The general limitations of the applicability of such a decoupled motion planning and control have already been mentioned in the Section 2. In addition, from planning perspective also limitations of this approach with respect to complex dynamic environments arise. However, this separation enables simplified planning for relevant use-cases like valet parking. The general path planning problem can be stated as follows: The planner has to compute a collision-free path from a given starting pose to a given final pose, complying constraints with respect to drivability (limited actuation systems, like a vehicle's bounded steering angle), comfort (bounded lateral acceleration and jerk), efficiency (length of the path and necessary actuation effort) and safety (distance to obstacles). Many surveys on different path planning approaches and algorithms have been published (see, for example, [2,5,50–52]).

In contrast to the state-of-the-art path tracking controllers, where it was necessary to establish a reasonable generalization of the input requirements via a careful study of state-of-the-art components, the generalization of the output requirements of path planning components is more straight forward and does not need an extensive consideration of specific state-of-the-art path planning algorithms. On a qualitative level, there are two different output formats provided by path planning systems: Hermite path data and analytic curve expressions.

Hermite path data consist of a set of consecutive way points (position $x, y \mapsto$ G0 Hermite data) and optional higher-order geometric information: tangent (position x, y + orientation $\psi \mapsto$ G1 Hermite data), curvature (position x, ys + orientation ψ + curvature κ: $\mapsto$ G2 Hermite data (cf. Figure 9)) and so on. Hermite path data give a hint about the qualitative course of a continuous path, assuming that the single samples do not skip a significant section of the path.

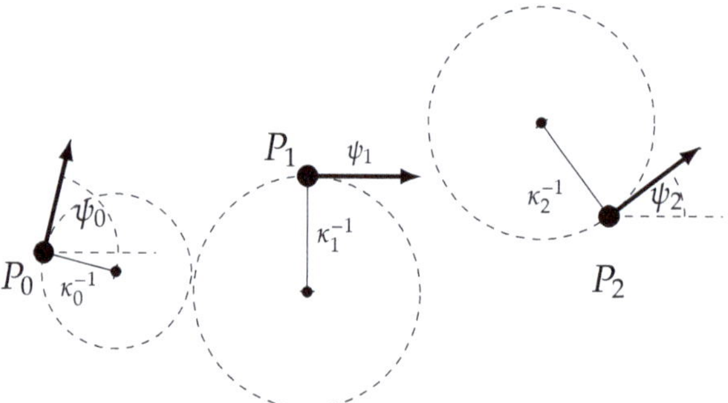

Figure 9. Exemplary G2 Hermite waypoint data, consisting of position P, orientation ψ and curvature κ.

Alternatively a path planning algorithm may provide the reference path in form of an analytic curve expression. The most commonly used analytic curve expressions are parametric curves. In Appendix A, an overview on the basics of parametric curves and commonly used parametric curves expressions in path planning are given.

The more accurate the shape of a single analytic curve shall be specified, the more complex representations (e.g., high-order polynomials) are required. In practice, this yields an increased computational complexity and corresponding numerical issues. Hence, it is far more practicable to represent the planned path not as one single analytic curve but as a sequence of aligned curve segments, in which each are defined by simple analytic curves, yielding a so-called spline. The application of a segmented path needs an adaption of the elementary path operations stated in Section 4. The determination of a tracking control error has to be managed in two stages:

1. Global: Identification for the respective path segment.
2. Local: Application of the actual error computation within the path segment (see Section 4).

The identification of the respective path segment can inflate to a complex issue, since the actual optimal solution of this task requires the application of the path operation (step 2) on each path segment. In order to reduce the computational effort for step 1, a practicable workaround is to aim for a sub-optimal solution. Such a solution can be obtained, for example by applying the path operation to a simplified version of the path, like for example a linear interpolation of the way point within the path segments. However, the discrepancy between the sub-optimal solution and the optimal one, may become significant especially for reference paths including sharp turns, loops or cusps.

In summary, the path representations provided by state-of-the-art path planning algorithms feature a huge diversity. There is no obvious link for a potential generalization similar to the error classification for path tracking. However, according to Section 5, the implementation of identified elementary path operations depends on the reference path representation. Therefore, it is reasonable to include an internal reference path representation into the interface, which, in general, may differ from a possible parametric path provided by the planner. Since it is straight forward to generate Hermite path data from a given parametric path, by path sampling, the generalized input format is defined as Hermite path data. The drawback of this definition is that the proposed interface may withdraw analytic curve expressions potentially provided by high-sophisticated planning components.

The choice for an internal path representation is a degree of freedom in the interface design. Appendix A gives important hints for the pros and cons of different curves. A common choice are polynomial splines. Such splines may be defined in different polynomial bases (e.g., monomial and Bernstein basis). The missing step to finalizing the wanted generic motion planning and control interface is the parametrization of the internal path with respect to given Hermite way point data. This is covered by the well investigate research field of interpolation. For sake of completeness, Appendix B surveys some state-of-the-art algorithms for the interpolation of Bezier splines, which serve as internal path representation for the exemplary implementation of the interface (cf. Section 8).

Figure 10 illustrates the intermediate result with respect to the input structure of the proposed generic interface.

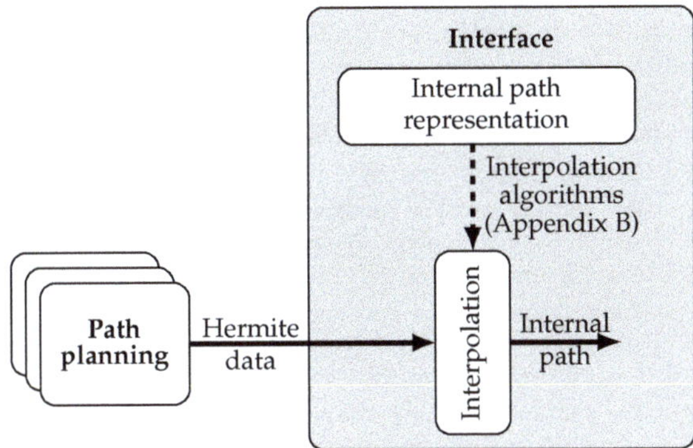

Figure 10. Interface concept to meet the input requirements defined by state-of-the-art reference path representations.

7. A Generic Path Planning and Tracking Interface

Combining the above considerations from path planning and path tracking side, the concept for the generic interface design shall now be summarized (see Figure 11). The three major steps in the implementation of such an interface are:

- Decision for an internal path representation.
- Implementation of corresponding Hermite waypoint data interpolation algorithms (see Appendix B), in order to accomplish input (planning) modularity with respect to different data types (G0, G1, ···).
- Implementation of corresponding error computation, based on the path operations discussed in Section 4 (path-line intersection, path-circle intersection and point projection), in order to accomplish output modularity (control).

The key design decision is the choice of an internal path representation. The internal path representation, on the one hand, has to be able to accurately describe a planned reference path. On the other hand the computational aspects of the required interpolation and path operation algorithms have to be considered. It is an important fact that this design decision impacts the final performance of the motion planning and control system, in terms of tracking performance (cf. Section 8) and computational effort.

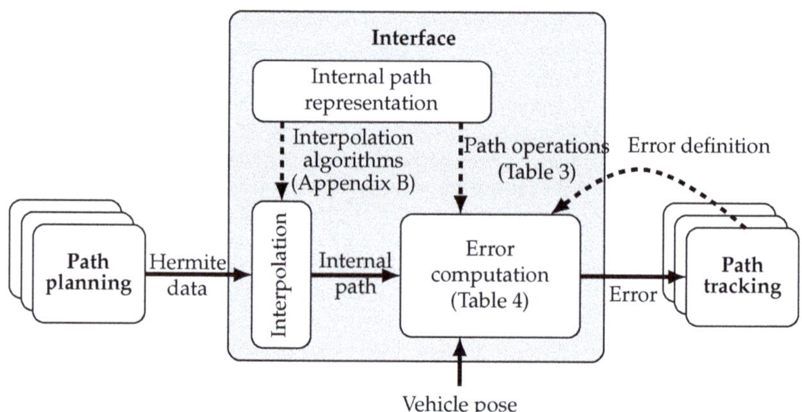

Figure 11. Interface for modular motion planning and control systems.

Such an interface is able to connect state-of-the-art path planning algorithms, which provide Hermite waypoint data, with state-of-the-art path tracking controller, which requires a lateral tracking error definition that is covered by the introduced error classifiers: vehicle reference, look-ahead and error orientation. Consequently, it enables the combination of a huge range of planning or tracking components without necessary interface modifications, which offers significant benefits in both simulation and operation:

- In simulation, it supports a straight-forward identification of an appropriate component set, based on iterative combination, simulation and evaluation, considering specific scenarios and corresponding KPIs. This is an important aspect of ODD-based AD function assembly. Furthermore, a specific error definition can be applied as common evaluation measure for a set of tracking controllers, which could not be compared on a quantitatively based on their different native error definition on a fair basis (cf. the example in Section 8).
- In operation, it enables the simultaneous execution of different software components as well as switching between different components. This on the one hand supports the design of AD-functions, for a set of varying ODDs extending the application range of Level 4 driving functions. On the other hand, it supports the application of redundant and fail-operational software in order to increase safety of an AD function.

In addition to these major benefits in simulation and operation, the application of the proposed interface offers additional possibilities in the AD function development. The resolution of the component interdependencies, which might impact the overall system performance, is a significant advantage in safety critical applications, like collision avoidance. Furthermore, as already mentioned in Sections 2 and 3.4, the tracking error computation can be executed also on a set of vehicle poses, generated by some motion prediction in order to support, e.g., MPC-based path tracking approaches. However, there might occur cases, which exclude the possibility of outsourcing the error computation to an interface component for some algorithmic reasons. In this case, the proposed interface still can operate as a reference path pre-processing unit, which may provide reference path sections of fixed quality, with respect to configurable requirements, e.g., equidistant sampling and path continuity, without any modifications of the planning components. Finally, the application of specific error definitions based on the current and predicted vehicle poses are also applicable and beneficial as risk indicators in threat and risk assessment (see for example [53,54]).

8. Exemplary Interface Application

In this final section, some of the aforementioned benefits and aspects of a generic interface shall be demonstrated in a practical use case. Starting point is a simplified but exemplary ODD-based AD function assembly problem. A path tracking controller shall be selected for application in an AD function for highway driving. This ODD arises a considerable number of specific scenarios. This example is restricted to an exemplary lane-change maneuver. Therefore, a standardized maneuver—the second half of a standardized double lane-change maneuver in [31]—is considered. The maneuver is performed at a constant speed of v = 72 km/h. Since no obstacles have to be considered in this maneuver a simple path planner is used. It uses cubic Bezier splines (cf. Appendix A) to plan a smooth trajectory within the defined maneuver corridor. The planned path is provided to the generic interface as G2 Hermite data, with a sample distance of 5 m. In order to show the impact of the internal path representation on the control performance, two different interface implementations are applied in simulation. The first one uses a simple linear path representation. The provided Hermite path data are interpolated linearly, resulting in piece-wise straight sections, respectively piece-wise linear orientation and curvature. Note that the linear interpolation of orientation and curvature do not represent the actual orientation and curvature of the resulting polygonal path, which would be piece-wise constant respectively zero, but serves as an improved approximation. The second interface implementation is based on piece-wise polynomial path sections an applies the interpolation algorithms described in Appendix B. The G2 Hermite waypoint data are interpolated with quintic Bezier splines in order to achieve a G2 continuous internal reference path.

An exemplary set of two state-of-the-art path tracking controllers is used with a fixed parametrization (see Tables 5 and 6), including the applied lateral tracking error definition according to the proposed classification in Section 3:

- Stanley ([25]):

$$\delta(t) = e_\psi - k_{\text{ag}} \times v^2 \times \kappa_{ref} \times \sin e_\psi + \arctan\left(\frac{k \times e_{\text{lat}}}{k_{\text{soft}} + v}\right) \tag{13}$$

- PurePursuit ([41]):

$$\delta(t) = \arctan\left(\frac{2 \times l \times e_{\text{lat}}}{d_{\text{lh}}^2}\right) \tag{14}$$

Table 5. Parametrization of Stanley tracking controller.

Parameter	Value
k	19
k_{soft}	1
k_{ag}	0.013
vehicle reference	front
look-ahead	no
error orientation	path

The two controllers shall just serve as an exemplary set of components for this demonstration. The performed evaluation can be extended straight forwardly for example to the entire set of tracking controllers listed in Table 2.

In order to compare the performance of different controllers, which, in general, apply different error definition, an additional lateral tracking error definition (vehicle reference: center of gravity, look-ahead: no, error orientation: heading) is used to provide a comparable error measure. The path planning system, the two exemplary interface implementations as well as the exemplary tracking controllers have been implemented in a Python-based software framework. This framework has been used to perform the simulations, which are discussed within the following section.

Table 6. Parametrization of PurePursuit tracking controller.

Parameter	Value
l	2.68 m
d_{lh}	10 m
vehicle reference	rear
look-ahead	path
error orientation	heading

8.1. Discussion

Figure 12 shows the results of a series of four simulations (all combinations of the two controllers and the two interfaces implementations). According to the top plots, both controllers in principle accomplish the required scenarios task and keep the vehicle inside the defined scenario corridor. Due to its look-ahead the lateral error of the PurePursuit controller increase earlier, resulting in an earlier initiation of the lane-change maneuver. Furthermore, this yields a more smooth maneuver with reduced steering effort due to the damping behavior of a look-ahead in path tracking (cf. Section 3.2 and [23]). The drawback of this property is the curve-cutting behavior of the final vehicle motion, which brings the vehicle to the borders of the scenario corridor several times. Considering the defined common error measure, this fact substantiates in a considerable ≈ 1 m) lateral offset of the vehicle's center of gravity with respect to the reference path. From this point of view, the Stanley controller shows superior tracking performance to the cost of an increased steering effort.

The comparison of the two different applied interfaces reveals the important fact that the internal path representation impacts the final tracking performance. The non-smooth linear reference path yields a non-smooth tracking error. The damping characteristic of the PurePursuit's look-ahead still ensures a satisfying steering command, which is almost equivalent to the corresponding steering command when applying the polynomial internal path. Contrary, the Stanley controller is affected through-out by this effect, yielding an unsatisfying jerky steering command. This consideration shows that in absence of an interface the Stanley controller is not applicable at all in this scenario in combination with the used planning system. Obviously, this is an illustrative edge-case example and the effect may be reduced when using a more dense sampling of the reference path, but, in general, this might be a fixed parameter of the planning system. The slight steering into the opposite direction of the Stanley controller results from an undesired property of the applied path interpolation algorithm, the so-called Runge's phenomenon.

In order to conclude this exemplary evaluation, based on the defined error measure lateral offset of the vehicle's center of gravity, the Stanley controller has to be favored in this scenario.

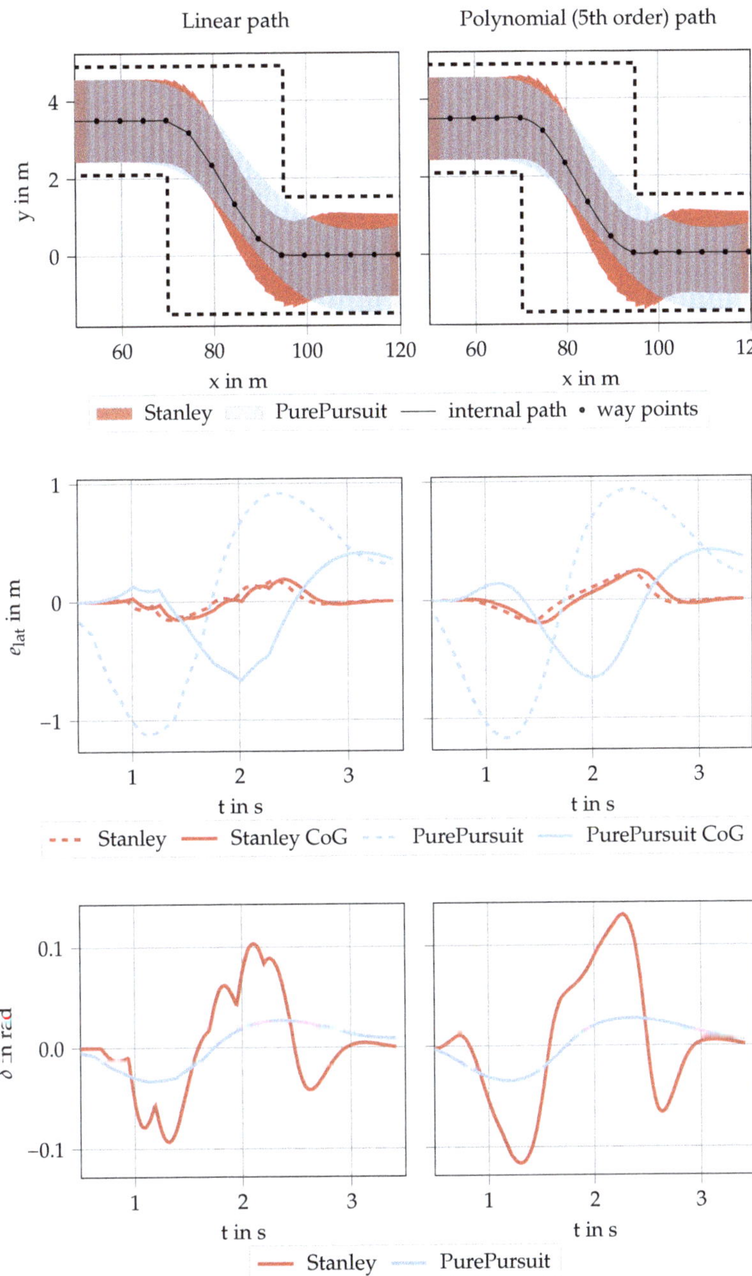

Figure 12. Simulation results: Lane-change maneuver [31] performed with two different path tracking controllers (Stanley [25] and PurePursuit [41]) using two different interface implementations (linear, respectively 5th order polynomial internal path representation). The plots on the top show the vehicle position, the plots in the middle illustrate the occurring lateral tracking errors (controller specific lateral tracking error + common error measure) and the bottom plots show the resulting steering commands computed by the two controllers.

9. Conclusions

Level 4 autonomous driving requires AD-functions, which are carefully matched to the specific ODD. The scientific state-of-the-art provides a tremendous amount of dedicated algorithms, which may be applicable for specific tasks of an AD-function in different ODDs. Hence, contrary to developing AD-functions from scratch for each ODD, it is beneficial to aim for a modular system architecture with generic interfaces, enabling fast combination and evaluation of sets of algorithms, without any code adaptions. This publication is dedicated to the design of such a generic interface between a local path planning and path tracking system. In order to ensure modularity with respect to different state-of-the-art path tracking controller, it contributes a classification of controllers based on the applied lateral tracking error definitions. Based on this classification the actual requirements from control side in the error computation are identified in terms of elementary path operations. This substantiates the advantage of including an internal reference path representation into the interface in order to resolve the interdependencies between the path planning and the path tracking system and, hence, to achieve the required input modularity (on planning side) and output modularity (on control side). With these building blocks, the publication contributes a generic interface concept and a solid theoretical basis for occurring design decisions in the implementation. Two exemplary implementations are finally applied in a demonstrative ODD-based AD assembly task. The application of such an interface offers significant advantages in simulation, like straight-forward combination, evaluation, and benchmarking of set of components in different scenarios, as well as in operation, as a key element for fail-operational and ODD-adaptive AD-function design. A concluding overview on the technical content of the publication is represented in Table 7.

Table 7. Overview on the proposed generic interface concept for path planning and tracking.

Approach	• Analysis of requirements based on state-of-the-art components • Identification of potentials for generalization • Definition of component independent interface tasks and requirements
Concept	• Interface internal continuous reference path representation (Section 6) • Generalization of tracking error definition based on classifiers ((11) and Section 3) • Implementation of elementary path operations (Table 3) based on the internal path representation (Section 4) • Generalized tracking error computation based on error definition and path operations (Table 4) • Parametrization of the internal path by Hermite interpolation of reference path data (Appendix B)
Benefits	• Arbitrary combination of state-of-the-art path planning and tracking algorithms • Avoidance of component independent input/output data processing within the components • Push modular ODD-based AD function design (iterative combination, simulation and evaluation of path planning and tracking components) • Support application of redundant and fail-operational software design to increase safety of an AD function (redundant operation and ODD-based switching of components)

Author Contributions: Conceptualization, J.R. and M.S.; methodology, J.R.; software, J.R.; validation, J.R. and M.S.; formal analysis, J.R. and M.S.; investigation, J.R.; writing—original draft preparation, J.R.; writing—review and editing, J.R., M.S. and D.W.; supervision, M.S. and D.W.; funding acquisition, D.W. All authors have read and agreed to the published version of the manuscript.

Funding: The authors would like to acknowledge the financial support within the COMET K2 Competence Centers for Excellent Technologies from the Austrian Federal Ministry for Climate Action (BMK), the Austrian Federal Ministry for Digital and Economic Affairs (BMDW), the Province of Styria (Dept. 12) and the Styrian Business Promotion Agency (SFG). The Austrian Research

Promotion Agency (FFG) has been authorized for the program management. Supported by TU Graz Open Access Publishing Fund.

Conflicts of Interest: The authors declare no conflict of interest.

Appendix A. Parametric Path

This section gives a theoretical basis on parametric path representation and exemplary concepts.

Appendix A.1. Basics

Analytic curve expressions can be given in terms of a implicit, explicit or parametric expression:

$$\text{implicit: } f(x,y) = 0, \quad \text{explicit: } y = f(x), \quad \text{parametric: } \gamma(\tau) = \begin{bmatrix} x(\tau) & y(\tau) \end{bmatrix}^T. \tag{A1}$$

In general, a parametric expression is most convenient, since it enables a global definition of arbitrary paths (from a starting point P_a to a target point P_b),

$$\gamma(\tau) = \begin{bmatrix} x(\tau) \\ y(\tau) \end{bmatrix} : \quad \gamma(\tau_a) = \begin{bmatrix} x(\tau_a) \\ y(\tau_a) \end{bmatrix} = P_a \quad \text{and} \quad \gamma(\tau_b) = \begin{bmatrix} x(\tau_b) \\ y(\tau_b) \end{bmatrix} = P_b, \tag{A2}$$

without any bijection issues as they occur for implicit and explicit expressions and which require path sectioning and transformation to local coordinate systems.

The curve parameter τ is a monotonically increasing parameter, related to the progress along the path. It is a degree of freedom and can be used for normalization,

$$\tau_a \stackrel{!}{=} 0, \quad \tau_b \stackrel{!}{=} 1, \tag{A3}$$

or to achieve a path length parametrization $\tau = s$ (also known as natural, arc-length or chord-length parametrization):

$$\frac{\partial s}{\partial \tau} = \sqrt{x'^2 + y'^2} \stackrel{!}{=} 1, \tag{A4}$$

holds. x' and y' are the derivatives with respect to the curve parameter $\frac{\partial x}{\partial \tau}$ and $\frac{\partial y}{\partial \tau}$. If the curve parameter in this case is interpreted as time ($\tau = t$), the curve is passed with a speed of $v = 1$ m/s. Therefore, this parametrization is also called unit-speed parametrization (see for example [55]). Orientation and curvature of a parametric path compute as:

$$\theta(\tau) = \arctan\left(\frac{y'}{x'}\right), \tag{A5}$$

$$\kappa(\tau) = \frac{\partial \theta}{\partial s} = \frac{\partial \theta}{\partial \tau}\frac{\partial \tau}{\partial s} = \frac{x'y'' - x''y'}{(x'^2 + y'^2)^{\frac{3}{2}}}, \tag{A6}$$

Appendix A.2. Clothoidal Path

A widely used curve in natural parametrization is the clothoid or Euler spiral. It is defined by a linear curvature with respect to arc length,

$$\kappa(s) = \sigma \times s, \tag{A7}$$

and is widely used as transition curve between straight and circular road segments in road network design to offer a good steering behavior. Therefore, for it is an important parametric curve candidate for path planning (see, for example, [5,56–59]). The consideration of clothoid x, y-coordinates by integration with respect to (A5) and (A6),

$$x(s) = \int_0^s \cos\left(\frac{\sigma\tau^2}{2}\right) d\tau, \quad y(s) = \int_0^s \sin\left(\frac{\sigma\tau^2}{2}\right) d\tau, \tag{A8}$$

reveals the main drawback of a natural curve parametrization: Whereas arc length-based curve parameters (curvature, ...) feature a simple form, the functions for the coordinates may be transcendental functions (in this case Fresnel-Integrals), which cannot be solved analytically. This property of curves in natural parametrization complicates a coordinate-based definition, like curve fitting or interpolation, which is much more straight forward for other curve parametrizations. Therefore, several approaches have been published, aiming at approximation of clothoid section with other parametric curves (see, for example, [60–62]) to overcome this drawback while maintaining the advantage of a bounded path curvature.

Appendix A.3. Polynomial Path

A widely-used and, hence, important parametric path is the polynomial path. It is defined by a set of coefficients,

$$\mathbf{a}_\mathrm{x}^T = \begin{bmatrix} a_{\mathrm{x},0} & \dots & a_{\mathrm{x},n} \end{bmatrix}, \quad \mathbf{a}_\mathrm{y}^T = \begin{bmatrix} a_{\mathrm{y},0} & \dots & a_{\mathrm{y},n} \end{bmatrix}, \tag{A9}$$

with respect to some set of polynomial basis functions—monomial basis functions,

$$\boldsymbol{\tau}_{\mathrm{m},n}^T = \begin{bmatrix} 1 & \tau & \dots & \tau^n \end{bmatrix}, \tag{A10}$$

in the most general case:

$$\gamma(\tau) = \begin{bmatrix} x(\tau) & y(\tau) \end{bmatrix}^T = \begin{bmatrix} \mathbf{a}_\mathrm{x}^T \boldsymbol{\tau}_{\mathrm{m},n} & \mathbf{a}_\mathrm{y}^T \boldsymbol{\tau}_{\mathrm{m},n} \end{bmatrix}^T. \tag{A11}$$

Polynomials in monomial basis can be efficiently evaluated using *Horner's scheme* [63], which features a minimum number of additions and multiplications. Due to possibly huge differences in the range of the single polynomial coefficients, however, numerical instabilities might occur. The Bernstein basis is a widely used alternative to the monomial basis. For a polynomial of order n, the basis functions,

$$\boldsymbol{\tau}_{\mathrm{b},n}^T = \begin{bmatrix} \tau_{\mathrm{b},n,0} & \tau_{\mathrm{b},n,1} & \dots & \tau_{\mathrm{b},n,n} \end{bmatrix}, \tag{A12}$$

compute as:

$$\tau_{\mathrm{b},n,i} = \binom{n}{i} \tau^i (1-\tau)^{n-i} \quad \text{with:} \quad i = [0, 1, \dots, n], \tag{A13}$$

where the relation

$$\sum_{i=0}^{n} \tau_{\mathrm{b}n,i}(\tau) \equiv 1 \quad \text{and:} \quad \tau \in (0,1), \tag{A14}$$

holds. A polynomial curve in Bernstein basis is known as Bezier curve:

$$\gamma(\tau) = \begin{bmatrix} x(\tau) & y(\tau) \end{bmatrix}^T = \begin{bmatrix} \mathbf{b}_\mathrm{x}^T \boldsymbol{\tau}_{\mathrm{b},n} & \mathbf{b}_\mathrm{y}^T \boldsymbol{\tau}_{\mathrm{b},n} \end{bmatrix}^T. \tag{A15}$$

The $(n+1)$ pairs of x, y-coefficients,

$$\begin{bmatrix} B_0 & \dots & B_n \end{bmatrix} = \begin{bmatrix} \mathbf{b}_\mathrm{x}^T \\ \mathbf{b}_\mathrm{y}^T \end{bmatrix}, \tag{A16}$$

the so-called control points, form a convex hull to the curve (see Figure A1) and, hence, offer a comprehensible geometric interpretation of the polynomial coefficients:

$$\gamma(\tau) = \begin{bmatrix} x(\tau) \\ y(\tau) \end{bmatrix} = \sum_{i=0}^{n} B_i \times \tau_{\mathrm{b},n,i}(t). \tag{A17}$$

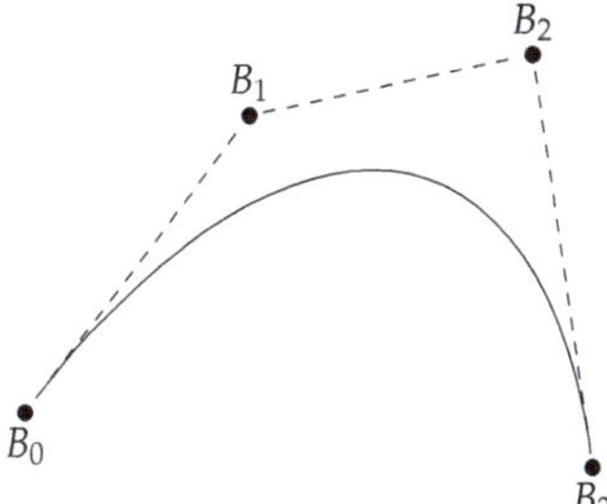

Figure A1. Exemplary cubic Bezier curve from B_0 to B_3 controlled by control B_1 and B_2.

Bezier curves play an important role in computer graphics, and also in path planning. With the *De Casteljau's algorithm* there exists a strong algorithm to evaluate a Bezier curve. It is not as efficient as the *Horner's scheme* but more numerically stable. As also the Bernstein basis finally consists of monomial terms, there exists a static transformation between these two bases:

$$\begin{bmatrix} A_{n,0}^T \\ \vdots \\ A_{n,n}^T \end{bmatrix} = \mathbf{M}_n \begin{bmatrix} B_{n,0}^T \\ \vdots \\ B_{n,n}^T \end{bmatrix} \tag{A18}$$

This transformation may be used to adaptively utilize the advantages of the different bases. Without proof the transformation matrix $\mathbf{M}_n$ can be computed with the following Hadamard product:

$$\mathbf{M}_n = \mathbf{Q}_n \circ \mathbf{P}_{L,-n}. \tag{A19}$$

With a polynomial coefficient matrix $\mathbf{Q}_n$ with switching signs with elements $q_{i,j}$:

$$q_{i,j} := \binom{n}{i}, \quad \text{for:} \quad i,j \in [0,\ldots,n] \tag{A20}$$

and lower triangular pascal matrix extension to negative coefficients $\mathbf{P}_{L,-n}$, which is for example,

$$\mathbf{P}_{L,-3} = \begin{bmatrix} 1 & 0 & 0 & 0 \\ 1 & -1 & 0 & 0 \\ -1 & 2 & -1 & 0 \\ 1 & -3 & 3 & 1 \end{bmatrix}, \tag{A21}$$

for $n = 3$. This relation offers a simple approach to assemble a transformation matrix for specific order of the polynomials, which has not been published yet to the knowledge of the authors.

In order to improve the quality of a polynomial approximations of specific curve sections, e.g., conic sections, while maintaining a low polynomial degree, there exists several generalizations (see, for example, [64]) like rational polynomial functions, B-splines and there combination: non-uniform rational B-splines (NURBS).

Appendix B. Hermite Data Interpolation

Path interpolation is an extensively researched topic and is considered as a special case of path smoothing (see [65]). In [65], a comprehensive overview on the state-of-the-art in path smoothing and interpolation is given. In [66], the mathematical properties of different interpolation functions are considered in very detail. This section shall not survey the complete state-of-the-art in path interpolation, but assemble basic definitions and a survey on interpolation algorithms for Bezier splines. The task of interpolation is to find a parametric curve expression for each path segment, which matches the given Gn Hermite data (see Figure A2), with $n = 0, 1, \cdots$.

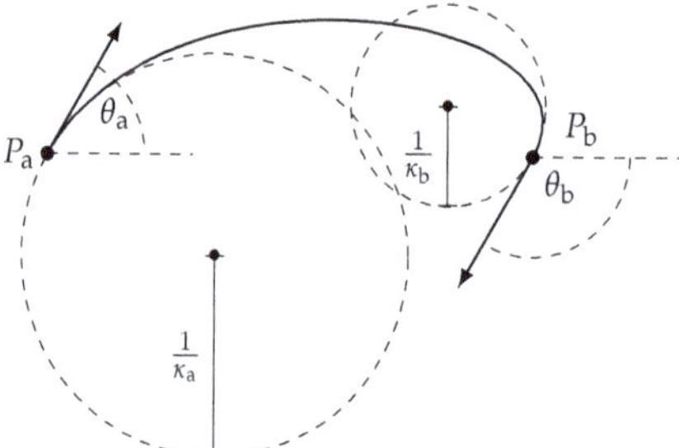

Figure A2. G2 Hermite data interpolation.

The number of matched geometric path information determine the minimal geometric continuity of the total path at the waypoints, i.e., the transition points between the single spline segments. If the path, furthermore, features continuous orientation (and curvature, ...) on the entire definition domain, the curve is called to have G1 (G2, ...) continuity. In contrast to this so-called geometric continuity, differential continuity refers to the derivatives of the curve with respect to the curve parameter τ. Differential continuity always implies geometric continuity. If the curve parameter τ is not applied also for longitudinal velocity planning, differential continuity is a too restrictive requirement, which does not provide any additional geometric features of the curve. However, in practice it is often more convenient to require differential continuity as it will be shown on a comprehensive example of a G2 Hermite interpolation with Bezier curves.

Consider a set of G2 Hermite way point data in two points P_a and P_b. This gives eight boundary conditions and, hence, a cubic Bezier spline (determined by four way points) should be sufficient to meet these conditions:

$$x_a, y_a, \theta_a, \kappa_a, x_b, y_b, \theta_b, \kappa_b \mapsto B_0, B_1, B_2, B_3, \tag{A22}$$

Due to non-linearity of tangent direction and curvature of a general parametric curve, the analysis of the existence and uniqueness of the solutions to this interpolation problem is rather complex. In fact, a cubic spline is not sufficient to interpolate arbitrary G2 Hermite data. Ref. [67] gives a proof that the necessary order for Hermite interpolation of arbitrary G2 data with a Bezier curves is $3 \leq n \leq 5$ depending on the given boundary conditions for tangent directions and curvatures. In [68], conditions for the G2 boundary conditions on the existence and uniqueness of a cubic curve solution are given, restricting on shape preserving curves. In summary, the Hermite interpolation of arbitrary G2 data with Bezier curves of fixed order, in general, requires quintic curves. According to [67], this, consequently, means that occasionally up to 4 additional degrees of freedom occur.

An alternative approach is to implicitly solve the problem, requiring differential continuity (C2 for this example). This enlarges the original set of eight boundary conditions to twelve,

$$x_a, y_a, x'_a, y'_a, x''_a, y''_a, x_b, y_b, x'_b, y'_b, x''_b, y''_b \mapsto B_0, B_1, B_2, B_3, B_4, B_5 \tag{A23}$$

and, consequently, directly requires a quintic Bezier curve (six control points). In [69,70], a parameter vector η has been proposed to utilize the additional degrees of freedom compared to the original problem for later curvature minimization. An extension of this concept for G3 Hermite interpolation has been published in [71,72].

A different common interpolation problem is that a specific geometric continuity between the single spline segments of a curve is required, but not (or not all of them) fixed to a given value. Following again an implicit approach over the differential continuity, a linear equation system can be set up to determine the coefficients of the curve segments. In order to obtain a fully determined equation system, additional global boundary conditions (initial conditions for the first segment and final conditions for the last segment) have to be specified. The necessary spline order n of the curve segments and the resulting number of additional global boundary conditions n_{BC} can be determined straight forwardly considering Cx input data at N way points and required Cy path continuity (with $x \leq y$):

$$\underbrace{2}_{\text{2D}} \cdot \overbrace{2(x+1)}^{\text{fixed local BCs}} \underbrace{(N-1)}_{\text{splines}} + \overbrace{2}^{\text{2D}} \underbrace{(y-x)}_{\text{free lokal BCs}} \overbrace{(N-2)}^{\text{joints}} + \underbrace{n_{\mathrm{BC}}}_{\text{global BCs}} \equiv \overbrace{2}^{\text{2D}} \underbrace{(N-1)}_{\text{splines}} \overbrace{(n+1)}^{\text{coeffs./spline}} \tag{A24}$$

The evaluation of this identity in N (the spline order should be independent of the number of way points) yields:

$$n_{\mathrm{BC}} = 2(y-x) \quad \text{and} \quad n = x+y+1. \tag{A25}$$

Table A1 lists the evaluation of (A25). These simple relations are quite useful in order to implement input data adaptive interpolation algorithm, but have not been published yet to the knowledge of the authors.

In the special case of $x = 0$ and y being an odd number and a homogeneous choice of the global boundary conditions, the resulting splines are called natural splines (see [66]). Such natural splines are characterized by the possibility to continuously extend them with polynomials of order $\frac{n-1}{2}$ outside of the nominal definition space, i.e., before the starting point P_0 respectively after the end point P_{N-1}.

While the splines, as proposed above, are simple to compute, they suffer from several drawbacks like an asymmetric shape in the case that the number of global boundary conditions is not a multiple of four. Furthermore, they are global splines, which means that appending additional way point data affects the entire spline. Another drawback of, in general, all polynomial approaches, is that, although a specific geometric or differential continuity is ensured at the spline transitions, there are, in general, no guarantees about the continuity within the single spline segments. In fact, for spline order $n \geq 3$ cusps (curvature discontinuities) may occur. [73] gives necessary and sufficient conditions for the existence of such cusps. To handle this problem, special types of more restrictive polynomials, which result in so-called regular curves (curves without cusps) have been proposed. For example, in [74] Bezier spirals (a spiral is a curve with monotonic curvature, i.e., without interior curvature extrema) are used to ensure specific continuity.

Table A1. Necessary spline order and number of free boundary conditions for given Hermite data and required differential path continuity.

Data	Continuity	Spline Order	Boundary Conditions
C0	C2	3	4
C1	C2	4	2
C0	C3	4	6
C1	C3	5	4

References

1. SAE. *Taxonomy and Definitions for Terms Related to Driving Automation Systems for On-Road Motor Vehicles*; SAE: Warrendale, PA, USA, 2018.
2. Yurtsever, E.; Lambert, J.; Carballo, A.; Takeda, K. A Survey of Autonomous Driving: Common Practices and Emerging Technologies. *IEEE Access* **2020**, *8*, 58443–58469. [CrossRef]
3. Ulbrich, S.; Reschka, A.; Rieken, J.; Ernst, S.; Bagschik, G.; Dierkes, F.; Nolte, M.; Maurer, M. Towards a functional system architecture for automated vehicles. *arXiv* **2017**, arXiv:1703.08557.
4. Siciliano, B.; Khatib, O. *Springer Handbook of Robotics*; Springer: Berlin/Heidelberg, Germany, 2007.

5. Paden, B.; Cap, M.; Yong, S.Z.; Yershov, D.; Frazzoli, E. A Survey of Motion Planning and Control Techniques for Self-Driving Urban Vehicles. *IEEE Trans. Intell. Veh.* **2016**, *1*, 33–55. [CrossRef]
6. Sorniotti, A.; Barber, P.; De Pinto, S. Path Tracking for Automated Driving: A Tutorial on Control System Formulations and Ongoing Research. In *Automated Driving*; Watzenig, D., Horn, M., Eds.; Springer: Cham, Switzerland, 2017; Chapter 5, pp. 71–140.
7. You, D.; Wang, H.; Yang, K. State-of-the-art and trends of autonomous driving technology. In Proceedings of the TEMS-ISIE 2018-1st Annual International Symposium on Innovation and Entrepreneurship of the IEEE Technology and Engineering Management Society, Beijing, China, 30 March–1 April 2018. [CrossRef]
8. Kato, S.; Tokunaga, S.; Maruyama, Y.; Maeda, S.; Hirabayashi, M.; Kitsukawa, Y.; Monrroy, A.; Ando, T.; Fujii, Y.; Azumi, T. Autoware on board: Enabling autonomous vehicles with embedded systems. In Proceedings of the 9th ACM/IEEE International Conference on Cyber-Physical Systems (ICCPS), Porto, Portugal, 11–13 April 2018; pp. 287–296.
9. Hamid, U.Z.A.; Saito, Y.; Rahman, M.A.A.; Zamzuri, H.; Raksincharoensak, P. A review on threat assessment, path planning and path tracking strategies for collision avoidance systems of autonomous vehicles. *Int. J. Veh. Auton. Syst.* **2018**, *14*, 134. [CrossRef]
10. Tang, J. Review: Analysis and Improvement of Traffic Alert and Collision Avoidance System. *IEEE Access* **2017**, *5*, 21419–21429. [CrossRef]
11. Werling, M.; Gröll, L. From flatness-based trajectory tracking to path following. In Proceedings of the 2009 IEEE Intelligent Vehicles Symposium, Xi'an, China, 3–5 June 2009; pp. 271–1275. [CrossRef]
12. De Luca, A.; Oriolo, G.; Samson, C. Feedback control of a nonholonomic car-like robot. In *Robot Motion Planning and Control*; Laumond, J.-P., Ed.; Lecture Notes in Control and Information Sciences; Springer: Berlin/Heidelberg, Germany, 1998; Volume 229, Chapter 4, pp. 171–253.
13. Snider, J.M. Automatic Steering Methods for Autonomous Automobile Path Tracking. In *Technical Report CMU-RI-TR-09-08*; Robotics Institute, Carnegie Mellon University: Pittsburgh, PA, USA, 2009.
14. Chatzikomis, C.; Sorniotti, A.; Gruber, P.; Zanchetta, M.; Willans, D.; Balcombe, B. Comparison of Path Tracking and Torque-Vectoring Controllers for Autonomous Electric Vehicles. *IEEE Trans. Intell. Veh.* **2018**, *3*, 559–570. [CrossRef]
15. Calzolari, D.; Schurmann, B.; Althoff, M. Comparison of trajectory tracking controllers for autonomous vehicles. In Proceedings of the IEEE Conference on Intelligent Transportation Systems, Proceedings, ITSC, Yokohama, Japan, 16–19 October 2018; pp. 1–8. [CrossRef]
16. Rouchon, P.; Fliess, M.; Levine, J.; Martin, P. Flatness, motion planning and trailer systems. In Proceedings of the 32nd Conference on Decision and Control, San Antonio, TX, USA, 15–17 December 1993; pp. 2700–2705.
17. Kritayakirana, K.; Gerdes, J.C. Using the centre of percussion to design a steering controller for an autonomous race car. *Veh. Syst. Dyn.* **2012**, *50*, 33–51. [CrossRef]
18. Peters, S.C.; Frazzoli, E.; Iagnemma, K. Differential flatness of a front-steered vehicle with tire force control. In Proceedings of the 2011 IEEE/RSJ International Conference on Intelligent Robots and Systems, San Francisco, CA, USA, 25–30 September 2011, pp. 298–304. [CrossRef]
19. Xu, S.; Peng, H. Design, Analysis, and Experiments of Preview Path Tracking Control for Autonomous Vehicles. *IEEE Trans. Intell. Transp. Syst.* **2020**, *21*, 48–58. [CrossRef]
20. Nestlinger, G.; Stolz, M. Bumpless Transfer for Convenient Lateral Car Control Handover. *IFAC-PapersOnLine* **2016**, *49*, 132–138. [CrossRef]
21. Broggi, A.; Bertozzi, M.; Fascioli, A.; Bianco, C.G.L.; Piazzi, A. The ARGO Autonomous Vehicle's Vision and Control Systems. *Int. J. Intell. Control Syst.* **1999**, *3*, 409–441.
22. Yuan, X.; Huang, G.; Shi, K. Improved Adaptive Path Following Control System for Autonomous Vehicle in Different Velocities. *IEEE Trans. Intell. Transp. Syst.* **2019**, *21*, 3247–3256. [CrossRef]
23. Guldner, J.; Sienel, W.; Tan, H.S.; Ackermann, J.; Patwardhan, S.; Bünte, T. Robust automatic steering control for look-down reference systems with front and rear sensors. *IEEE Trans. Control Syst. Technol.* **1999**, *7*, 2–11. [CrossRef]
24. Hu, C.; Wang, R.; Yan, F.; Chen, N. Should the Desired Heading in Path Following of Autonomous Vehicles be the Tangent Direction of the Desired Path? *IEEE Trans. Intell. Transp. Syst.* **2015**, *16*, 3084–3094. [CrossRef]
25. Hoffmann, G.M.; Tomlin, C.J.; Montemerlo, M.; Thrun, S. Autonomous automobile trajectory tracking for off-road driving: Controller design, experimental validation and racing. In Proceedings of the American Control Conference, New York, NY, USA, 9–13 July 2007; pp. 2296–2301. [CrossRef]
26. Kolb, J.K.; Nitzsche, G.; Wagner, S. A simple yet efficient Path Tracking Controller for Autonomous Trucks. *IFAC-PapersOnLine* **2019**, *52*, 307–312. [CrossRef]
27. Sun, C.; Zhang, X.; Xi, L.; Tian, Y. Design of a path-tracking steering controller for autonomous vehicles. *Energies* **2018**, *11*, 1451. [CrossRef]
28. Sun, C.; Zhang, X.; Zhou, Q.; Tian, Y. A Model Predictive Controller with Switched Tracking Error for Autonomous Vehicle Path Tracking. *IEEE Access* **2019**, *7*, 53103–53114. [CrossRef]
29. Chen, C.; Jia, Y.; Shu, M.; Wang, Y. Hierarchical Adaptive Path-Tracking Control for Autonomous Vehicles. *IEEE Trans. Intell. Transp. Syst.* **2015**, *16*, 2900–2912. [CrossRef]
30. Hu, C.; Chen, Y.; Wang, J.; Member, S. Fuzzy Observer-Based Transitional Path-Tracking Control for Autonomous Vehicles. *IEEE Trans. Intell. Transp. Syst.* **2020**, 1–11. [CrossRef]

31. Bruschetta, M.; Picotti, E.; Mion, E.; Chen, Y.; Beghi, A.; Minen, D. A Nonlinear Model Predictive Control based Virtual Driver for high performance driving. In Proceedings of the CCTA 2019-3rd IEEE Conference on Control Technology and Applications, Hong Kong, China, 19–21 August 2019; pp. 9–14. [CrossRef]
32. Zhang, C.; Hu, J.; Qiu, J.; Yang, W.; Sun, H.; Chen, Q. A Novel Fuzzy Observer-Based Steering Control Approach for Path Tracking in Autonomous Vehicles. *IEEE Trans. Fuzzy Syst.* **2019**, *27*, 278–290. [CrossRef]
33. Hiraoka, T.; Nishihara, O.; Kumamoto, H. Automatic path-tracking controller of a four-wheel steering vehicle. *Veh. Syst. Dyn.* **2009**, *47*, 1205–1227. [CrossRef]
34. Tieber, K. Motion planning and control of a people mover. Master's Thesis, Graz University of Technology, Graz, Austria, 2019.
35. Samson, C. Path Following And Time-Varying Feedback Stabilization of a Wheeled Mobile Robot. In Proceedings of the Second International Conference on Automation, Robotics and Computer Vision, Singapore, 16–18 September 1992; vol.3.
36. Dominguez, S.; Ali, A.; Garcia, G.; Martinet, P. Comparison of lateral controllers for autonomous vehicle : Experimental results. In Proceedings of the IEEE Conference on Intelligent Transportation Systems (ITSC), Rio de Janeiro, Brazil, 1–4 November 2016; pp. 1418–1423. [CrossRef]
37. Solea, R.; Nunes, U. Trajectory planning with velocity planner for fully-automated passenger vehicles. In Proceedings of the IEEE Conference on Intelligent Transportation Systems (ITSC), Toronto, ON, Canada, 17–20 September 2006; pp. 474–480. [CrossRef]
38. Ackermann, J.; Guldner, J.; Sienel, W.; Steinhauser, R. Linear and Nonlinear Controller Design for Robust Automatic Steering. *IEEE Trans. Control Syst. Technol.* **1995**, *3*, 132–143. [CrossRef]
39. Elkaim, G.H.; Connors, J.; Nagle, J. The overbot: An off-road autonomous ground vehicle testbed. In Proceedings of the Institute of Navigation-19th International Technical Meeting of the Satellite Division (ION GNSS 2006), Fort Worth, TX, USA, 26–29 September 2006; Volume 3, pp. 1449–1456.
40. Sentouh, C.; Chevrel, P.; Mars, F.; Claveau, F. A sensorimotor driver model for steering control. In Proceedings of the IEEE International Conference on Systems, Man and Cybernetics, San Antonio, TX, USA, 11–14 October 2009; pp. 2462–2467. [CrossRef]
41. Coulter, R.C. Implementation of the Pure Pursuit Path Tracking Algorithm. In *Technical Report CMU-RI-TR-92-01*; Carnegie Mellon University: Pittsburgh, PA, USA, 1992.
42. Cibooglu, M.; Karapinar, U.; Soylemez, M.T. Hybrid controller approach for an autonomous ground vehicle path tracking problem. In Proceedings of the 2017 25th Mediterranean Conference on Control and Automation (MED 2017), Valletta, Malta, 3–6 July 2017; Volume 8, pp. 583–588. [CrossRef]
43. Spencer, M.R. Polynomial Real Root Finding in Bernstein Form Polynomial Real Root Finding in Bernstein Form. Ph.D Thesis, Birgham Young University-Provo, Provo, UT, USA, 1994.
44. McNamee, J.M.; Pan, V.Y. Efficient polynomial root-refiners: A survey and new record efficiency estimates. *Comput. Math. Appl.* **2012**, *63*, 239–254. [CrossRef]
45. Meek, D.S.; Walton, D.J. A note on finding clothoids. *J. Comput. Appl. Math.* **2004**, *170*, 433–453. [CrossRef]
46. Ma, Y.L.; Hewitt, W.T. Point inversion and projection for NURBS curve: Control polygon approach. In Proceedings of the Theory and Practice of Computer Graphics (TPCG), Birmingham, UK, 5 June 2003; Volume 20, pp. 113–120. [CrossRef]
47. Chen, X.D.; Zhou, Y.; Shu, Z.; Su, H.; Paul, J.C.; Improved Algebraic Algorithm On Point Projection For Bézier Curves. In Proceedings of the Second International Multi-Symposiums on Computer and Computational Sciences (IMSCCS 2007), Iowa City, IA, USA, 13–15 August 2007.
48. Song, H.C.; Xu, X.; Shi, K.L.; Yong, J.H. Projecting points onto planar parametric curves by local biarc approximation. *Comput. Graph.* **2014**, *38*, 183–190. [CrossRef]
49. Hu, S.M.; Wallner, J. A second order algorithm for orthogonal projection onto curves and surfaces. *Comput. Aided Geom. Des.* **2005**, *22*, 251–260. [CrossRef]
50. Latombe, J.C. *Robot Motion Planning*; Springer: Boston, MA, USA, 1991.
51. Lavalle, S.M. *Planning Algorithms*; Cambridge University Press: Cambridge, UK, 2006.
52. González, D.; Pérez, J.; Milanés, V.; Nashashibi, F. A Review of Motion Planning Techniques for Automated Vehicles. *IEEE Trans. Intell. Transp. Syst.* **2016**, *17*, 1135–1145. [CrossRef]
53. Li, Y.; Zheng, Y.; Morys, B.; Pan, S.; Wang, J.; Li, K. Threat Assessment Techniques in Intelligent Vehicles: A Comparative Survey. *IEEE Intell. Transp. Syst. Mag.* **2020**. [CrossRef]
54. Lefèvre, S.; Vasquez, D.; Laugier, C. A survey on motion prediction and risk assessment for intelligent vehicles. *ROBOMECH J.* **2014**, *1*, 1–14. [CrossRef]
55. Choe, R.; Puig-Navarro, J.; Cichella, V.; Xargay, E.; Hovakimyan, N. Trajectory Generation Using Spatial Pythagorean Hodograph Bézier Curves. In Proceedings of the AIAA Guidance, Navigation, and Control Conference, Kissimee, FL, USA, 5–9 January 2015; pp. 1–32. [CrossRef]
56. Scheuer, A.; Laugier, C. Planning Sub-Optimal and Continuous-Curvature Paths for Car-Like Robots. In Proceedings of the 1998 IEEE/RSJ International Conference on Intelligent Robots and Systems. Innovations in Theory, Practice and Applications (Cat. No.98CH36190), Victoria, BC, Canada, 17 October 1998; Volume 1, pp. 25–31. doi: 10.1109/IROS.1998.724591. [CrossRef]
57. Gim, S.; Adouane, L.; Lee, S.; Dérutin, J.P. Clothoids Composition Method for Smooth Path Generation of Car-Like Vehicle Navigation. *J. Intell. Robot. Syst. Theory Appl.* **2017**, *88*, 129–146. [CrossRef]
58. Bertolazzi, E.; Frego, M. On the G 2 Hermite Interpolation Problem with clothoids. *J. Comput. Appl. Math.* **2018**, *341*, 99–116. [CrossRef]

59. Meek, D.S.; Walton, D.J. Clothoid Spline Transition Spirals. *Math. Comput.* **1992**, *59*, 117. [CrossRef]
60. Vázquez-Méndez, M.E.; Casal, G. Clothoid computation: A simple and efficient numerical algorithm. *J. Surv. Eng.* **2016**, *142*, 1–9, [CrossRef]
61. Chen, Y.; Cai, Y.; Zheng, J.; Thalmann, D. Accurate and Efficient Approximation of Clothoids Using Bézier Curves for Path Planning. *IEEE Trans. Robot.* **2017**, *33*, 1242–1247. [CrossRef]
62. Montés, N.; Herraez, A.; Armesto, L.; Tornero, J. Real-time clothoid approximation by rational bezier curves. In Proceedings of the IEEE International Conference on Robotics and Automation, Pasadena, CA, USA, 19–23 May 2008; pp. 2246–2251. [CrossRef]
63. Horner, W.G. A new method of solving numerical equations of all orders, by continuous approximation. *Philos. Trans. R. Soc. Lond.* **1819**, *109*, 308–335.
64. Piegl, L.; Tiller, W. *The NURBS Book*; Springer: Berlin/Heidelberg, Germany, 1995.
65. Ravankar, A.; Ravankar, A.A.; Kobayashi, Y.; Hoshino, Y.; Peng, C.C. Path smoothing techniques in robot navigation: State-of-the-art, current and future challenges. *Sensors* **2018**, *18*, 3170. [CrossRef]
66. Bojanov, B.D. *Spline Functions and Multivariate Interpolations*; Springer Netherlands Imprint Springer: Dordrecht, The Newtherland, 1993.
67. Schaback, R. Optimal Geometric Hermite Interpolation of Curves. In *Mathematical Methods for Curves and Surfaces II*; Daehlen, M., Lyche, T., Schumaker, L.L., Eds.; Vanderbilt University Press: Nashville, TN, US, 1998; pp. 1–12.
68. Krajnc, M. Interpolation scheme for planar cubic G 2 spline curves. *Acta Appl. Math.* **2011**, *113*, 129–143. [CrossRef]
69. Piazzi, A.; Guarino Lo Bianco, C. Quintic G/sup 2/-splines for trajectory planning of autonomous vehicles. In Proceedins of the IEEE Intelligent Vehicles Symposium 2000, Dearborn, MI, USA, 3–5 October 2000; pp. 198–203. [CrossRef]
70. Piazzi, A.; Lo Bianco, C.G.; Bertozzi, M.; Fascioli, A.; Broggi, A. Quintic G2-Splines for the Iterative Steering of Vision-Based Autonomous Vehicles. *IEEE Trans. Intell. Transp. Syst.* **2002**, *3*, 27–36. [CrossRef]
71. Piazzi, A.; Romano, M.; Lo Bianco, C.G. G3-splines for the path planning of wheeled mobile robots. In Proceedings of the European Control Conference (ECC 2003), Cambridge, UK, 1–4 September 2003; pp. 1845–1850. [CrossRef]
72. Piazzi, A.; Lo Bianco, C.; Romano, M. n3-Splines for the Smooth Path Generation of Wheeled Mobile Robots. *IEEE Trans. Robot.* **2007**, *23*, 1089–1095. [CrossRef]
73. Manocha, D.; Canny, J.F. Detecting cusps and inflection points in curves. *Comput. Aided Geom. Des.* **1992**, *9*, 1–24. [CrossRef]
74. Walton, D.J.; Meek, D.S. Cubic Bezier Spiral Segments for Planar G2 Curve Design. In Proceedings of the 7th International Conference on Computer Graphics, Virtual Reality, Visualisation and Interaction in Africa, Franschhoek, South Africa, 21–23 June 2010; pp. 21–26. [CrossRef]

 electronics

MDPI

Article

Integrated Comfort-Adaptive Cruise and Semi-Active Suspension Control for an Autonomous Vehicle: An LPV Approach

Gia Quoc Bao Tran [1], Thanh-Phong Pham [2,*], Olivier Sename [1], Eduarda Costa [1] and Peter Gaspar [3]

[1] CNRS, Grenoble INP, GIPSA-Lab, Univ. Grenoble Alpes, 38000 Grenoble, France; gia-quoc-bao.tran@grenoble-inp.org (G.Q.B.T.); olivier.sename@grenoble-inp.fr (O.S.); eduarda-karoliny.costa@grenoble-inp.org (E.C.)

[2] Faculty of Electrical and Electronic Engineering, The University of Danang—University of Technology and Education, Danang 550000, Vietnam

[3] Systems and Control Laboratory, Institute for Computer Science and Control, Hungarian Academy of Sciences, Kende u. 13-17, H-1111 Budapest, Hungary; gaspar.peter@sztaki.mta.hu

* Correspondence: ptphong@ute.udn.vn

Abstract: This paper presents an integrated linear parameter-varying (LPV) control approach of an autonomous vehicle with an objective to guarantee driving comfort, consisting of cruise and semi-active suspension control. First, the vehicle longitudinal and vertical dynamics (equipped with a semi-active suspension system) are presented and written into LPV state-space representations. The reference speed is calculated online from the estimated road type and the desired comfort level (characterized by the frequency weighted vertical acceleration defined in the ISO 2631 norm) using precomputed polynomial functions. Then, concerning cruise control, an LPV $\mathcal{H}_2$ controller using a linear matrix inequality (LMI) based polytopic approach combined with the compensation of the estimated disturbance forces is developed to track the comfort-oriented reference speed. To further enhance passengers' comfort, a decentralized LPV $\mathcal{H}_2$ controller for the semi-active suspension system is proposed, minimizing the effect of the road profile variations. The interaction with cruise control is achieved by the vehicle's actual speed being a scheduling parameter for suspension control. To assess the strategy's performance, simulations are conducted using a realistic nonlinear vehicle model validated from experimental data. The simulation results demonstrate the proposed approach's capability to improve driving comfort.

Keywords: autonomous vehicle; advanced driver-assistance system; LPV approach; robust control; cruise control; semi-active suspension control; passenger comfort

Citation: Tran, G.Q.B.; Pham, T.-P.; Sename, O.; Costa, E.; Gaspar, P. Integrated Comfort-Adaptive Cruise and Semi-Active Suspension Control for an Autonomous Vehicle: An LPV Approach. *Electronics* **2021**, *10*, 813. https://doi.org/10.3390/electronics10070813

Academic Editor: Umar Zakir Abdul Hamid

Received: 23 February 2021
Accepted: 25 March 2021
Published: 30 March 2021

1. Introduction

Autonomous vehicles always remain an interesting research topic thanks to their numerous advantages, including collision avoidance and fuel consumption reduction capabilities, satisfying traffic safety and environmental objectives.

There has been a considerable amount of research work conducted on either cruise or suspension control of autonomous vehicles. Cruise control refers to the control of the vehicle speed, which is related to longitudinal dynamics, for multiple purposes such as collision avoidance [1,2]. For this, different control strategies (optimal, robust, linear parameter-varying (LPV), etc.) have been proposed [3–7]. Recently, cruise control has been linked to a comfort objective [8–10], which extends the field to the coordination between longitudinal and vertical controllers.

Indeed, the suspension system is a key subsystem that allows us to improve the driving comfort and road-holding performance of the vehicle [11,12]. This is thanks to its remarkable ability to limit the vertical oscillations of the vehicle body caused by road displacements at the four wheels. From recent years, it is known that the semi-active suspension system provides better performance than the passive one while being less energy-consuming than the active one [12]. Existing work on semi-active suspension control includes model predictive control or state feedback with various observers, from robust, LPV, to unified ones [11–15].

However, there has not been much work combining cruise and suspension control into an integrated problem, considering their interaction. Besides, very few studies do consider the driving comfort level in a cruise control problem. To improve driving comfort, a potential strategy is to relate the speeds at which the vehicle should travel to the desired comfort level and w.r.t specific road profiles. Such speed values are determined using criteria formed by examining the human body, including which range of frequency is most absorbed by humans. Our group has conducted a study [16] into relating the vehicle speed with the comfort level measured using the ISO 2631 standard [17] and the international roughness index (IRI) [18] for each road type from A to D (defined in [19]). Recent research about road profile estimation using adaptive observers allows us to detect which road type the vehicle is traveling on [20], thus enabling this strategy.

The purpose of this paper is to bring further results and to introduce a comfort-oriented strategy of the integrated cruise–suspension control of an autonomous vehicle. There has been some existing work combining these problems [9,21]. In the latter, the coupling between longitudinal and vertical motions is considered but not the comfort objective. The work [9] requires too many assumptions and much information from the environment (therefore being challenging to embed in reality). Therefore, this work proposes a more realistic approach, handling unknown inputs using a robust LPV control approach. We analyze both longitudinal and vertical dynamics and their interaction through the road displacement at each of the four wheels. The $\mathcal{H}_2$ condition is used as it is suited for the type of noise we are faced with in this suspension control case where one of the sensors is an accelerometer. For cases where the variation of a specific parameter(s) significantly affects the system, we model the parameter(s) into an LPV problem, which is solved as a set of linear matrix inequalities (LMIs). We also show how driving comfort is evaluated by measuring the vertical acceleration transmitted to passengers, from which we propose a way to relate the current speed to comfort level using the ISO 2631 standard. This allows us to determine which speed the vehicle should travel at in order to guarantee that the acceleration felt by one passenger does not exceed a predefined value. Combining the cruise and suspension controllers with a comfort-oriented reference speed generation algorithm leads to the proposed integrated comfort-oriented vehicle control. The integrated control scheme is then tested using simulations on a realistic nonlinear vehicle model validated from experimental data.

This paper is organized as follows. In Section 2, we present the vehicle longitudinal and vertical dynamics (quarter-car model) then the integrated dynamics model. The general scheme of the strategy is presented in Section 3, which consists of comfort-guaranteeing speed calculation (described in detail in Section 4) and integrated cruise–suspension control (discussed in Section 5). Finally, simulation results are presented in Section 6, which shows the effectiveness of our strategy.

2. Vehicle Dynamics Modeling

This section introduces and discusses the vehicle longitudinal and vertical dynamics considered in this paper. The integrated full-vehicle model is also briefly presented.

2.1. Longitudinal Dynamics

In this part, we present the vehicle longitudinal dynamics and the corresponding system's LPV state-space representation. First, let us make some assumptions for the longitudinal dynamics system:

- The vehicle mass is considered to be time-varying. It is measured online thanks to multiple built-in sensors that detect the additional load (the mass of the empty vehicle is the known nominal mass). This is the most crucial assumption as it allows for gain-scheduling based on mass. The vehicle speed is also directly measurable using a speedometer;

- The road slope is known/estimated in real time thanks to algorithms such as in [22–24]. Such an assumption allows us to implement road slope compensation using a feed-forward term in the cruise control input.

Suppose we have a vehicle of mass m traveling at the speed of v, as shown in Figure 1. Let F be the longitudinal control force on the vehicle, and F_d the total disturbance force.

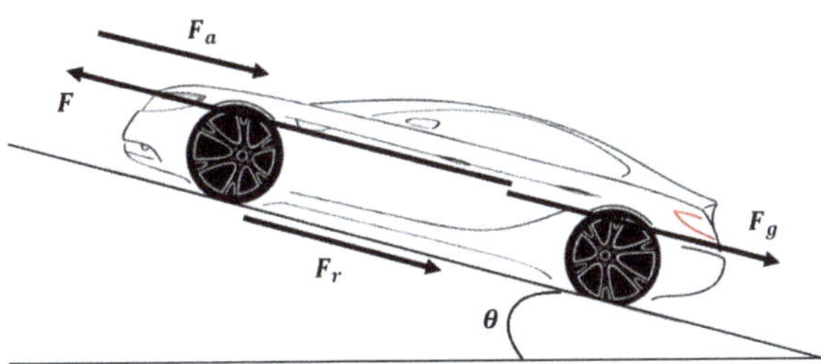

Figure 1. Longitudinal forces on the vehicle.

We have the following equation of motion [3]:

$$m\dot{v} = F - F_d. \tag{1}$$

The disturbance force F_d consists of three components: The rolling friction supposed to have a constant value, the drag by gravity supposing the road slope θ to be sufficiently small (between $\pm 10^\circ$ which is a realistic assumption for real roads), and the aerodynamic drag that adds nonlinearity to the system, respectively:

$$F_r = mgC_r cos(\theta) \approx mgC_r, \tag{2}$$

$$F_g = mgsin(\theta) \approx mg\theta, \tag{3}$$

$$F_a = \frac{1}{2}C_v D_a S v^2, \tag{4}$$

where C_r is the rolling friction coefficient, C_v is the aerodynamic drag coefficient, D_a is the air's density, and S denotes the vehicle's frontal area. The disturbance force thus has the following equation:

$$F_d = mgC_r + mg\theta + \frac{1}{2}C_v D_a S v^2. \tag{5}$$

Finally, the vehicle's motion equation is formulated as:

$$m\dot{v} = F - mgC_r - mg\theta - \frac{1}{2}C_v D_a S v^2. \tag{6}$$

The input force F is composed of two parts:

$$F = F_{ff} + F_l, \tag{7}$$

where $F_{ff} = mg\hat{C}_r + mg\hat{\theta}$ is the feed-forward term that compensates for the rolling friction and the road slope and F_l is the feedback term of the longitudinal control force. Here $\hat{C}_r$ is an estimated nominal value for C_r (constant) and $\hat{\theta}$ is the road slope estimated in real time by the methods in [22–24]. As this compensation is inexact, i.e., $\hat{C}_r \neq C_r$ and $\hat{\theta} \neq \theta$, all the uncertainty is modeled by Δw_l, where Δ is a constant bound and w_l is white noise ($|w_l| \leq 1$).

The system is then written in the LPV form, with $x_l = v$ being the state variable, $u_l = F_l$ being the cruise control input, $y_l = v$ being the measured output, and $\rho_l = \begin{bmatrix}\rho_{l1} & \rho_{l2}\end{bmatrix}^\top = \begin{bmatrix}1/m & v\end{bmatrix}^\top$ being the varying parameter of the longitudinal control case, as:

$$\Sigma_l(\rho_l) : \begin{cases} \dot{x}_l = A_l(\rho_l)x_l + B_{l1}w_l + B_{l2}(\rho_l)u_l \\ y_l = C_l x_l + D_{l1}w_l + D_{l2}u_l, \end{cases} \tag{8}$$

where:

$$A_l(\rho_l) = \left[-\frac{1}{2}C_v D_a S \rho_{l1}\rho_{l2}\right], \quad B_{l1} = [-g\Delta], \quad B_{l2}(\rho_l) = [\rho_{l1}], \quad C_l = [1], \quad D_{l1} = [0], \quad D_{l2} = [0].$$

Vehicle longitudinal dynamics parameters are given in Table 1.

Table 1. Longitudinal dynamics parameters.

Symbol	Value	SI Unit	Parameter Name
m	1400	kg	Vehicle mass
C_r	0.01	-	Rolling friction coefficient
C_v	0.32	-	Aerodynamic drag coefficient
D_a	1.3	kg/m^3	Density of air
S	2.4	m^2	Vehicle frontal area
g	9.8	m/s^2	Gravitational acceleration
τ	0.1	s	Actuator time constant
σ	0.2	s	Communication delay in the vehicle
v_{wind}	12	km/h	Average wind speed

2.2. Vertical Dynamics

The suspension control design is carried out using the quarter-car suspension system [11]. Indeed, this model is simple enough to catch the comfort objective w.r.t the bounce motion and to cope with the requirements about reducing the complexity of an embedded controller. For pitch/roll control, a full-vehicle model would be needed, which is not the case here.

We use the quarter-car model with a semi-active magneto-rheological (MR) suspension system to model the vehicle vertical dynamics, as shown in Figure 2. This consists of the sprung mass m_s, the unsprung mass m_{us}, and the suspension components positioned between them, including a spring element with stiffness k_s and the damper part. Let us denote z_s and z_{us} as the sprung and unsprung mass' displacement, respectively.

From Newton's second law of motion, we obtain:

$$\begin{cases} m_s\ddot{z}_s = -F_{spring} - F_{damper} \\ m_{us}\ddot{z}_{us} = F_{spring} + F_{damper} - F_{tire}, \end{cases} \tag{9}$$

where $F_{spring} = k_s(z_s - z_{us})$ is the spring force and $F_{tire} = k_t(z_{us} - z_r)$ is the tire force.

Concerning the damper force F_{damper}, two models are considered:

- A control-oriented model as given below:

$$F_{damper} = \underbrace{k_0(z_s - z_{us}) + c_0(\dot{z}_s - \dot{z}_{us})}_{F_{passive}} + F_v, \tag{10}$$

where F_v is the control input;

- A simulation model as given below and shown in Figure 3:

$$F_{damper} = k_0(z_s - z_{us}) + c_0(\dot{z}_s - \dot{z}_{us}) + I \cdot f_c \cdot tanh(k_1(z_s - z_{us}) + c_1(\dot{z}_s - \dot{z}_{us})), \quad (11)$$

where c_0, k_0, c_1, and k_1 are constant parameters and I is the applied current. In order to design the controller, the controlled part in (11) is defined as $F_v = I \cdot f_c \cdot tanh(k_1(z_s - z_{us}) + c_1(\dot{z}_s - \dot{z}_{us}))$.

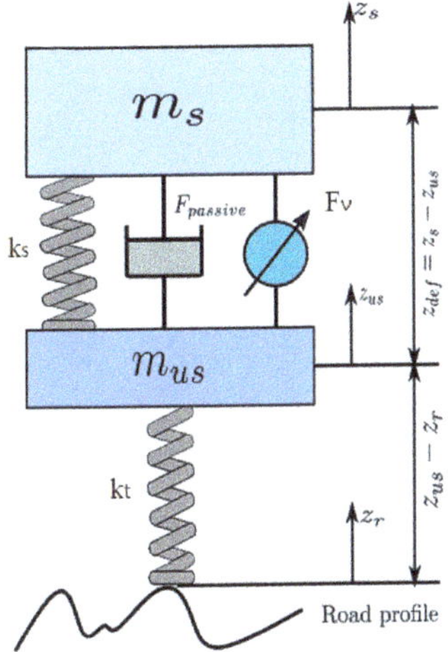

Figure 2. The quarter-car model for an illustration of vehicle vertical dynamics.

In Figure 3, the considered MR damper force – deflection velocity ($\dot{z}_{def} = \dot{z}_s - \dot{z}_{us}$) characteristic is shown, from the MR damper available at ITESM, Mexico (refer to [25]).

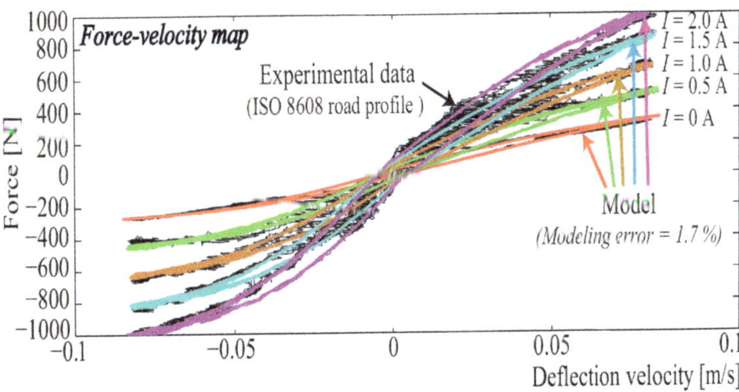

Figure 3. Magneto-rheological damper force–deflection velocity characteristic [25].

The model in (10) is used to design the controller. Then, the nonlinear model (11) is used as the inverse model to simulate the suspension controlled system for the full car model presented later.

Remark 1: The controller to be designed in this paper is applied to the semi-active suspension system using the clipped strategy as used in [26]. Then, the control input current to be applied to the MR damper is computed from the clipped controlled damper force and given the deflection ($z_{def} = z_s - z_{us}$) and the deflection velocity $\dot{z}_{def}$.

To link with longitudinal dynamics, we here consider benefiting from some knowledge of the road displacement input model z_r, which is related to the current vehicle speed according to [27] as:

$$\dot{z}_r + a \cdot v \cdot z_r = b \cdot v \cdot w_v, \tag{12}$$

where w_v is white noise, and a and b are coefficients that depend on the road type according to International Organization for Standardization (ISO) classification [19].

Remark 2: Using a road profile model is indeed possible since the information on the type of road profile may be obtained using some adaptive road profile estimator, as proposed in [20], or a frequency-wise approach [27].

From (9) and (12), by selecting the system states as $x_v = \begin{bmatrix} z_s & \dot{z}_s & z_{us} & \dot{z}_{us} & z_r \end{bmatrix}^\top \in \mathbb{R}^5$, the measured variables $y_v = \begin{bmatrix} \ddot{z}_s & z_s - z_{us} \end{bmatrix}^\top \in \mathbb{R}^2$, the control input $u_v = F_v$, and by choosing the scheduling variable $\rho_v = v$ to link with longitudinal dynamics, the extended quarter-car system can be written in the LPV form as:

$$\Sigma_v(\rho_v) : \begin{cases} \dot{x}_v = A_v(\rho_v)x_v + B_{v1}(\rho_v)w_v + B_{v2}u_v \\ y_v \ = C_{v2}x_v + D_{v21}w_v + D_{v22}u_v, \end{cases} \tag{13}$$

where:

$$A_v(\rho_v) = \begin{bmatrix} 0 & 1 & 0 & 0 & 0 \\ -\frac{k}{m_s} & -\frac{c_0}{m_s} & \frac{k}{m_s} & \frac{c_0}{m_s} & 0 \\ 0 & 0 & 0 & 1 & 0 \\ \frac{k}{m_{us}} & \frac{c_0}{m_{us}} & -\frac{k+k_t}{m_{us}} & -\frac{c_0}{m_{us}} & \frac{k_t}{m_{us}} \\ 0 & 0 & 0 & 0 & -a \cdot \rho_v \end{bmatrix}, \quad B_{v1}(\rho_v) = \begin{bmatrix} 0 \\ 0 \\ 0 \\ 0 \\ b \cdot \rho_v \end{bmatrix}, \quad B_{v2} = \begin{bmatrix} 0 \\ -\frac{1}{m_s} \\ 0 \\ \frac{1}{m_{us}} \\ 0 \end{bmatrix},$$

$$C_{v2} = \begin{bmatrix} -\frac{k}{m_s} & -\frac{c_0}{m_s} & \frac{k}{m_s} & \frac{c_0}{m_s} & 0 \\ 1 & 0 & -1 & 0 & 0 \end{bmatrix}, \quad D_{v21} = \begin{bmatrix} 0 \\ 0 \end{bmatrix}, \quad D_{v22} = \begin{bmatrix} -\frac{1}{m_s} \\ 0 \end{bmatrix},$$

where $k = k_s + k_0$. In this work, the coefficients a and b are consistent with those of a road profile of type B in [19].

Vehicle vertical dynamics parameters are given in Table 2.

Table 2. Vertical dynamics parameters.

Symbol	Value	SI Unit	Parameter Name
m_s	315	kg	Sprung mass
m_{us}	37.5	kg	Unsprung mass
c_0	3000	Ns/m	Viscous damping coefficient
$k = k_s + k_0$	29,500	N/m	Spring and damper stiffness
k_t	208,000	N/m	Tire stiffness

2.3. *Full-Vehicle Dynamics*

In this paper, the full-vehicle model presented in [28,29] is used for simulation and validation purposes. This model and its parameters have been validated on a real Renault Megane vehicle (thanks to M. Basset, from the MIAM research team). For illustration, the model is presented in Figure 4, but interested readers should refer to [28] for more details.

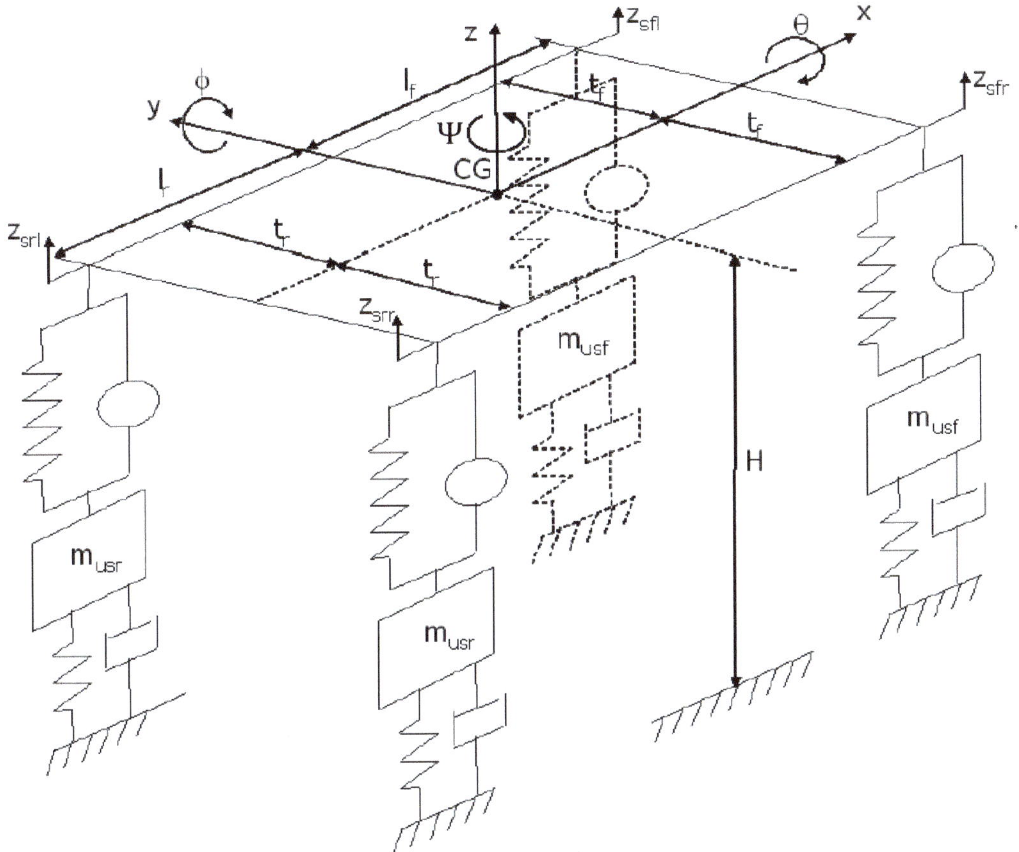

Figure 4. Full-vehicle model.

Note that the main interest in using the full nonlinear vehicle model is that it allows us to consider a nonlinear load transfer, fast nonlinear dynamics entering the tire force description, and consequently, in the global chassis dynamic. It reproduces the longitudinal (x), lateral (y), vertical (z), roll (θ), pitch (ϕ), and yaw (ψ) dynamics of the chassis. It also models the vertical and rotational motions of the wheels ($z_{us_{ij}}$ and ω_{ij} respectively), the slip ratios (λ_{ij}), and the center of gravity side slip angle (β_{cog}) dynamics, as a function of the tires and suspensions forces.

3. Integrated Cruise—Suspension LPV Control of an Autonomous Vehicle for Comfort: Structure and Objectives

The proposed strategy is illustrated in Figure 5.

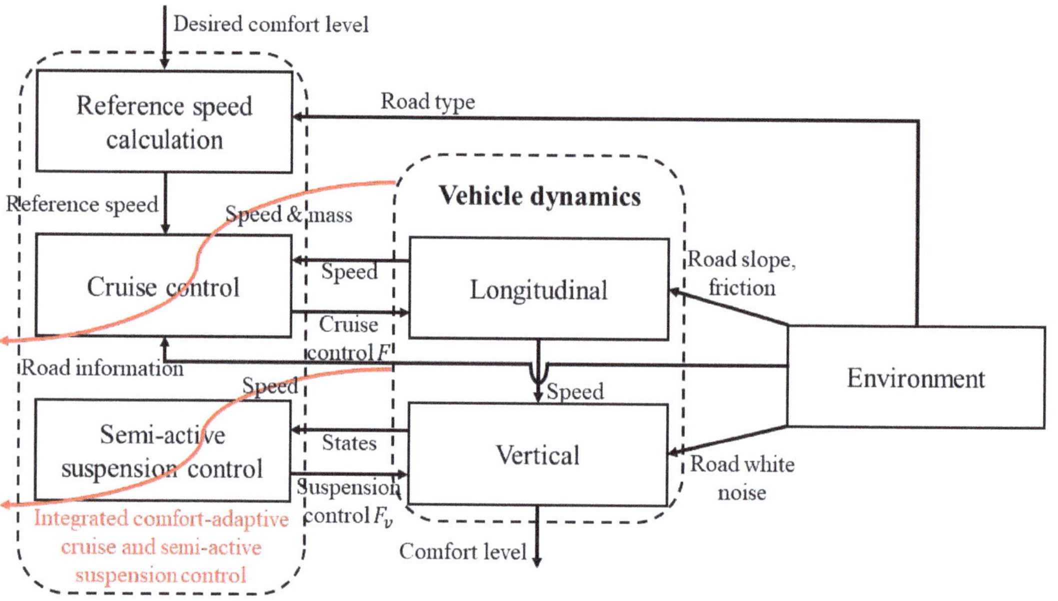

Figure 5. Integrated cruise—suspension linear parameter-varying (LPV) control for comfort.

Our strategy consists of three main parts closely connected to each other and the full-vehicle dynamics. Note that the vehicle speed connects the longitudinal and vertical dynamics due to the relationship (12).

The road type is assumed to be known/estimated in real time thanks to algorithms such as in [20]. This is the condition that enables the making of the proposed reference speed generation strategy, which gives us suitable speed values based on the road profile and comfort objective. In the reference speed calculation part, given the estimated road type and the desired comfort level specified by the driver/passenger(s), a suitable reference speed value is determined so as to guarantee this level. How we quantify driving comfort and calculate the reference speed is presented in Section 4.

In the cruise control part, given the calculated reference speed value, the cruise controller drives the vehicle speed to track this value. This uses not only the feedback measured by the speedometer but also road information such as road slope in order to compensate for this, providing a smoother response. How we design this part is discussed in Section 5.2.

In the semi-active suspension control design method, a semi-active suspension control strategy is used to further improve driving comfort. How we design this part is discussed in Section 5.3.

Combining the three mentioned parts constitutes what we propose in this paper as the integrated cruise–suspension control of an autonomous vehicle with a comfort objective.

4. Comfort-Oriented Reference Speed Calculation

4.1. Comfort Evaluation Using the ISO 2631 Standard

First, the road types are characterized by the ISO norm [19]. In Figure 6, we examine the road displacement profiles of types from A to D described in the ISO standard, with the vehicle's speed being 15 m/s. As shown in [30], such profiles do change w.r.t the speed as we considered in the modeling step (see (12)).

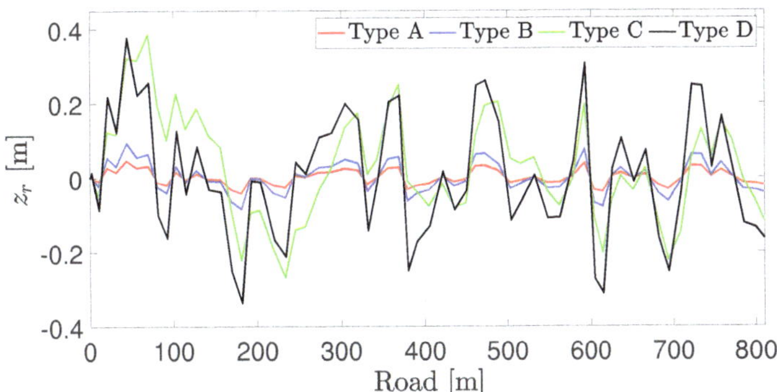

Figure 6. Road displacement profiles of road types A–D for a given speed.

This displacement is then transmitted to the passengers through the vehicle vertical dynamics. What affects driving comfort is the acceleration felt by passengers, as analyzed in the ISO 2631 norm [17]. In order to characterize human comfort, i.e., the effect of exposure to vibration, a filter is applied on the sprung mass acceleration [31]. This filter's transfer function is:

$$W_{ISO}(s) = \frac{81.89s^3 + 796.6s^2 + 1937s + 0.1446}{s^4 + 80s^3 + 2264s^2 + 7172s + 21196}. \tag{14}$$

Driving comfort is then assessed according to the following scale of the ISO 2631 standard using the root mean square (RMS) value of the vertical acceleration [17] (see Table 3).

Table 3. Vertical acceleration RMS (root mean square) value and comfort level.

RMS Value of Acceleration	Comfort Level
Less than 0.315 m/s^2	Not uncomfortable
0.315–0.63 m/s^2	A little uncomfortable
0.5–1 m/s^2	Fairly uncomfortable
0.8–1.6 m/s^2	Uncomfortable
1.25–2.5 m/s^2	Very uncomfortable
Greater than 2 m/s^2	Extremely uncomfortable

4.2. Modeling of Vehicle Speed—Comfort Interaction

It is known that, as the vehicle travels on certain different road types with the same speed and vice versa, it experiences different road displacement profiles. Thus, the felt human comfort varies according to the vehicle speed [30,32].

Our objective is to propose a comfort-oriented reference speed profile to link the comfort level to the vehicle's speed. This is carried out using a vertical vehicle model performing simulation with different speed values and computing the comfort criterion. This allows us first to evaluate the human comfort (for the RMS of the vertical acceleration) as seen in Figure 7.

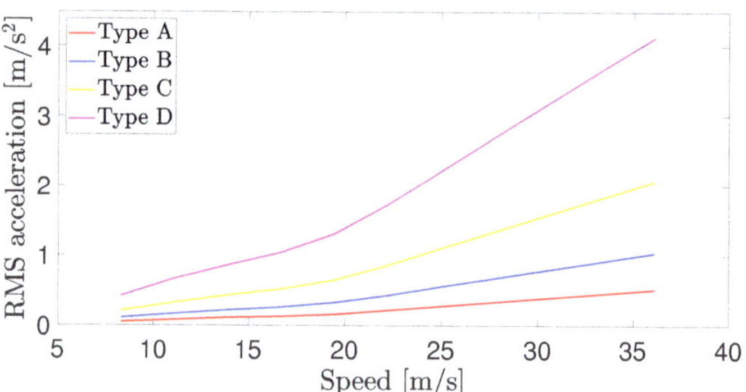

Figure 7. Comfort evaluation for different road types.

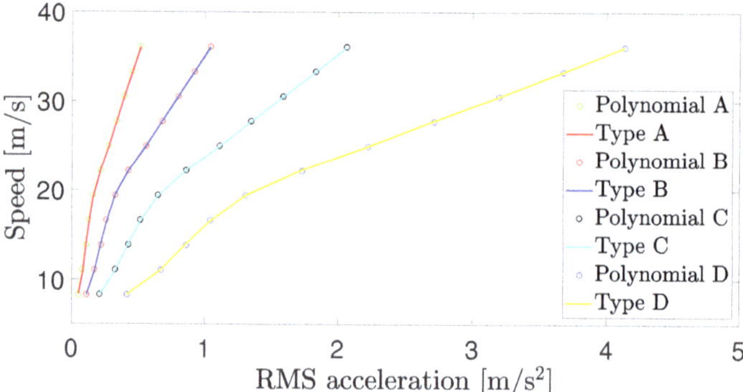

Figure 8. Polynomial functions: Speed vs. comfort level for different road types.

Then, following our previous study in [16], we define the comfort-oriented reference speed values from a polynomial fitting method, which are illustrated in Figure 8 and Table 4.

Table 4. Calculation of comfort-oriented reference speed values.

Road Type	Comfort-Oriented Reference Speed Values $v_{ref}(x)$ Where x is RMS Acceleration
A	$-281058.82x^7 + 616932.65x^6 - 553287.74x^5 + 259269.77x^4 - 67006.34x^3$ $+9126.1x^2 - 490.82x + 17.03$
B	$-2918.46x^7 + 12566.60x^6 - 22133.93x^5 + 20420.08x^4 - 10438.88x^3$ $+2839.03x^2 - 319.40x + 20.36$
C	$-20.58x^7 + 176.82x^6 - 620.95x^5 + 1140.85x^4 - 1158.9x^3$ $+623.06x^2 - 134.38x + 17.84$
D	$-0.19x^7 + 3.22x^6 - 22.33x^5 + 81.06x^4 - 162.92x^3$ $+174.09x^2 - 76.47x + 19.57$

These polynomials are precomputed and programmed into the autonomous vehicle's computer. In practice, the vehicle detects the current road type by performing the road estimation algorithm as assumed. Then from a given desired comfort level, characterized by a given RMS acceleration, the corresponding reference speed is calculated thanks to these polynomials.

5. LPV $\mathcal{H}_2$ Cruise and Semi-Active Suspension Control

In this section, both longitudinal and vertical controllers are synthesized using the $\mathcal{H}_2$ control approach for LPV systems. A short background of the control approach is presented herein, followed by the two design applications.

5.1. Preliminaries on LPV $\mathcal{H}_2$ Controller Design

Throughout this paper, we design two LPV controllers for cruise and semi-active suspension control. This part presents briefly the LPV approach including the optimization problem to be solved [28]. An LPV system is represented as:

$$\Sigma(\rho): \begin{bmatrix} \dot{x} \\ z \\ y \end{bmatrix} = \begin{bmatrix} A(\rho) & B_1(\rho) & B_2(\rho) \\ C_1(\rho) & D_{11}(\rho) & D_{12}(\rho) \\ C_2(\rho) & D_{21}(\rho) & D_{22}(\rho) \end{bmatrix} \begin{bmatrix} x \\ w \\ u \end{bmatrix}, \quad (15)$$

where x is the state, z is the controlled output, y is the measured output, w is the disturbance, u is the control input, and $\rho = \begin{bmatrix} \rho_1 & \rho_2 & \dots & \rho_N \end{bmatrix}^\top \in \Omega$ is the vector of varying parameters (Ω is a convex set). The assumptions on ρ are:

- ρ varies in the set of continously differentiable parameter curves and is known or measurable;
- ρ is bounded, i.e., $\rho_j \in [\underline{\rho_j}, \overline{\rho_j}], \forall j$;
- The system matrices $A(\cdot)$, etc. are continuous on Ω.

The vector of parameters evolves inside a polytope represented by 2^N vertices ω_i, as:

$$\rho \in \mathbf{Co}\{\omega_1, \dots, \omega_Z\}. \quad (16)$$

It is then written as the convex combination:

$$\rho = \sum_{i=1}^{2^N} \alpha_i \omega_i, \qquad \alpha_i \geq 0, \qquad \sum_{i=1}^{2^N} \alpha_i = 1, \quad (17)$$

where the vertices are defined by a vector $\omega_i = [\nu_{i1}, \dots, \nu_{iN}]$ where ν_{ij} equals $\underline{\rho_j}$ or $\overline{\rho_j}$.

Therefore, we consider a polytopic model of the LPV system above, represented as:

$$\Sigma(\rho) = \sum_{i=1}^{2^N} \alpha_i(\rho) \begin{bmatrix} A(\omega_i) & B(\omega_i) \\ C(\omega_i) & D(\omega_i) \end{bmatrix}, \qquad \alpha_i(\rho) \geq 0, \qquad \sum_{i=1}^{2^N} \alpha_i(\rho) = 1, \quad (18)$$

where $\begin{bmatrix} A(\omega_i) & B(\omega_i) \\ C(\omega_i) & D(\omega_i) \end{bmatrix}$ is the linear time-invariant (LTI) system corresponding to one of the system's 2^N vertices.

An LPV controller has the following structure:

$$K(\rho): \begin{bmatrix} \dot{x}_c \\ u \end{bmatrix} = \begin{bmatrix} A_c(\rho) & B_c(\rho) \\ C_c(\rho) & D_c(\rho) \end{bmatrix} \begin{bmatrix} x_c \\ y \end{bmatrix}. \quad (19)$$

Solving for an LPV controller using the $\mathcal{H}_2$ condition is here carried out using the polytopic approach so computing the controllers K_i $\forall i$, at each vertex of the parameter polytope, such that a single, global performance γ_2 is minimized. For a given parameter value ρ, the controller is then determined as:

$$K(\rho) = \sum_{i=1}^{2^N} \alpha_i(\rho) K_i, \qquad \alpha_i(\rho) \geq 0, \qquad \sum_{i=1}^{2^N} \alpha_i(\rho) = 1. \quad (20)$$

Proposition 1. *A dynamical output-feedback controller $K(\rho)$ (19) that solves the control problem is obtained by solving the following LMIs in ($\mathbf{X}(\rho)$, $\mathbf{Y}(\rho)$, $\widetilde{\mathbf{A}}(\rho)$, $\widetilde{\mathbf{B}}(\rho)$, $\widetilde{\mathbf{C}}(\rho)$, and $\widetilde{\mathbf{D}}(\rho)$) at the 2^N vertices ω_i of the polytope, while minimizing γ_2:*

$$\begin{array}{l} \begin{bmatrix} M_{11} & (*)^\top & (*)^\top \\ M_{21} & M_{22} & (*)^\top \\ M_{31} & M_{32} & M_{33} \end{bmatrix} \prec 0, \quad \forall i, \\ \begin{bmatrix} N_{11} & (*)^\top & (*)^\top \\ N_{21} & N_{22} & (*)^\top \\ N_{31} & N_{32} & N_{33} \end{bmatrix} \succ 0, \quad \forall i, \\ Trace(\mathbf{Z}) < \gamma_2, \end{array} \tag{21}$$

where:

$$\begin{array}{rcl} M_{11} & = & A(\omega_i)\mathbf{X}(\omega_i) + \mathbf{X}(\omega_i)A(\omega_i)^\top + B_2\widetilde{\mathbf{C}}(\omega_i) + \widetilde{\mathbf{C}}(\omega_i)^\top B_2^\top, \\ M_{21} & = & \widetilde{\mathbf{A}}(\omega_i) + A(\omega_i)^\top + C_2^\top \widetilde{\mathbf{D}}(\omega_i)^\top B_2^\top, \\ M_{22} & = & \mathbf{Y}(\omega_i)A(\omega_i) + A(\omega_i)^\top \mathbf{Y}(\omega_i) + \widetilde{\mathbf{B}}(\omega_i)C_2 + C_2^\top \widetilde{\mathbf{B}}(\omega_i)^\top, \\ M_{31} & = & B_1(\omega_i)^\top + D_{21}(\omega_i)^\top \widetilde{\mathbf{D}}(\omega_i)^\top B_2^\top, \\ M_{32} & = & B_1(\omega_i)^\top \mathbf{Y}(\omega_i) + D_{21}(\omega_i)^\top \widetilde{\mathbf{B}}(\omega_i)^\top, \\ M_{33} & = & -I_{n_u}, \\ N_{11} & = & X(\omega_i), \\ N_{21} & = & I_n, \\ N_{22} & = & Y(\omega_i), \\ N_{31} & = & C_1(\omega_i)\mathbf{X}(\omega_i) + D_{12}(\omega_i)\widetilde{\mathbf{C}}(\omega_i), \\ N_{32} & = & C_1(\omega_i) + D_{12}(\omega_i)\widetilde{\mathbf{D}}(\omega_i)C_2, \\ N_{33} & = & \mathbf{Z}. \end{array}$$

Then, the reconstruction of the controller K_i is obtained by the following equivalent transformation:

$$\left\{ \begin{array}{rcl} D_c(\omega_i) & = & \widetilde{\mathbf{D}}(\omega_i) \\ C_c(\omega_i) & = & (\widetilde{\mathbf{C}}(\omega_i) - D_c(\omega_i)C_2(\omega_i)\mathbf{X}(\omega_i))M(\omega_i)^{-\top} \\ B_c(\omega_i) & = & N(\omega_i)^{-1}(\widetilde{\mathbf{B}}(\omega_i) - \mathbf{Y}(\omega_i)B_2(\omega_i)D_c(\omega_i)) \\ A_c(\omega_i) & = & N(\omega_i)^{-1}(\widetilde{\mathbf{A}}(\omega_i) - \mathbf{Y}(\omega_i)A(\omega_i)\mathbf{X}(\omega_i) - \mathbf{Y}(\omega_i)B_2(\omega_i)D_c(\omega_i)C_2(\omega_i)\mathbf{X}(\omega_i) \\ & - & N(\omega_i)B_c(\omega_i)C_2(\omega_i)\mathbf{X}(\omega_i) - \mathbf{Y}(\omega_i)B_2(\omega_i)C_c(\omega_i)M(\omega_i)^\top)M(\omega_i)^{-\top}, \end{array} \right. \tag{22}$$

where $M(\omega_i)$ and $N(\omega_i)$ are defined such that $M(\omega_i)N(\omega_i)^\top = I_n - X(\omega_i)Y(\omega_i)$ (that can be solved through a singular value decomposition plus a Cholesky factorization).

5.2. Application of the LPV $\mathcal{H}_2$ Approach to Cruise Control

5.2.1. Cruise Controller Design

The approach above is here applied to the LPV longitudinal model presented before in (8) as:

$$\Sigma_l(\rho_l) : \left\{ \begin{array}{l} \dot{x}_l = A_l(\rho_l)x_l + B_{l1}w_l + B_{l2}(\rho_l)u_l \\ y_l = C_l x_l + D_{l1}w_l + D_{l2}u_l, \end{array} \right. \tag{23}$$

where ρ_l includes the vehicle mass and speed.

To use the polytopic approach, the control input matrix has to be independent of the scheduling parameter. Therefore, following the method in [33], the system is extended with the following filter at the input variable:

$$W_f : \begin{bmatrix} \dot{x}_f \\ u_l \end{bmatrix} = \begin{bmatrix} A_f & B_f \\ C_f & 0 \end{bmatrix} \begin{bmatrix} x_f \\ u_f \end{bmatrix}, \tag{24}$$

where A_f, B_f, and C_f are constant matrices. Here, we choose $A_f = -1/\tau_f$, $B_f = 1/\tau_f$, and $C_f = 1$ where τ_f is a small constant. To synthesize the controller, we define the generalized system denoted $\Sigma_{gl}(\rho_l)$ (see in Figure 9) consisting of the extended state-space representation with a parameter-independent control input:

$$\begin{cases} \begin{bmatrix} \dot{x}_l \\ \dot{x}_f \end{bmatrix} = \begin{bmatrix} A_l(\rho_l) & B_{l2}(\rho_l)C_f \\ 0 & A_f \end{bmatrix} \begin{bmatrix} x_l \\ x_f \end{bmatrix} + \begin{bmatrix} B_{l1} \\ 0 \end{bmatrix} w_l + \begin{bmatrix} 0 \\ B_f \end{bmatrix} u_f \\ y_l = \begin{bmatrix} C_l & D_{l2}C_f \end{bmatrix} \begin{bmatrix} x_l \\ x_f \end{bmatrix} + D_{l1}w_l, \end{cases} \tag{25}$$

and the following weighting functions in order to ensure tracking performances and to cope with the actuator limitations:

$$W_e = \frac{0.5s+2}{s+0.0002}, \quad W_u = \frac{1}{100}, \quad W_d = 0.01. \tag{26}$$

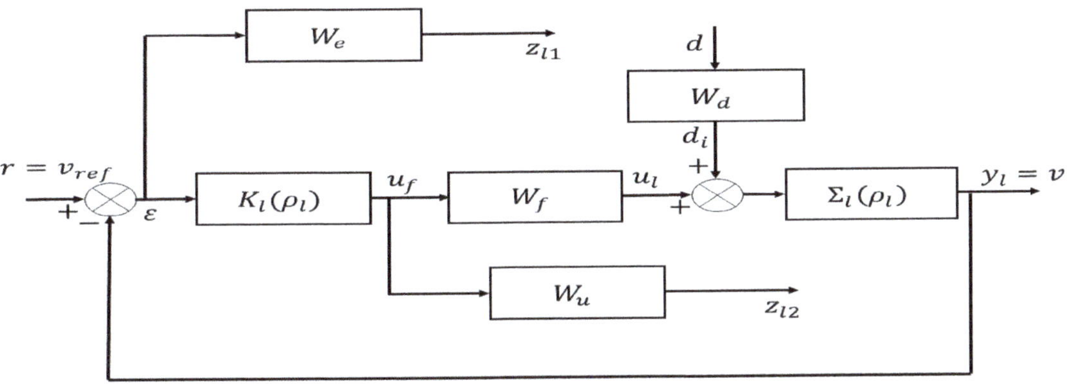

Figure 9. LPV cruise control generalized system $\Sigma_{gl}(\rho_l)$.

According to [33], since the parameter dependence is affine and since the scheduling parameter ρ_l varies in a polytope of four vertices, i.e., $\rho_{l1} \in [\underline{\rho_{l1}}, \overline{\rho_{l1}}]$ and $\rho_{l2} \in [\underline{\rho_{l2}}, \overline{\rho_{l2}}]$, the generalized system $\Sigma_{gl}(\rho_l)$ in Figure 9 can be expressed as a polytopic system composed of four vertices:

$$\Sigma_{gl}(\rho_l) = \sum_{i=1}^{4} \alpha_{l_i}(\rho_l)\Sigma_{gl_i}, \quad \alpha_{l_i}(\rho_l) \geq 0, \quad \sum_{i=1}^{4} \alpha_{l_i}(\rho_l) = 1, \tag{27}$$

where $\Sigma_{gl_1} = \Sigma_{gl}(\underline{\rho_{l1}}, \underline{\rho_{l2}})$, $\Sigma_{gl_2} = \Sigma_{gl}(\underline{\rho_{l1}}, \overline{\rho_{l2}})$, $\Sigma_{gl_3} = \Sigma_{gl}(\overline{\rho_{l1}}, \underline{\rho_{l2}})$, and $\Sigma_{gl_4} = \Sigma_{gl}(\overline{\rho_{l1}}, \overline{\rho_{l2}})$. Solving the LMIs in Proposition 1, the LPV controller $K_l(\rho_l)$ with the scheme as shown in Figure 9 is defined as:

$$K_l(\rho_l) : \begin{bmatrix} \dot{x}_{cl} \\ u_f \end{bmatrix} = \begin{bmatrix} A_{cl}(\rho_l) & B_{cl}(\rho_l) \\ C_{cl}(\rho_l) & D_{cl}(\rho_l) \end{bmatrix} \begin{bmatrix} x_{cl} \\ \epsilon \end{bmatrix}, \tag{28}$$

where $\epsilon = r - y_l$ denotes the tracking error where r is the reference. The controller $K_l(\rho_l)$ can be transformed into a convex interpolation as follows:

$$K_l(\rho_l) = \sum_{i=1}^{4} \alpha_{l_i}(\rho_l) \begin{bmatrix} A_{cl_i} & B_{cl_i} \\ C_{cl_i} & D_{cl_i} \end{bmatrix}, \quad \alpha_{l_i}(\rho_l) \geq 0, \quad \sum_{i=1}^{4} \alpha_{l_i}(\rho_l) = 1. \tag{29}$$

5.2.2. Cruise Control Simulation

We verify by simulation that the vehicle speed can track a given reference value in the presence of disturbance and noise in the form of inexact disturbance force compensation.

We see that the tracking performance is guaranteed, and tracking is achieved after a few hundreds of meters (see Figure 10), with a longitudinal control force smaller than 4000 N (see Figure 11).

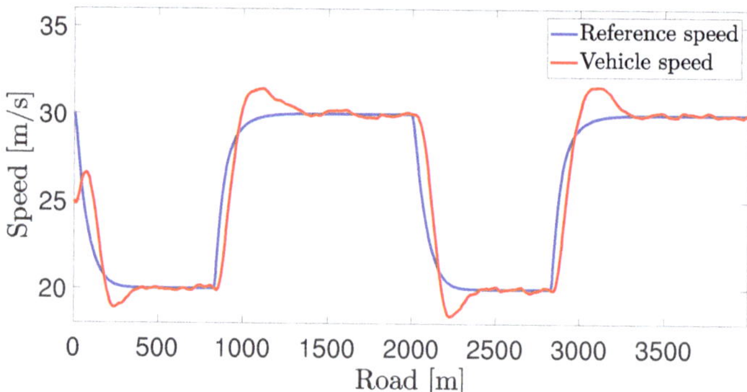

Figure 10. Reference and real speed.

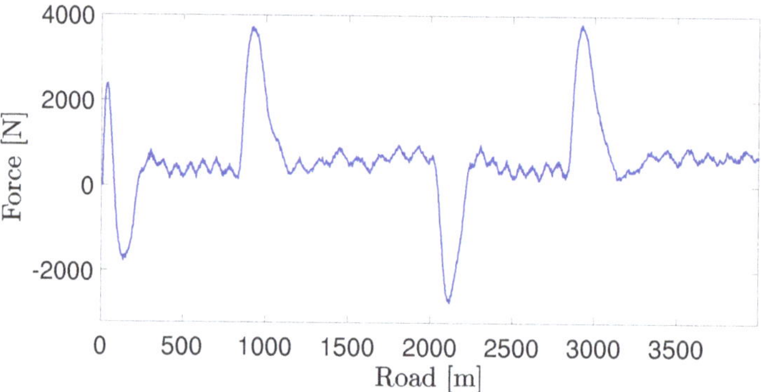

Figure 11. Longitudinal control force.

5.3. Application of the LPV $\mathcal{H}_2$ Approach to Semi-Active Suspension Control

5.3.1. Semi-Active Suspension Controller Design

First, let us recall the LPV system where the extended quarter-car system is described in (13) as:

$$\Sigma_v(\rho_v) : \begin{cases} \dot{x}_v = A_v(\rho_v)x_v + B_{v1}(\rho_v)w_v + B_{v2}u_v \\ y_v \ = C_{v2}x_v + D_{v21}w_v + D_{v22}u_v, \end{cases} \tag{30}$$

where ρ_v is the vehicle speed.

In order to guarantee the comfort and road holding objectives, we define the performance output vector as $z_v = \begin{bmatrix} z_{us} & \ddot{z}_s \end{bmatrix}^\top \in \mathbb{R}^2$. To synthesize the controller, we define the generalized system denoted $\Sigma_{gv}(\rho_v)$ (see in Figure 12) consisting of:

$$\begin{cases} \dot{x}_v = A_v(\rho_v)x_v + B_{v1}(\rho_v)w_v + B_{v2}u_v \\ z_v = C_{v1}x_v + D_{v11}w_v + D_{v12}u_v \\ y_v \ = C_{v2}x_v + D_{v21}w_v + D_{v22}u_v, \end{cases} \tag{31}$$

where:

$$C_{v1} = \begin{bmatrix} 0 & 0 & 1 & 0 & 0 \\ -\frac{k}{m_s} & -\frac{c_0}{m_s} & \frac{k}{m_s} & \frac{c_0}{m_s} & 0 \end{bmatrix}, \quad D_{v11} = \begin{bmatrix} 0 \\ 0 \end{bmatrix}, \quad D_{v12} = \begin{bmatrix} 0 \\ -\frac{1}{m_s} \end{bmatrix},$$

and the parameter-dependent weighting function $W_{\ddot{z}_s}(\rho_v)$ and the weighting function $W_{z_{us}}$ shaped in order to reduce the amplification of the sprung mass acceleration $\ddot{z}_s$ and unsprung mass displacement z_{us} depending on the vehicle speed; W_w and W_n model white noise (w_v) and measurement noise, respectively. These weighting functions can be chosen as:

$$W_{\ddot{z}_s}(\rho_v) = \rho_v \cdot k_{\ddot{z}_s} \cdot \frac{s^2+2\zeta_{11}\Omega_{11}s+\Omega_{11}^2}{s^2+2\zeta_{12}\Omega_{12}s+\Omega_{12}^2}, \quad W_{z_{us}} = k_{z_{us}} \cdot \frac{s^2+2\zeta_{21}\Omega_{21}s+\Omega_{21}^2}{s^2+2\zeta_{22}\Omega_{22}s+\Omega_{22}^2}, \quad W_w = \frac{0.5s+0.1}{s+0.001}, \quad W_n = 10^{-3}. \tag{32}$$

Remark 3: The parameters in the weighting functions are chosen following our previous studies where a Genetic Algorithm is applied to find these parameters optimizing multiple objectives: Passenger's comfort and road holding (safety). Refer to [34] for more details.

According to [33], since the parameter dependence is affine and since the scheduling parameter ρ_v varies in a polytope of two vertices, i.e., $\rho_v \in [\underline{\rho_v}, \overline{\rho_v}]$, the generalized system $\Sigma_{gv}(\rho_v)$ in Figure 12 can be expressed as a polytopic system composed of two vertices:

$$\Sigma_{gv}(\rho_v) = \sum_{i=1}^{2} \alpha_{v_i}(\rho_v)\Sigma_{gv_i}, \quad \alpha_{v_i}(\rho_v) \geq 0, \quad \sum_{i=1}^{2} \alpha_{v_i}(\rho_v) = 1, \tag{33}$$

where $\Sigma_{gv_1} = \Sigma_{gv}(\underline{\rho_v})$ and $\Sigma_{gv_2} = \Sigma_{gv}(\overline{\rho_v})$. Solving the LMIs in Proposition 1, the LPV controller $K_v(\rho_v)$ with the scheme as shown in Figure 12 is defined as:

$$K_v(\rho_v) : \begin{bmatrix} \dot{x}_{cv} \\ u_v \end{bmatrix} = \begin{bmatrix} A_{cv}(\rho_v) & B_{cv}(\rho_v) \\ C_{cv}(\rho_v) & D_{cv}(\rho_v) \end{bmatrix} \begin{bmatrix} x_{cv} \\ y_v \end{bmatrix}. \tag{34}$$

The controller $K_v(\rho_v)$ can be transformed into a convex interpolation as follows:

$$K_v(\rho_v) = \sum_{i=1}^{2} \alpha_{v_i}(\rho_v) \begin{bmatrix} A_{cv_i} & B_{cv_i} \\ C_{cv_i} & D_{cv_i} \end{bmatrix}, \quad \alpha_{v_i}(\rho_v) \geq 0, \quad \sum_{i=1}^{2} \alpha_{v_i}(\rho_v) = 1. \tag{35}$$

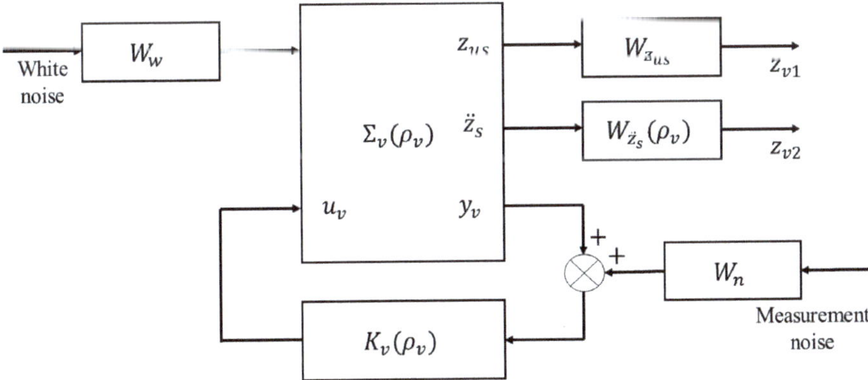

Figure 12. LPV semi-active suspension control generalized system $\Sigma_{gv}(\rho_v)$.

5.3.2. Frequency-Domain Analysis

The Bode diagrams given in Figure 13 show the system frequency performance according to the varying ρ_v. Compared with the passive suspension, the closed-loop system provides efficient vibration mitigation (attenuation) in the whole frequency range of 10^{-3}–10^4 Hz. The effects of road vibrations on the performance output (sprung mass acceleration, which is directly linked to driving comfort) are shown, which are effectively attenuated for both vertices.

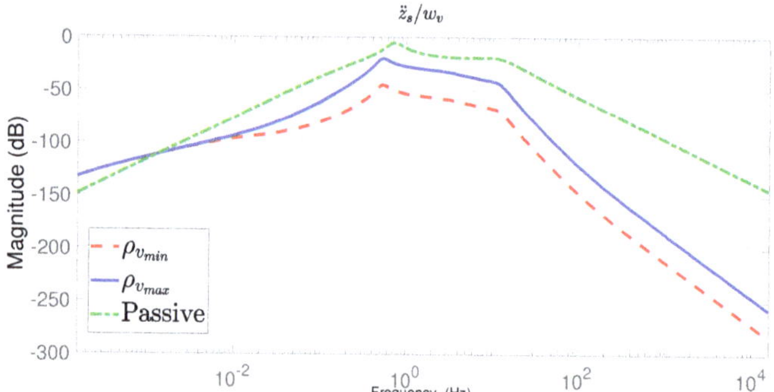

Figure 13. Bode diagram (from w_v to $\ddot{z}_s$) of the closed-loop systems corresponding to the vertices of $\rho_{v_{min}}$ and $\rho_{v_{max}}$, and of the passive system.

5.3.3. Semi-Active Suspension Control Simulation

In order to demonstrate the performance of the LPV $\mathcal{H}_2$ approach, simulation in the time domain is performed in this part. In this simulation, the sprung mass acceleration at one corner of the vehicle is considered. The simulation scenario is as follows:

- The vehicle speed rises from its minimum (10 m/s) to maximum value (35 m/s);
- The ISO road profile (type B) is used (shown in Figure 14).

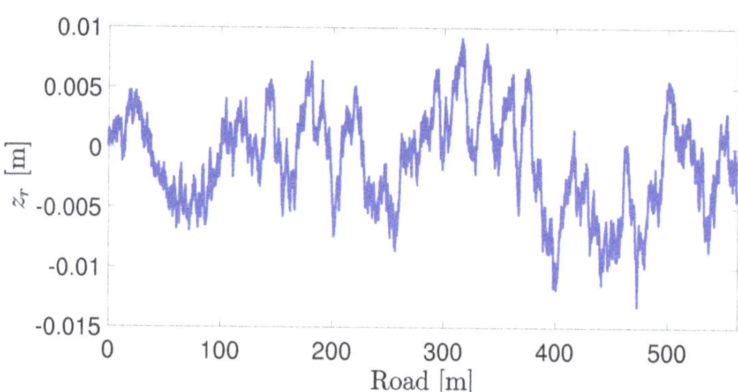

Figure 14. Road input at the front right corner.

From Figures 15 and 16, it can be seen that the LPV $\mathcal{H}_2$ control for the semi-active suspension provides better driving comfort than the passive one.

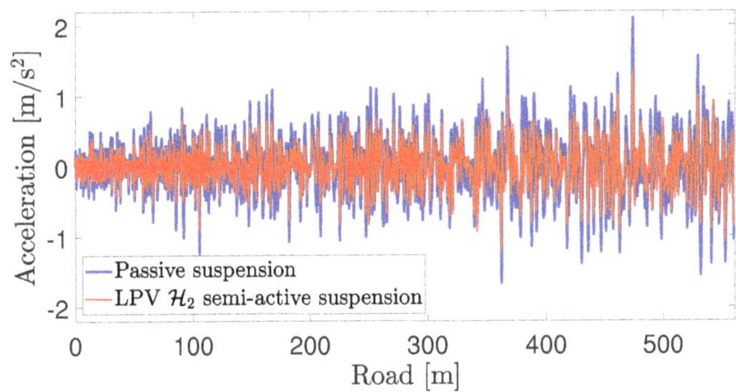

Figure 15. Sprung mass acceleration at the front right corner (filtered by (14)).

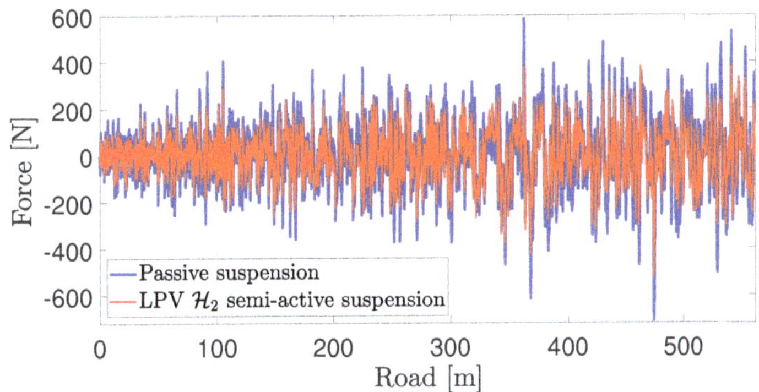

Figure 16. Damper force at the front right corner.

6. Simulation of the Integrated Control Strategy

In this part, we perform simulations using the full-vehicle model presented in [29], following the scheme presented in Figure 17. It should be noted that the parameters used for simulations are chosen according to a real Megane vehicle. Because of this, the speed is limited to 10–35 m/s. This vehicle is equipped with four independent semi-active suspension systems controlled with a sampling period of 0.005 s. Since we perform simulations with a full-vehicle model, there is a varying time delay of L/v (where L is the distance between the front and rear wheels, i.e., $L = l_f + l_r$ in Figure 4) in the road profile z_r at the rear wheels compared to the front wheels. Driving comfort is evaluated using the RMS value of the acceleration at the vehicle's center filtered by (14). We combine the two control strategies into an integrative case where the reference speed varies according to the road type and the desired comfort level to guarantee driving comfort. In this part, we perform simulations with various road types and desired RMS acceleration values to test the reference speed generation and integrated cruise-suspension vehicle control strategies.

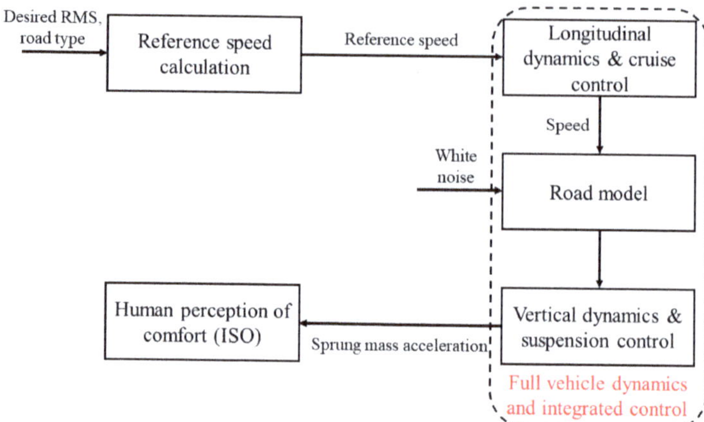

Figure 17. Simulation scheme.

6.1. Simulation Scenario 1

The road profile inputs are chosen according to the road model (12) for the given road types, where the input is white noise. The total simulation time is 54 s. The simulation scenario is as follows (see Figures 18 and 19):

- At 18 s (around 550 m), the desired comfort level (characterized by the given RMS acceleration) decreases from 0.4 to 0.3 m/s^2;
- At 36 s (around 1000 m), the road type (characterized by the estimated road roughness) changes from A to B.

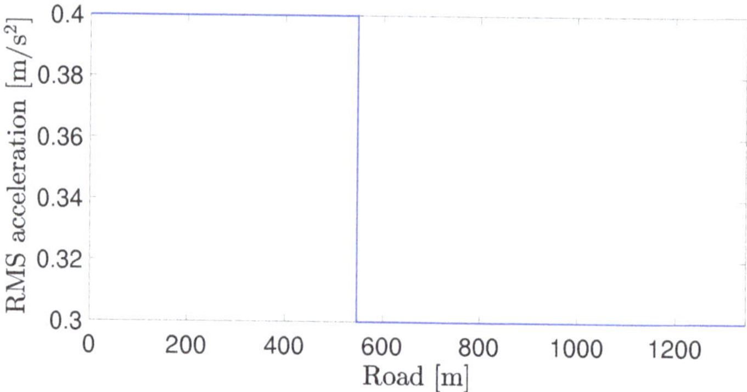

Figure 18. Desired comfort level.

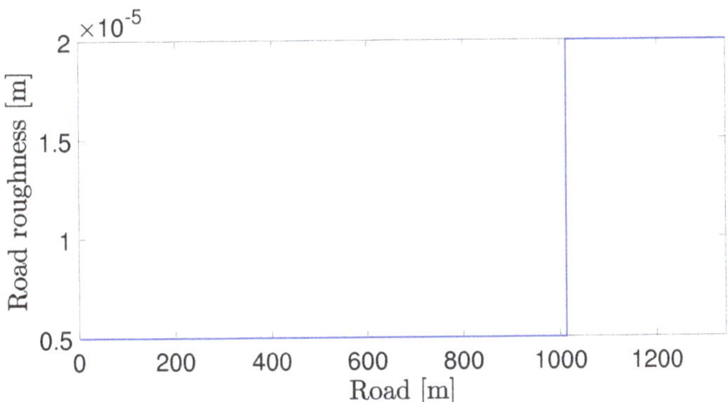

Figure 19. Road roughness.

We see from Figure 20 that each time the road type or the desired RMS value changes, a new reference speed is calculated, and the cruise control effectively tracks this value. The resulting road displacement z_r significantly increases after 36 s (around 1000 m) due to the change in road type (only the displacement at the front right corner of the vehicle is shown in Figure 21).

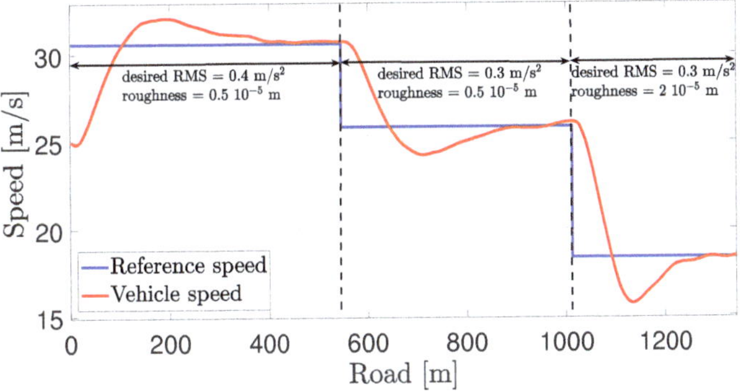

Figure 20. Resulting reference and vehicle speed.

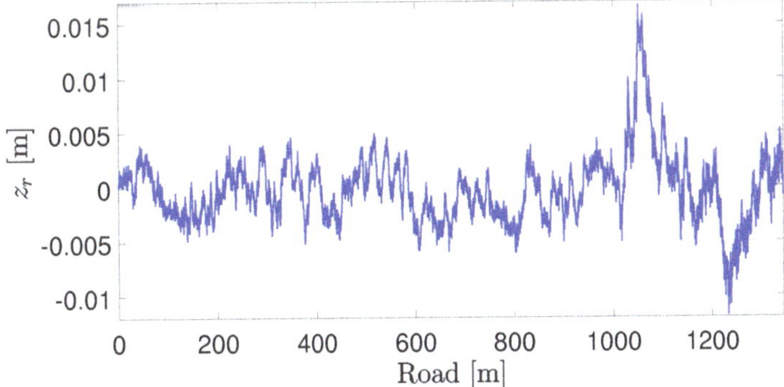

Figure 21. Resulting road input at the front right corner.

The resulting accelerations are shown in Figure 22, whose RMS values are 1.2360 m/s^2 for the passive suspension case and 0.4301 m/s^2 for the LPV $\mathcal{H}_2$ semi-active suspension case, which correspond to the comfort level of "uncomfortable" and "a little uncomfortable", respectively, according to Table 3. These results show that the latter further improves driving comfort by limiting the acceleration transmitted to passengers. Finally, the resulting deflections are shown in Figure 23, from which we can see that semi-active suspension leads to smaller deflection values compared to passive suspension.

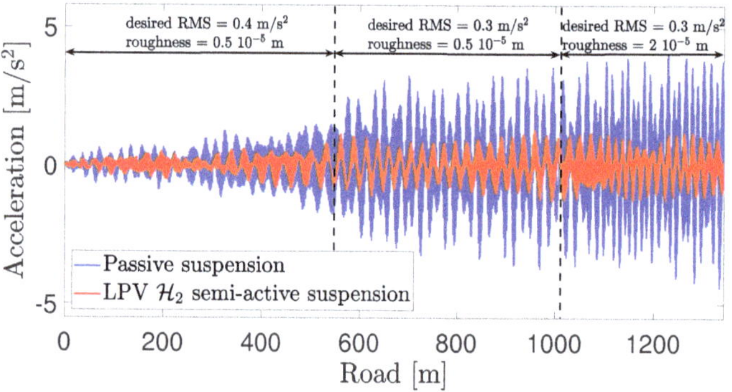

Figure 22. Resulting acceleration felt by passengers (filtered by (14)).

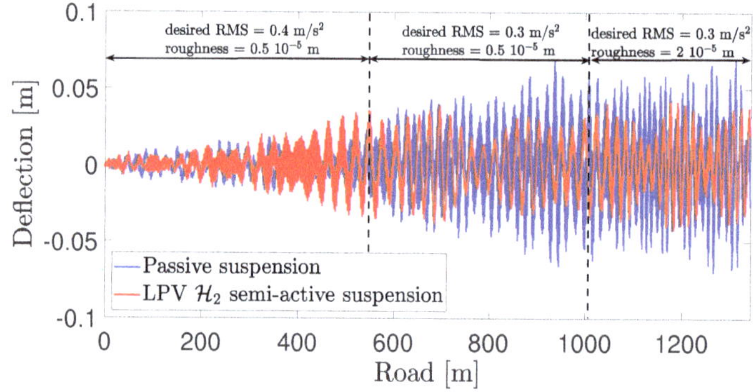

Figure 23. Resulting deflection at the front right corner.

6.2. Simulation Scenario 2

The second simulation scenario is used to assess the robustness of the proposed approach w.r.t uncertainty on the sprung mass and viscous damping coefficients at the corners. This scenario is designed by adding the following uncertainty into the first one. The uncertain parameters are shown in Table 5.

Table 5. Uncertain parameters.

Uncertain Parameter	Value
Sprung mass m_s at each corner	315 + 2.5% kg
Viscous damping coefficients c_{fl} and c_{fr} at the front corners	3000 + 10% Ns/m
Viscous damping coefficients c_{rl} and c_{rr} at the rear corners	6000 + 10% Ns/m

The total simulation time is 60 s. The simulation scenario is as follows (see Figures 24 and 25):

- At 20 s (around 160 m), the desired comfort level (characterized by the given RMS acceleration) increases from 0.2 to 0.3 m/s^2;
- At 40 s (around 360 m), the road type (characterized by the estimated road roughness) changes from B to A.

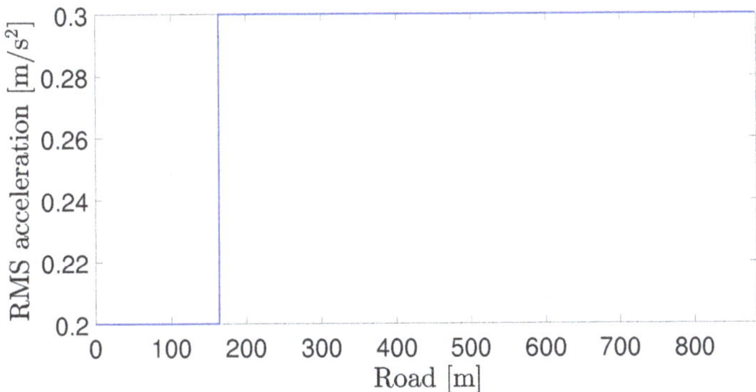

Figure 24. Desired comfort level.

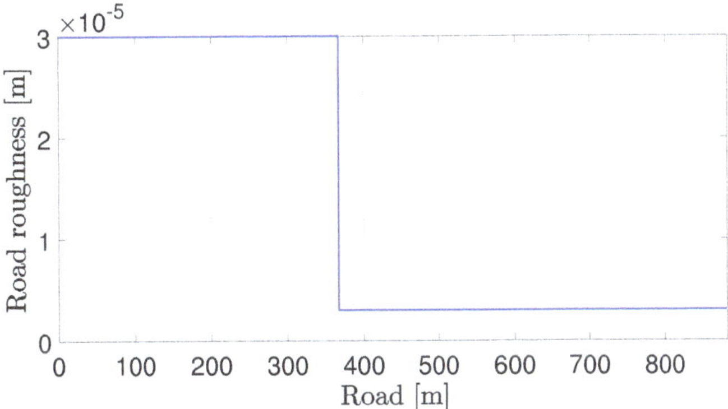

Figure 25. Road roughness.

Again, we see from Figure 26 that the new reference speed is calculated, and the cruise control effectively tracks this value. The resulting road displacement z_r significantly decreases after 40 s (around 360 m) due to the change in road type (only the displacement at the front right corner of the vehicle is shown in Figure 27).

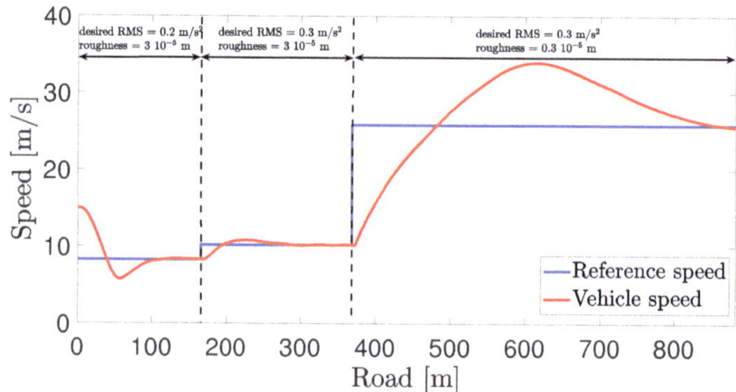

Figure 26. Resulting reference and vehicle speed.

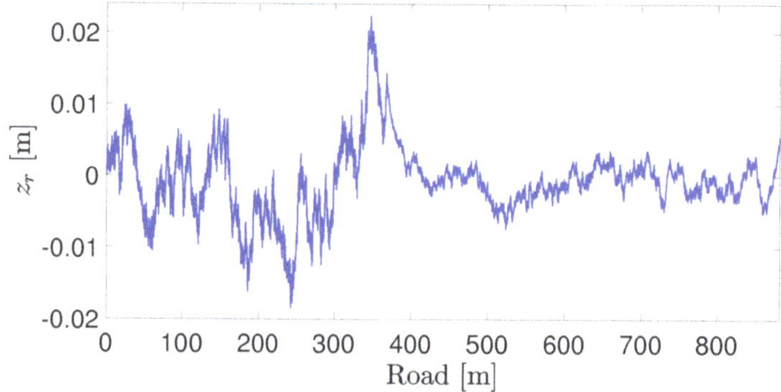

Figure 27. Resulting road input at the front right corner.

The resulting accelerations are shown in Figure 28, whose RMS values are 1.1947 $\mathrm{m/s^2}$ for the passive suspension case and 0.6916 $\mathrm{m/s^2}$ for the LPV $\mathcal{H}_2$ semi-active suspension case, which correspond to the comfort level of "uncomfortable" and "fairly uncomfortable", respectively, according to Table 3. Finally, the resulting deflections are shown in Figure 29, from which we can see that again, semi-active suspension leads to smaller deflection values compared to passive suspension. These results show that the proposed method is robust enough w.r.t the uncertainty.

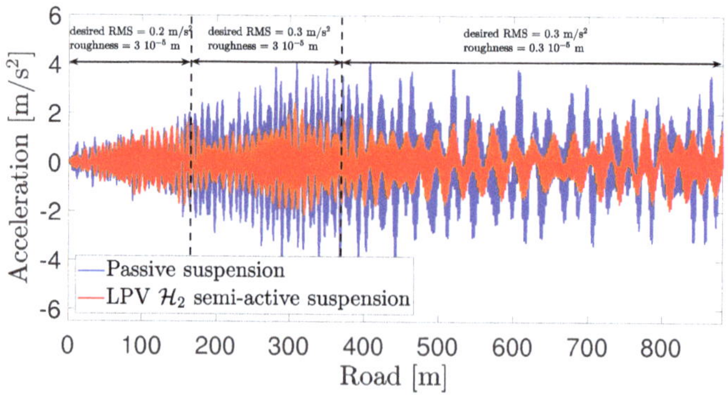

Figure 28. Resulting acceleration felt by passengers (filtered by (14)).

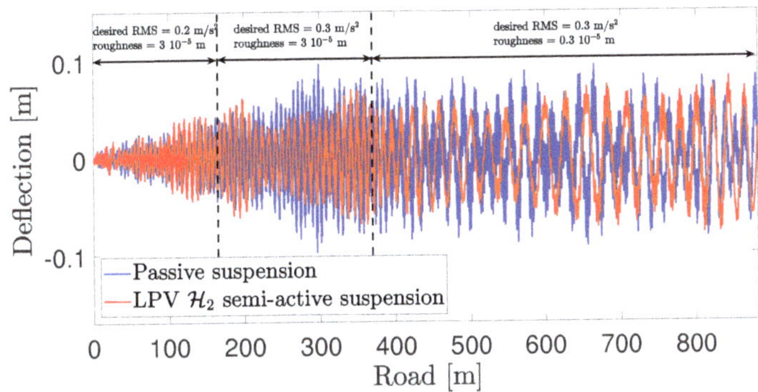

Figure 29. Resulting deflection at the front right corner.

6.3. Comparison of Comfort Performances: Passive vs. Semi-Active Suspension

Table 6 summarizes the comfort evaluation for the two performed simulation scenarios.

Table 6. RMS acceleration (m/s^2).

Simulation Scenario	Passive Suspension	Semi-Active Suspension
Scenario 1	1.2360	0.4301
Scenario 2	1.1947	0.6916

Clearly, the LPV semi-active suspension control outperforms the passive one, and it allows for an efficient coupling between the longitudinal and vertical dynamics.

7. Conclusions

This work presented an integrated strategy for comfort-oriented vehicle cruise and suspension control with a robust/LPV approach in the $\mathcal{H}_2$ framework. We related driving comfort (quantified using the filtered sprung mass acceleration) and road type with the vehicle speed to obtain comfort-guaranteeing reference speed functions. We then designed a cruise and a semi-active suspension controller respectively for vehicle longitudinal and vertical dynamics, which are linked to each other through the vehicle speed as a scheduling parameter. These controllers were tested by performing simulations, first independently and then in an integrated framework using a nonlinear full-vehicle model validated from real data. Results showed that the vehicle was capable of finding a speed value guaranteeing comfort and tracking this value thanks to cruise control, while semi-active suspension control provided further enhancement of comfort level. Indeed, the integrated control approach was adapted to the comfort requirement and vehicle speed. It is worth mentioning that we relied on basic assumptions and a reasonable amount of knowledge of the environment, making this strategy realistic.

Author Contributions: Conceptualization, O.S. and P.G.; Methodology, G.Q.B.T., T.-P.P., and E.C.; Validation, O.S.; Formal analysis, G.Q.B.T. and T.-P.P.; Writing-original draft preparation, G.Q.B.T. and T.-P.P.; Writing-review and editing, O.S.; Supervision, O.S. and P.G.; Project administration, O.S. All authors have read and agreed to the published version of the manuscript.

Funding: This research received no external funding.

Conflicts of Interest: The authors declare no conflict of interest.

Abbreviations

The following abbreviations are used in this manuscript:

IRI	international roughness index
ISO	International Organization for Standardization
LMI	linear matrix inequality
MR	magneto-rheological
LPV	linear parameter-varying
LTI	linear time-invariant
RMS	root mean square

References

1. Hedrick, J.K.; Tomizuka, M.; Varaiya, P. Control Issues in Automated Highway Systems. *IEEE Control. Syst. Mag.* **1994**, *14*, 21–32.
2. Ioannou, P.A.; Chien, C.C. Autonomous Intelligent Cruise Control. *IEEE Trans. Veh. Technol.* **1993**, *42*, 657–672. [CrossRef]
3. Gáspár, P.; Szabó, Z.; Bokor, J.; Nemeth, B. *Robust Control Design for Active Driver Assistance Systems*; Springer: Berlin/Heidelberg, Germany, 2016; Volume 10, pp. 978–983.
4. Rajamani, R.; Zhu, C. Semi-autonomous Adaptive Cruise Control Systems. *IEEE Trans. Veh. Technol.* **2002**, *51*, 1186–1192. [CrossRef]
5. Kayacan, E. Multiobjective H_∞ Control for String Stability of Cooperative Adaptive Cruise Control Systems. *IEEE Trans. Intell. Veh.* **2017**, *2*, 52–61. [CrossRef]
6. Németh, B.; Gáspár, P. LPV-based Control Design of Vehicle Platoon Considering Road Inclinations. *IFAC Proc. Vol.* **2011**, *44*, 3837–3842. [CrossRef]
7. Öncü, S.; Ploeg, J.; Van de Wouw, N.; Nijmeijer, H. Cooperative Adaptive Cruise Control: Network-aware Analysis of String Stability. *IEEE Trans. Intell. Transp. Syst.* **2014**, *15*, 1527–1537. [CrossRef]
8. Du, Y.; Liu, C.; Li, Y. Velocity Control Strategies to Improve Automated Vehicle Driving Comfort. *IEEE Intell. Transp. Syst. Mag.* **2018**, *10*, 8–18. [CrossRef]
9. Wu, J.; Zhou, H.; Liu, Z.; Gu, M. Ride Comfort Optimization via Speed Planning and Preview Semi-active Suspension Control for Autonomous Vehicles on Uneven Roads. *IEEE Trans. Veh. Technol.* **2020**, *69*, 8343–8355. [CrossRef]
10. Tran, G.Q.B.; Sename, O.; Gaspar, P.; Nemeth, B.; Costa, E. Adaptive Speed Control of an Autonomous Vehicle with a Comfort Objective. In Proceedings of the VSDIA 2020—20th International Conference on Vehicle System Dynamics, Identification and Anomalies, Budapest, Hungary, 9–11 November 2020.
11. Savaresi, S.M.; Poussot-Vassal, C.; Spelta, C.; Sename, O.; Dugard, L. *Semi-Active Suspension Control Design for Vehicles*; Elsevier: Amsterdam, The Netherlands, 2010.
12. Poussot-Vassal, C.; Spelta, C.; Sename, O.; Savaresi, S.M.; Dugard, L. Survey and Performance Evaluation on Some Automotive Semi-active Suspension Control Methods: A Comparative Study on a Single-corner Model. *Annu. Rev. Control.* **2012**, *36*, 148–160. [CrossRef]
13. Murali Madhavan Rathai, K. Synthesis and Real-Time Implementation of Parameterized NMPC Schemes for Automotive Semi-Active Suspension Systems. Ph.D. Thesis, Université Grenoble Alpes, Grenoble Alpes, France, 2020.
14. Pham, T.P.; Sename, O.; Dugard, L. Unified $\mathcal{H}_\infty$ Observer for a Class of Nonlinear Lipschitz Systems: Application to a Real ER Automotive Suspension. *IEEE Control. Syst. Lett.* **2019**, *3*, 817–822. [CrossRef]
15. Poussot-Vassal, C.; Sename, O.; Dugard, L.; Gáspár, P.; Szabó, Z.; Bokor, J. A New Semi-active Suspension Control Strategy through LPV Technique. *Control. Eng. Pract.* **2008**, *16*, 1519–1534. [CrossRef]
16. Costa, E.; Pham, T.P.; Sename, O.; Tran, G.Q.B.; Do, T.T.; Gaspar, P. Definition of a Reference Speed of an Autonomous Vehicle with a Comfort Objective. In Proceedings of the VSDIA 2020—20th International Conference on Vehicle System Dynamics, Identification and Anomalies, Budapest, Hungary, 9–11 November 2020.
17. International Organization for Standardization. ISO 2631-1:1997, Mechanical Vibration and Shock—Evaluation of Human Exposure to Whole Body Vibration—Part 1: General Requirement. 1997. Available online: https://www.iso.org/standard/7612.html (accessed on 20 February 2021).
18. Pawar, P.R.; Mathew, A.T.; Saraf, M. IRI (International Roughness Index): An Indicator of Vehicle Response. *Mater. Today Proc.* **2018**, *5*, 11738–11750. [CrossRef]
19. International Organization for Standardization. ISO 8608:2016, Mechanical Vibration—Road Surface Profiles—Reporting of Measured Data. 2016. Available online: https://www.iso.org/standard/71202.html (accessed on 20 February 2021).
20. Tudón-Martínez, J.C.; Fergani, S.; Sename, O.; Martinez, J.J.; Morales-Menendez, R.; Dugard, L. Adaptive Road Profile Estimation in Semiactive Car Suspensions. *IEEE Trans. Control. Syst. Technol.* **2015**, *23*, 2293–2305. [CrossRef]
21. Basargan, H.; Mihály, A.; Gáspár, P.; Sename, O. Adaptive Semi-Active Suspension and Cruise Control through LPV Technique. *Appl. Sci.* **2021**, *11*, 290. [CrossRef]
22. Vahidi, A.; Stefanopoulou, A.; Peng, H. Recursive Least Squares with Forgetting for Online Estimation of Vehicle Mass and Road Grade: Theory and Experiments. *Veh. Syst. Dyn.* **2005**, *43*, 31–55. [CrossRef]

23. Yin, Z.; Dai, Q.; Guo, H.; Chen, H.; Chao, L. Estimation Road Slope and Longitudinal Velocity for Four-wheel Drive Vehicle. *IFAC-PapersOnLine* **2018**, *51*, 572–577. [CrossRef]
24. Li, B.; Zhang, J.; Du, H.; Li, W. Two-layer Structure Based Adaptive Estimation for Vehicle Mass and Road Slope under Longitudinal Motion. *Measurement* **2017**, *95*, 439–455. [CrossRef]
25. Tudón-Martínez, J.C.; Fergani, S.; Varrier, S.; Sename, O.; Dugard, L.; Morales-Menendez, R.; Ramírez-Mendoza, R. Road Adaptive Semi-Active Suspension in an Automotive Vehicle using an LPV Controller. *IFAC Proc. Vol.* **2013**, *46*, 231–236. [CrossRef]
26. Soheib, F.; Sename, O.; Dugard, L. An LPV/H∞ Integrated Vehicle Dynamic Controller. *IEEE Trans. Veh. Technol.* **2015**, *65*, 1880–1889. [CrossRef]
27. Unger, A.; Schimmack, F.; Lohmann, B.; Schwarz, R. Application of LQ-based Semi-active Suspension Control in a Vehicle. *Control. Eng. Pract.* **2013**, *21*, 1841–1850. [CrossRef]
28. Poussot-Vassal, C. Robust LPV Multivariable Automotive Global Chassis Control. Ph.D. Thesis, Institut National Polytechnique de Grenoble-INPG, Grenoble, France, 2008.
29. Poussot-Vassal, C.; Sename, O.; Dugard, L.; Savaresi, S.M. Vehicle Dynamic Stability Improvements through Gain-scheduled Steering and Braking Control. *Veh. Syst. Dyn.* **2011**, *49*, 1597–1621. [CrossRef]
30. Loprencipe, G.; Zoccali, P.; Cantisani, G. Effects of Vehicular Speed on the Assessment of Pavement Road Roughness. *Appl. Sci.* **2019**, *9*, 1783. [CrossRef]
31. Zuo, L.; Nayfeh, S. Low Order Continuous-time Filters for Approximation of the ISO 2631-1 Human Vibration Sensitivity Weightings. *J. Sound Vib.* **2003**, *265*, 459–465. [CrossRef]
32. Ahlin, K.; Granlund, N.J. Relating Road Roughness and Vehicle Speeds to Human Whole Body Vibration and Exposure Limits. *Int. J. Pavement Eng.* **2002**, *3*, 207–216. [CrossRef]
33. Apkarian, P.; Gahinet, P.; Becker, G. Self-scheduled H∞ Control of Linear Parameter-varying Systems: A Design Example. *Automatica* **1995**, *31*, 1251–1261. [CrossRef]
34. Do, A.L.; Sename, O.; Dugard, L.; Soualmi, B. Multi-objective Optimization by Genetic Algorithms in H_∞/LPV Control of Semi-active Suspension. *IFAC Proc. Vol.* **2011**, *44*, 7162–7167. [CrossRef]

Motivated by this, the EU-H2020 funded project ESRIUM [2] strives for the high-level goal of increased safety and resource efficiency of transport on European roads. In doing this, its key innovation is a digital map of road surface damage and road wear. This shall help reduce both the number of road works and the associated problems by using new digital services for managing the traffic, as well as by controlling the utilization of the road. The road wear map will contain unique information for the road operators to enhance the road maintenance planning, as well as to provide route and driving recommendations to CAVs and connected vehicles. These recommendations will be used for adjusting the driving path (in-lane and between lanes), which shall help with gradual degradation of the road surface, thereby reducing the regular maintenance actions.

The development of CAVs and infrastructure-assisted automated driving functions have attracted increasing attention in recent years, as summarized in [3,4]. Particularly, lane change and merge maneuvers present a challenge for vehicle automation. The authors in [5] provided a survey in this area, including vehicle positioning systems, communication systems, control systems and system validation. The energy saving potentials of CAVs, on the other hand, have been highlighted in [6]. The conservative assessment in [6] shows 3% energy saving in highway driving in presence of static environment information, and 10% energy saving in arterial driving if traffic signals are provided via vehicle-to-infrastructure (V2I) communications. However, the effect of automated vehicles (AVs) on mixed traffic might be negative. The authors in [7] used real traffic data together with the traffic simulation tool simulation of urban mobility (SUMO) [8] to accomplish the traffic efficiency analysis of mixed traffic, and they claimed that maximum traffic flow rate reduces with higher penetration rate of AVs in mixed traffic. Another important application for CAVs is truck platooning and many problems remain unsolved [9].

In addition to perception and control, trajectory planning is an indispensable function to realize high level of autonomous driving [10]. The configuration space of AVs or CAVs on a 2D plane comprises three dimensions: two dimensions indicating the vehicle's position (x, y), and one dimension indicating the vehicle's heading θ. However, the time dimension or velocity dimension has to be added, if differential constraints and vehicle dynamics are considered. Quite a lot of work tries to tackle this high dimensional problem using the so-called spatiotemporal planning. McNaughton and et al. proposed conformal spatiotemporal lattice to represent search space for structured environments [11]. However, the high dimension of the state space makes the lattice expensive to be repeatedly constructed and searched in a dynamical environment. Recently, Sun and et al. attempted to address this problem by using intelligent driver model (IDM) as a velocity feedback policy [12]. Therefore, the number of dimensions of the lattice planner is reduced.

To balance the trade-off between computation time and quality of the trajectory planning result, a framework of combining graph search and optimization was proposed in [13] and has been augmented recently. In [14,15], an A* search is used to find a rough reference spatiotemporal trajectory along with a collision-free driving corridor, which is subsequently refined by optimization considering safety and dynamical constraints. Moreover, [16] proposed a different approach to identify the spatiotemporal constraints (collision-free driving corridors) using set-based reachability analysis. By combining their approach with optimization-based spatiotemporal planning, arbitrary traffic scenarios can be solved.

Wheel wander, which is related to road surface damage, is of interest for infrastructure designers. Gungor analyzed the literature about wheel wander and its impact on the pavement [17]. According to this report, wheel wander was defined as uncertainty of the lateral position of wheel loads on a lane. The lateral position of the wheels of human driven cars is not uniformly distributed over the whole lane but can be modeled as a normal distribution with non-zero mean and known standard deviation. It was shown in simulation, that wheel wander reduces rutting in comparison to the case without wheel wander.

CAVs, especially heavy-duty trucks in platoon formations, are per design expected to have no wheel wander, and without any measure will lead to more road damage in the from of rutting of the road surface. This is, in fact, very different than the effects of human driven trucks on the road networks, and can cause infrastructure maintenance issues as the penetration of such vehicles increase. In the recent paper [18], this problem was also described and some counter measures were analyzed. Particularly, optimization strategies using V2I communication are proposed for a desired lane off-set the vehicles should drive.

On another strand of work, Bouchihati [19] analyzed the impact of truck platooning on pavement structure on Dutch motorways. It is stated in this publication that automated truck platoons will lead to substantial higher damage on the pavement as result of rut development and fatigue. As a potential solution, smart lanes were proposed. Accordingly, sensors can be used to measure the positions of all previous passed truck platoons and the optimum lateral position can be communicated to the leading truck who will then adjust its lateral position according to the given recommendation. With this method, an optimal use of the road surface can be achieved.

Inspired by these, the EU project ESRIUM investigates infrastructure assisted routing and driving recommendations utilizing C-ITS communications. In this respect, specially designed ADAS functions (i.e., lateral and longitudinal tracking controllers and a trajectory planner) are being developed with capabilities to adapt their behavior according to specific routing and driving recommendations received in the form of infrastructure to vehicle information message (IVIM). The current paper presents the specific use-cases to demonstrate the utilization of this approach for reduction in road rutting (due to reduced wheel wander of CAVs), and to initiate informed lane changes to avoid potential hazard situations. The paper further describes the developed C-ITS assisted ADAS functions together with their verification results utilizing a simulation framework.

The paper is organized as follows: First, we introduce the two use cases of the C-ITS assisted routing and driving recommendation in Section 2. In Section 3, the simulation framework is introduced. Next, the ADAS functions are defined in Section 4. Simulation implementation and verification results are finally given in Section 5 followed by the conclusions and outlook in Section 6.

2. Descriptions of the Use Case Scenario and Performance Indicators

In this paper, two use case scenarios of ESRIUM are described. The first one is an in-lane off-set recommendation for connected automated vehicles (CAVs) and the second is a strategic lane change and lane utilization information for CAVs. In both cases, the ego vehicle receives the information via the C-ITS message type IVIM [20].

In the in-lane off-set recommendation scenario, as sketched in Figure 1, the ego vehicle shall drive in automated mode (SAE Level-3 equivalent Motorway Chauffeur combining adaptive cruise control (ACC) and lane keeping assist (LKA) driving functions) in a detection zone when it receives an IVIM containing a recommended lane-off-set information. This C-ITS routing recommendation in the form of a IVIM comes from an infrastructure road-side unit (RSU) and is received by an on-board unit (OBU) and the linked automated driving functions on the ego vehicle interpret it. The detection zone defines the region, where the vehicle is expected to be in a bi-lateral communication link with the RSU. Before entering the relevance zone (this is the zone, where the recommended action by the IVIM needs to be implemented), the ego vehicle adapts the typical LKA task of tracking the center of the existing lane and transitions to driving along the same lane with the given in-lane off-set. The ego vehicle is expected to drive throughout the relevance zone with the recommended in-lane off-set in case traffic conditions permit it. Immediately after leaving the relevance zone, the ego vehicle is expected to follow the default centerline tracking task unless otherwise recommended.

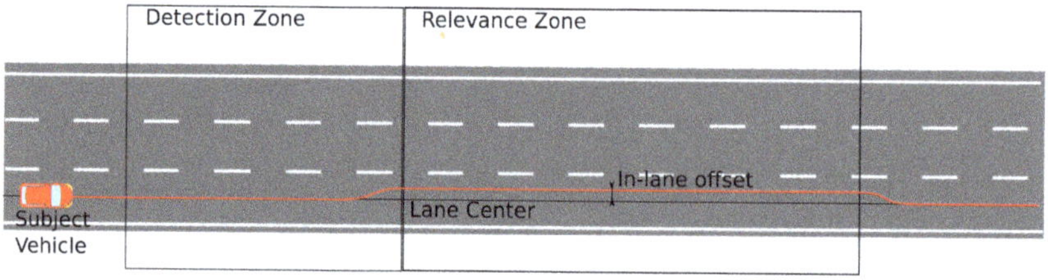

Figure 1. In-lane off-set recommendation scenario description.

In the lane change recommendation scenario (see Figure 2) , the ego vehicle shall drive in automated mode (SAE Level-3 equivalent Motorway Chauffeur combining ACC and LKA driving functions) in a detection zone when it receives an IVIM containing a set of three relevance zones with instructions to avoid the rightmost lane, because of road damage. When the ego vehicle is driving on the rightmost lane in relevance zone 1, it will change the lane to next lane if traffic allows it, and in relevance zone 2 it will continue driving on the remaining lanes and avoid the rightmost lane. In relevance zone 3, the rightmost lane is cleared and the ego vehicle uses the most appropriate lane according to the traffic situation.

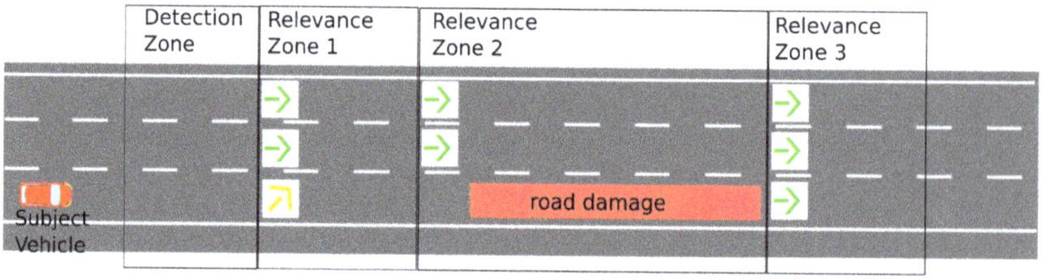

Figure 2. Lane change recommendation scenario description (see Table 1 for the definitions of the pictogram symbols)

3. Simulation Environment and Setup

3.1. Driving Functions and Vehicle Dynamics

For the development of the driving functions, a co-simulation environment based on Matlab/Simulink and the vehicle dynamics software IPG Carmaker was utilized (see Figure 3a). This simulation environment was preferred since it is well suited for rapid-prototyping purposes. As depicted in the figure, all driving functions in terms of trajectory planner, as well as longitudinal and lateral tracking controllers were developed and implemented in Matlab/Simulink. Those include the vehicle state, an object list and lane marking information. The vehicle state includes actuation signals (steering angle, brake and gas pedal), velocities and accelerations. For the object list, no sensor dynamics were assumed considering only occlusion effects. Taking into account limitations of a prospective real world demonstration of the presented implementations, lateral vehicle guidance was based on lane marking information only. For that purpose, a polynomial lane marking model was recreated in simulation, which describes lane boundaries as third order polynomials and related polynomial domains.

The routing recommendations based on IVIM messages were implemented as preset messages. Future demonstrations will be performed on sections of the Austrian motorway A2. Therefore, the corresponding UHDmaps(®) [21] in the OpenDRIVE format was used in the CarMaker driving scenario.

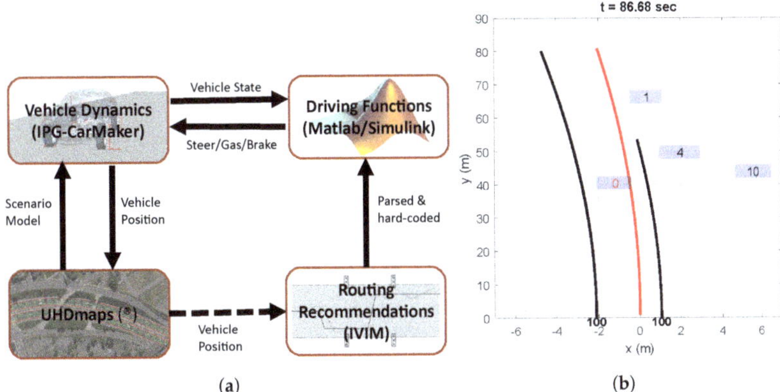

Figure 3. Simulation environment, setup, and ego vehicle aligned visualization. (**a**) Co-simulation architecture utilizing Matlab/Simulink and IPG Carmaker. (**b**) Visualization of objects (blue boxes) and assigned ID, lane markings and quality of detection (black), and ego prediction (red).

Object list and lane marking inputs to the driving function were visualized in an ego-centered view online (Figure 3b), i.e., in parallel with the simulation. This allowed to validate driving function inputs and outputs.

3.2. C-ITS Message Structure and Simulation Implementation

In the final implementation, the recommendations for lane off-set and lane change will be sent using the IVIM messages. IVIM information will be broadcast according to TS 103 301 V2.1.1 and ISO 19321: 2020 standard. The automated vehicle container of the broadcast IVIM message will include the pictograms of the recommended action. In the case of lane change recommendations, one of the four categories of recommendations associated with corresponding pictograms according to the ISO 14823:2017 "Intelligent transport systems-Graphic data dictionary", as indicated in Table 1, will be used.

Table 1. IVIM lane choice recommendation according to the ISO 14823:2017 code.

Recommendation	ISO14823 Code	Pictogram
Keep the current lane	13,660	↓
Move to the left lane	13,661	↙
Move to the right lane	13,662	↘
Lane closed	13,669	✕

The lane off-set recommendation will also be broadcast using the free text option in the automated vehicle container of the IVIM message. The free text will contain the signage followed by the off-set in centimeters. In this setting "+" sign indicates an off-set to the right of the lane center whereas "−" sign indicates an off-set to the left relative to the lane center. The example information content of the lane off-set IVIM message is given in Table 2.

Table 2. IVIM lane off-set messages information structure.

AV-Container Part	Detection-Zone Ids	Relevance-Zone Ids	Applicable Lane	±Offset [cm]
1	1	11	1	−20
2	1	13	2	10
⋮	⋮	⋮	⋮	⋮

In the scope of the simulation based development of the driving functions, the implementation of sending and receiving of the IVIM communications is not considered. Although parsing of the message standard structure described above is not part of this first simulations. For both use case scenarios a simple emulation subsystem was developed. Its aim is to check if the ego vehicle is inside the detection or relevance zones and to provide the trajectory planner with the necessary information. In the case of the first scenario (i.e., in-lane off-set recommendation), this information is the desired off-set for each lane in a give relevance zone. For the lane change recommendation scenario, on the other hand, the specific information that is provided to the driving functions includes one detection zone and three relevance zones. Therefore, the developed emulation block checks and interprets whether the ego vehicle is inside one of these areas and consequently specifies the desired lane to the trajectory planner. The specific implementation of the IVIM parser will be conducted as part of the real vehicle tests. The ego vehicle has to be equipped with a GNSS-receiver and compares its position with the definitions of the detection and relevance zones to evaluate if it is inside of a zone.

4. AD Functions Development

The AD functions presented in this article are discussed in the well known sense–plan–act scheme. As mentioned in Section 3, the sensing task is delegated to CarMaker, which provides all the required ego vehicle states, as well as obstacle and road information. The sensors are therefore modeled as low-fidelity ideal sensors providing object lists within specified cones representing their respective field of view. The planning task is accomplished by a rule-based trajectory planner adapted to utilize C-ITS messages, as described below in Section 4.1.

4.1. Planning: Rule-Based Trajectory Planner

The planning task is accomplished by a rule-based trajectory planner (TP) that was developed for structured environments like highways with well defined lane boundaries. It uses a finite state machine and a set of discrete decisions to trigger lane changes or keep the vehicle on its current lane. By default, the ego vehicle drives in the middle of the rightmost lane. If a slower vehicle prevents the ego vehicle from reaching its desired cruising speed, and the target lane is not occupied, a Bézier curve is planned to perform a lane change.

The value of the Bézier curve at a tunable look-ahead distance is then used as the desired lateral off-set for the underlying lateral controller. As can be seen from Figure 4, this off-set is defined with respect to the center of the current lane. Therefore, the TP needs to handle a change of reference during the lane change maneuver. This allows future deployment of the proposed algorithm to a demonstrator vehicle which utilizes only a vision sensor for lane marking based in-lane localization.

Together with emulations blocks for parsing the IVIM (see Section 3.2), the TP is able to generate the input signals for the lateral and longitudinal controllers to accomplish the described scenarios in Section 2.

Bézier Based Path Planning Approach

Path planning relies on a Bézier curve approach [22] that was enhanced to handle specific limitations that are common in real world applications. Apart from using HD maps, currently available sensors lack the ability of providing trustworthy lane information, specifically the lane width, besides the current lane. To overcome this issue, the planned path consists of two quadratic Bézier curves connected via a straight segment as shown in Figure 4.

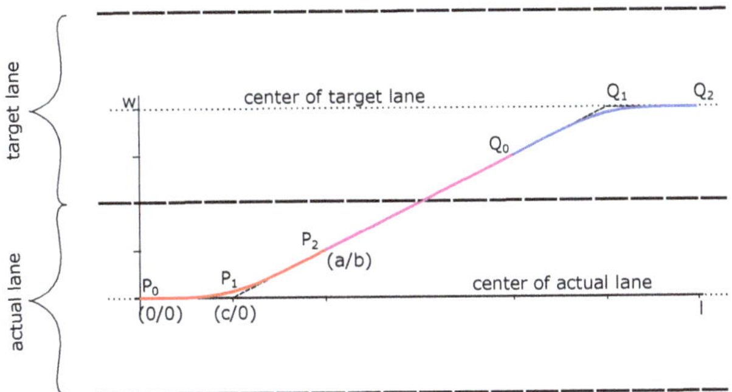

Figure 4. Lane change with two quadratic Bézier curves and a linear segment.

The Bézier curves provide a smooth transition to initiate and finish the lane change, while the straight segment can be shortened or extended during the lane change maneuver according to the width of the target lane.

The path as shown in Figure 4 can then be written as a spline

$$S(\tau) = \begin{cases} \Gamma_1(\tau) & 0 \le \tau \le 1, \\ \Gamma_2(\tau) & 1 \le \tau \le 2, \\ \Gamma_3(\tau) & 2 \le \tau \le 3. \end{cases} \tag{1}$$

with three segments Γ_i $(i = 0, 1, 2)$ and the path parameter τ. Considering a quadratic Bézier curve

$$\mathcal{C}(\tau, C_0, C_1, C_2) = C_0(1-\tau)^2 + 2C_1(1-\tau)\tau + C_2\tau^2, \quad \tau \in [0, 1] \tag{2}$$

with control points C_i $(i = 0, 1, 2)$, the spline segments Γ_1 and Γ_3 were chosen as

$$\Gamma_1(\tau) := \mathcal{C}(\tau, P_0, P_1, P_2) \quad \text{and} \quad \Gamma_3(\tau) := \mathcal{C}(\tau, Q_0, Q_1, Q_2). \tag{3}$$

For the terminal points of $S(\tau)$, we can immediately state from Figure 4:

$$P_0 = (0, 0) \quad \text{and} \quad Q_2 = (l, w). \tag{4}$$

To achieve a comfortable lane change, the length l of the lane change is chosen according to the current vehicle speed, while the lane change off-set w is pre-defined by the widths of the current and target lane:

$$w = \frac{w_{\text{actual}} + w_{\text{target}}}{2}. \tag{5}$$

Since Γ_3 is skew-symmetric to Γ_1, the relations

$$Q_0 = Q_2 - P_2 \quad \text{and} \quad Q_1 = Q_2 - P_1 \tag{6}$$

hold, leaving only $P_1 := (c, 0)$ and $P_2 := (a, b)$ undetermined. The straight segment Γ_2 shown in magenta connects the points P_2 and Q_0, therefore

$$\Gamma_2(\tau) = (Q_0 - P_2)\tau + P_2 = (Q_2 - 2P_2)\tau + P_2. \tag{7}$$

Claiming geometric continuity at the transition from Γ_1 to Γ_2 (and, therefore, from Γ_2 to Γ_3), the relation

$$\frac{b}{a-c} = \frac{w-2b}{l-2a}. \tag{8}$$

must be fulfilled. According to Figure 4, it is reasonable to restrict $b \leq \frac{1}{2}w_{\text{actual}}$, which is usually fulfilled by setting b to half the width of the ego vehicle. Then, introducing $f_c = \frac{c}{a}$ as a tunable parameter, (8) can be solved for

$$a = \frac{bl}{w + 2bf_c - wf_c}. \tag{9}$$

To conclude the path planning approach, the control points are

$$\begin{aligned} &P_0 = (0,0), \quad P_1 = (c,0), \quad P_2 = (a,b), \\ &Q_0 = (l-a, w-b), \quad Q_1 = (l-a/2, w), \quad Q_2 = (l,w). \end{aligned} \tag{10}$$

As mentioned, b was set to half the ego vehicle width, while f_c was determined empirically considering the trade-off between the slope and continuity of the spline S. For the simulations presented in Section 5, $b = 0.9$ and $f_c = 0.5$ were used.

4.2. Actuation: Lateral and Longitudinal Control

For tracking the lateral and longitudinal references from the trajectory planner, two proven controller implementations were reused. The lateral controller implements a state-feedback control law based on [23], modified according to [24] as motivated from a demonstrator vehicle application. It makes use of the reference path from the trajectory planner up to the second geometric continuity, i.e., lateral error, heading error, and path curvature. The actuated signal is the steering wheel angle. Although, the dynamics of a steer-by-wire actuator are considered.

The longitudinal controller is implemented as a time-discrete PI (proportional-integral) controller with an anti-wind-up measure providing the control signal p_k at time step k according to

$$p_k = k_{\mathrm{P}} e_k + \alpha_k \tag{11}$$

$$\alpha_k = k_{\mathrm{I}} e_k + \max(-100, \min(\alpha_{k-1}, 100)). \tag{12}$$

Here, e_k is the acceleration error, k_{P} and k_{I} are the proportional and integral gain and p_k is the commanded brake pedal position for $p_k < 0$ and the throttle position for $p_k > 0$. The anti-wind-up measure is implemented via max() and min() functions ensuring $p_k \in [-100, 100]$. To ensure bumpless activation of the longitudinal controller, α_0 is initialized according to the current pedal positions.

Both controllers execute with a sample time of 20 ms.

5. Simulation Results and Analysis

The described use case scenarios in Section 2 were simulated on a straight section of the A2 motorway without other traffic elements. The vehicle speed was set to 130 km/h for the longitudinal tracking controller.

Figure 5 shows the results of the in-lane off-set recommendation scenario. For this simulation a detection zone with a length of 200 m was chosen and a relevance zone with a length of 1000 m and the in-lane off-set recommendation is only valid for the rightmost lane. In Figure 5a, the steering wheel angle during the maneuver is depicted. Moreover, the lateral acceleration in Figure 5b, the yaw rate in Figure 5c, and the desired and driven off-sets in Figure 5d are shown, respectively.

According to the simulated scenario, at approximately 2.5 s the ego vehicle starts the transition to the desired in-lane off-set. In this example the vehicle reaches its desired steady-state off-set value of −0.2 m at approximately 7 s. The total time it took to reach the set off-set value is, therefore, about 4.5 s, which matches the in the controller defined value. According to the designed scenario, the ego vehicle drives through the relevance zone (RZ) with the desired off-set and leaves it at about second 35 returning back to zero off-set value.

As seen from the results in Figure 5, the developed enhanced driving functions are quite effective in achieving the desired off-set value. The difference between the desired and achieved off-set values are quite small except a small overshoot. The overshoot in the lateral position was smaller than 2 cm in this case. The small oscillation in the middle point of the maneuver (which should not be existing at all, at about 21 s) was caused by the CarMaker itself and is believed to be a modeling bug, as it varied when the vehicle model was changed, but could not be avoided completely.

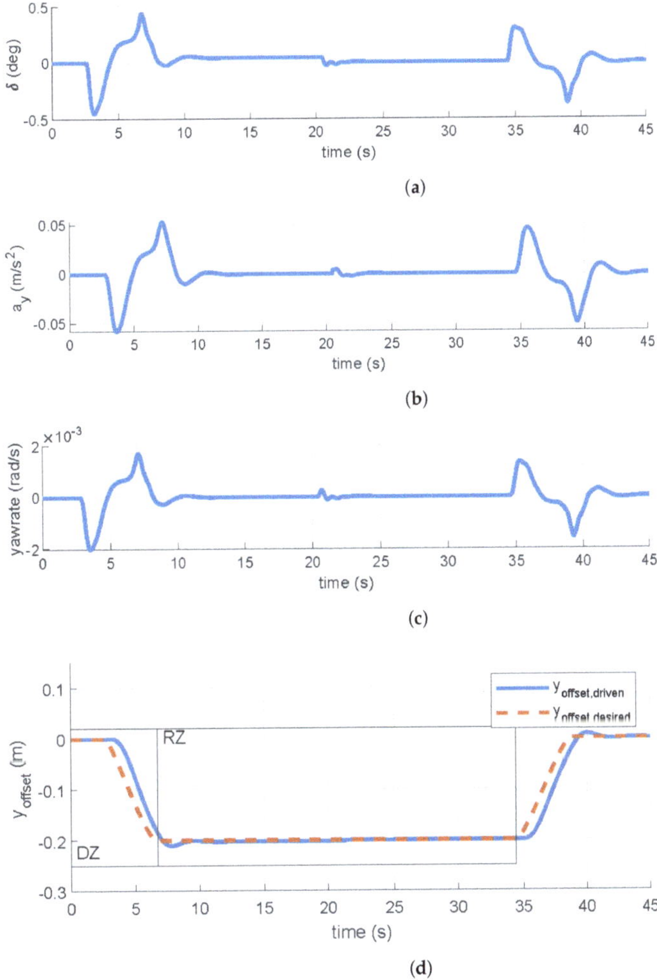

Figure 5. Result of the in-lane off-set recommendation scenario: (**a**) steering wheel angle, (**b**) lateral acceleration, (**c**) yaw rate, and (**d**) lane off-set.

Figure 6 shows the use case scenario simulation results for the strategic lane change and lane utilization recommendation for CAVs. Figure 6a shows the steering wheel angle variation during this example recommended lane change maneuver. Furthermore, the lateral acceleration in Figure 6b, the yaw rate in Figure 6c, and, finally, the vehicle lateral position relative to the desired lane in Figure 6d are shown, respectively. According to this example scenario, the ego vehicle enters the relevance zone 1 (RZ1) at about 6.6 s and the desired lane, therefore, switches from 0 to 1. Note that in the TP the rightmost lane is indexed with 0 and the index is increased while going towards the left side, which is different to the numbering convention in the IVIM. Since there is no hindering traffic in the scenerio, the TP initiates immediately a lane change maneuver to lane 1, where it stays while passing the relevance zones 1 and 2 (RZ1 and RZ2). In relevance zone 3 (RZ3) the TP computes a lane change maneuver back to the original starting lane (i.e., lane 0). In this use case as well, the recommended maneuvers were conducted effectively with little overshoot. It can be observed that the total lane change maneuver was achieved with less than 5 s transition times between steady state driving positions.

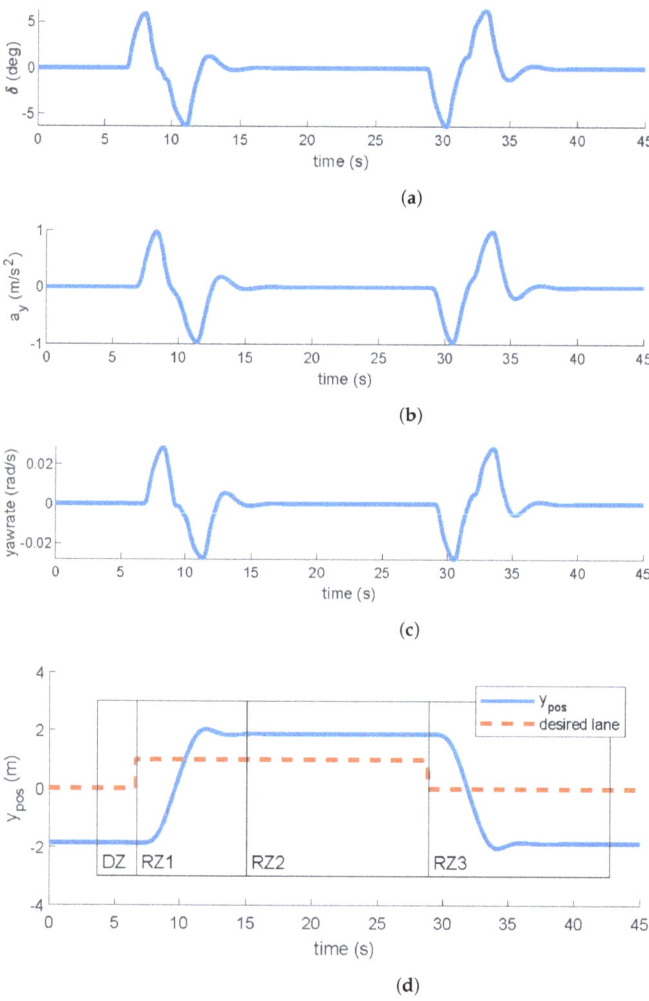

Figure 6. Result of the lane change recommendation scenario: (**a**) steering wheel angle, (**b**) lateral acceleration, (**c**) yaw rate, (**d**) lateral vehicle position.

As a further simulation use case example, we implemented also a combined in-lane off-set and lane change recommendations, as seen in Figure 7. In this scenario, the ego vehicle starts in the right lane and first receives a in-lane off-set recommendation and than consecutively a lane change to the left lane followed by reverse maneuvers to return to the right lane lateral 0-off-set position. In this case, the in-lane off-set values were varied with specific values {−1 m, −0.5 m, −0.2 m, −0.1 m, 0 m, 0.1 m, 0.2 m, 0.5 m, 1 m} and the resulting trajectories were displayed in Figure 7 as a superimposed plot. The results indicate that complex maneuvers combining the two specific recommendations can also be achieved with efficacy.

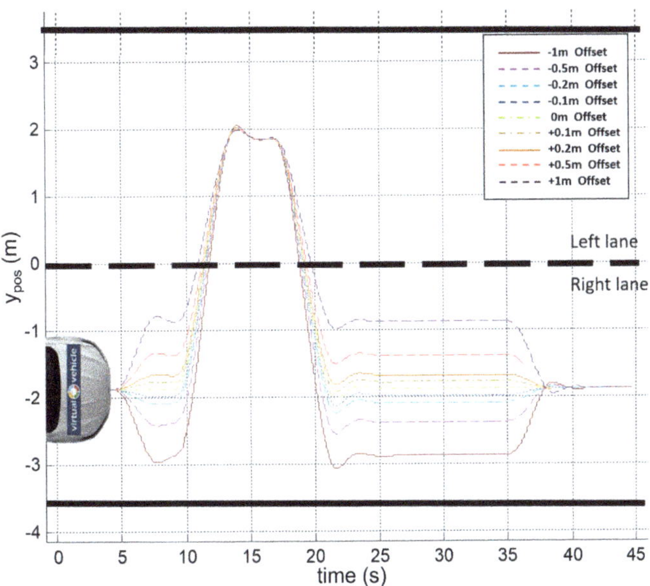

Figure 7. Combined in-lane off-set and lane change recommendations with varying lateral off-set values. The lane boundaries were shown to aid the visualization.

6. Conclusions and Outlook

Given that the number of automated vehicles increase rapidly, their behavior on the motorways can have abnormal ramifications on the road surface wear. Consider, for example, automated truck platoons driving along same road stretch, which will surely lead to frequent rutting. Motivated by this, we introduced in this paper a new model for operation of automated driving functions, namely by using specific infrastructure routing recommendations for enhancing or adapting their behavior.

The recommendations are in the form of specific lane-change and in-lane off-set suggestions. It is envisaged that from the collective behavior of the automated vehicles (and connected vehicles in general) with a capability to adapt their behavior according to the routing recommendations will benefit both the road operators and the vehicle owners. These vehicles will, on the one hand, reduce rutting of the surface, as well as unbalanced wear of the road surface, especially when their penetration rates increase. On the other hand, such vehicles will avoid rutted or damaged road sections and blocked lanes thanks to the routing recommendations, providing a safer and more comfortable ride. This is an idea being developed within the recently started EU-H2020 funded project ESRIUM, where the complete value chain for implementing this basic idea is being developed to investigate its feasibility.

In this paper we described the specific ADAS/AD functions including longitudinal and lateral tracking controllers, as well as a rule based trajectory planner, which are utilized to achieve the routing recommendations. We also introduced the specific use cases that will be implemented in this context, and the structure of the routing recommendations based on the IVIM message standard. The corresponding ADAS/AD functions were developed in a high-fidelity simulation framework based on Matlab/Simulink and Carmaker software with the view of deploying the same functions in a test vehicle for real-life implementation. The ADAS/AD functions were demonstrated for representative use cases to verify their basic operation. In the future extensions, the same driving functions will be imported to an automated drive demonstrator vehicle to implement and realize the demonstration of the routing recommendations in real-life use cases on a public road. It is planned to conduct this demonstration during the progress of the ESRIUM project.

Author Contributions: Conceptualization, M.R., G.N. and S.S.; methodology, M.R., G.N. and S.S.; software, M.R. and G.N.; validation, M.R., G.N. and S.S.; formal analysis, M.R., G.N. and S.S.; investigation, M.R., K.T. and S.S.; data curation, M.R and G.N.; writing—original draft preparation, M.R., G.N., K.T. and S.S.; writing—review and editing, M.R., G.N., K.T. and S.S.; visualization, M.R., G.N. and S.S.; supervision, S.S.; project administration, S.S.; All authors have read and agreed to the published version of the manuscript.

Funding: This work was conducted in the scope of the ESRIUM Project, which has received funding from the European Union Agency for the Space Programme under the European Union's Horizon 2020 research and innovation programme and under grant agreement No 101004181. The content of this paper reflects only the authors' view. Neither the European Commission nor the EUSPA is responsible for any use that may be made of the information it contains.

Acknowledgments: Virtual Vehicle Research GmbH has received funding within COMET Competence Centers for Excellent Technologies from the Austrian Federal Ministry for Climate Action, the Austrian Federal Ministry for Digital and Economic Affairs, the Province of Styria (Dept. 12) and the Styrian Business Promotion Agency (SFG). The Austrian Research Promotion Agency (FFG) has been authorised for the programme management.

Conflicts of Interest: The authors declare no conflict of interest. The content of this paper reflects only the authors' view. Neither the European Commission nor the EUSPA is responsible for any use that may be made of the information it contains. The funders had no role in design of the study, in the collection, analyses, or interpretation of data; in the writing of the manuscript, or in the decision to publish the results.

References

1. SAE-J3016. *Surface Vehicle Recommended Practice–(R) Taxonomy and Definitions for Terms Related to Driving Automation Systems for On-Road Motor Vehicles*; Standard SAE J3016:APR2021; Society of Automotive Engineers: Warrendale, PA, USA, 2018.
2. ESRIUM. EGNSS-Enabled Smart Road Infrastructure Usage and Maintenance for Increased Energy Efficiency and Safety on European Road Networks. Available online: https://esrium.eu/ (accessed on 14 July 2021).
3. Pendleton, S.; Andersen, H.; Du, X.; Shen, X.; Meghjani, M.; Eng, Y.; Rus, D.; Ang, M. Perception, Planning, Control, and Coordination for Autonomous Vehicles. *Machines* **2017**, *5*, 6. [CrossRef]
4. Guanetti, J.; Kim, Y.; Borrelli, F. Control of connected and automated vehicles: State of the art and future challenges. *Annu. Rev. Control* **2018**, *45*, 18–40. [CrossRef]
5. Bevly, D.; Cao, X.; Gordon, M.; Ozbilgin, G.; Kari, D.; Nelson, B.; Woodruff, J.; Barth, M.; Murray, C.; Kurt, A.; et al. Lane Change and Merge Maneuvers for Connected and Automated Vehicles: A Survey. *IEEE Trans. Intell. Veh.* **2016**, *1*, 105–120. [CrossRef]
6. Vahidi, A.; Sciarretta, A. Energy saving potentials of connected and automated vehicles. *Transp. Res. Part C Emerg. Technol.* **2018**, *95*, 822–843. [CrossRef]
7. Berrazouane, M.; Tong, K.; Solmaz, S.; Kiers, M.; Erhart, J. Analysis and Initial Observations on Varying Penetration Rates of Automated Vehicles in Mixed Traffic Flow utilizing SUMO. In Proceedings of the 2019 IEEE International Conference on Connected Vehicles and Expo (ICCVE), Graz, Austria, 4–8 November 2019; pp. 1–7.
8. Lopez, P.A.; Behrisch, M.; Bieker-Walz, L.; Erdmann, J.; Flötteröd, Y.P.; Hilbrich, R.; Lücken, L.; Rummel, J.; Wagner, P.; Wießner, E. Microscopic Traffic Simulation using SUMO. In Proceedings of the 21st IEEE International Conference on Intelligent Transportation Systems, Maui, HI, USA, 4–7 November 2018.
9. Bhoopalam, A.K.; Agatz, N.; Zuidwijk, R. Planning of truck platoons: A literature review and directions for future research. *Transp. Res. Part B Methodol.* **2018**, *107*, 212–228. [CrossRef]
10. Tong, K.; Ajanovic, Z.; Stettinger, G. Overview of tools supporting planning for automated driving. In Proceedings of the 2020 IEEE 23rd International Conference on Intelligent Transportation Systems (ITSC), Rhodes, Greece, 20–23 September 2020; pp. 1–8.
11. McNaughton, M.; Urmson, C.; Dolan, J.M.; Lee, J.W. Motion planning for autonomous driving with a conformal spatiotemporal lattice. In Proceedings of the 2011 IEEE International Conference on Robotics and Automation, Shanghai, China, 9–13 May 2011; pp. 4889–4895. [CrossRef]

12. Sun, K.; Schlotfeldt, B.; Chaves, S.; Martin, P.; Mandhyan, G.; Kumar, V. Feedback Enhanced Motion Planning for Autonomous Vehicles. In Proceedings of the 2020 IEEE/RSJ International Conference on Intelligent Robots and Systems (IROS), Las Vegas, NV, USA, 24 October–24 January 2020; pp. 2126–2133. [CrossRef]
13. Hubmann, C.; Aeberhard, M.; Stiller, C. A generic driving strategy for urban environments. In Proceedings of the 2016 IEEE 19th International Conference on Intelligent Transportation Systems (ITSC), Rio de Janeiro, Brazil, 1–4 November 2016; pp. 1010–1016. [CrossRef]
14. Zhang, T.; Fu, M.; Song, W.; Yang, Y.; Wang, M. Trajectory Planning Based on Spatio-Temporal Map with Collision Avoidance Guaranteed by Safety Strip. *IEEE Trans. Intell. Transp. Syst.* **2020**, 1–14. [CrossRef]
15. Xin, L.; Kong, Y.; Li, S.E.; Chen, J.; Guan, Y.; Tomizuka, M.; Cheng, B. Enable faster and smoother spatio-temporal trajectory planning for autonomous vehicles in constrained dynamic environment. *Proc. Inst. Mech. Eng. Part D J. Automob. Eng.* **2021**, *235*, 1101–1112. [CrossRef]
16. Schäfer, L.; Manzinger, S.; Althoff, M. Computation of Solution Spaces for Optimization-based Trajectory Planning. *IEEE Trans. Intell. Veh.* **2021**. [CrossRef]
17. Gungor, O.E. *A Literature Review on Wheel Wander*; Technical Report; Illinois Asphalt Pavement Association: Springfield, MO, USA, 2018.
18. Gungor, O.E.; Al-Qadi, I.L. All for one: Centralized optimization of truck platoons to improve roadway infrastructure sustainability. *Transp. Res. Part C Emerg. Technol.* **2020**, *114*, 84–98. [CrossRef]
19. el Bouchihati, M. The Impact of Truck Platooning on the Pavement Structure of Dutch Motorways. Master's Thesis, Delft University of Technology, Delft, The Netherlands, 2020.
20. ISO. *ISO/TS 19321:2020 Intelligent Transport Systems—Cooperative ITS—Dictionary of In-Vehicle Information (IVI) Data Structures*; ISO: Geneva, Switzerland, 2020.
21. Tihanyi, V.; Tettamanti, T.; Csonthó, M.; Eichberger, A.; Ficzere, D.; Gangel, K.; Hörmann, L.B.; Klaffenböck, M.A.; Knauder, C.; Luley, P.; et al. Motorway Measurement Campaign to Support R&D Activities in the Field of Automated Driving Technologies. *Sensors* **2021**, *21*, 2169. [CrossRef] [PubMed]
22. Choi, J.W.; Curry, R.; Elkaim, G. Path Planning Based on Bézier Curve for Autonomous Ground Vehicles. In Proceedings of the Advances in Electrical and Electronics Engineering—IAENG Special Edition of the World Congress on Engineering and Computer Science 2008, San Francisco, CA, USA, 22–24 October 2008; pp. 158–166. [CrossRef]
23. Nestlinger, G.; Stolz, M. Bumpless Transfer for Convenient Lateral Car Control Handover. *IFAC-PapersOnLine* **2016**, *49*, 132–138. [CrossRef]
24. Solmaz, S.; Nestlinger, G.; Stettinger, G. Compensation of Sensor and Actuator Imperfections for Lane-Keeping Control Using a Kalman Filter Predictor. *SAE Int. J. Connect. Autom. Veh.* **2021**, *4*, 97–106. [CrossRef]

Article

Automatically Learning Formal Models from Autonomous Driving Software

Yuvaraj Selvaraj [1,2,*,†], Ashfaq Farooqui [2,†], Ghazaleh Panahandeh [1], Wolfgang Ahrendt [3] and Martin Fabian [2]

1 Zenseact, 417 56 Gothenburg, Sweden; ghazaleh.panahandeh@zenseact.com
2 Department of Electrical Engineering, Chalmers University of Technology, 412 96 Gothenburg, Sweden; ashfaqf@chalmers.se (A.F.); fabian@chalmers.se (M.F.)
3 Department of Computer Science and Engineering, Chalmers University of Technology, 412 96 Gothenburg, Sweden; ahrendt@chalmers.se
* Correspondence: yuvaraj.selvaraj@zenseact.com
† These authors contributed equally to this work.

Abstract: The correctness of autonomous driving software is of utmost importance, as incorrect behavior may have catastrophic consequences. Formal model-based engineering techniques can help guarantee correctness and thereby allow the safe deployment of autonomous vehicles. However, challenges exist for widespread industrial adoption of formal methods. One of these challenges is the model construction problem. Manual construction of formal models is time-consuming, error-prone, and intractable for large systems. Automating model construction would be a big step towards widespread industrial adoption of formal methods for system development, re-engineering, and reverse engineering. This article applies *active learning* techniques to obtain formal models of an existing (under development) autonomous driving software module implemented in MATLAB. This demonstrates the feasibility of automated learning for automotive industrial use. Additionally, practical challenges in applying automata learning, and possible directions for integrating automata learning into the automotive software development workflow, are discussed.

Keywords: autonomous driving; active learning; formal methods; model-based engineering; automata learning

Citation: Selvaraj, Y.; Farooqui, A.; Panahandeh, G.; Ahrendt, W.; Fabian, M. Automatically Learning Formal Models from Autonomous Driving Software. *Electronics* **2022**, *11*, 643. https://doi.org/10.3390/electronics11040643

Academic Editors: Chuan Hu, Umar Zakir Abdul Hamid and Argyrios Zolotas

Received: 28 December 2021
Accepted: 15 February 2022
Published: 18 February 2022

1. Introduction

In recent years, the global automotive industry has made significant progress towards the development of autonomous vehicles. Such vehicles potentially have several benefits including the reduction of traffic accidents and increased traffic safety [1]. However, these are highly complex and safety critical systems, for which correct behavior is paramount, as incorrect behavior can have catastrophic consequences. Ensuring safety of autonomous vehicles is a multi-disciplinary challenge, where software design and development processes play a crucial role. A strong emphasis is placed on updating current engineering practices to create an end-to-end design, verification, and validation process that integrates all safety concerns into a unified approach [2].

Automotive software engineering is faced with several challenges that include non-technical aspects (organization, strategic processes, etc.) and technical aspects such as the need for new methodologies that combine traditional control theory and discrete event systems, quality assurance for reliability, etc. [3,4]. Model-based engineering techniques can address some of the challenges and help tackle the complexity in developing dependable automotive software [5–8].

An autonomous vehicle consists of several software and hardware components that interact to solve different tasks. Software in a modern car typically consists of hundreds of thousands of lines of code deployed over several distributed units developed by different suppliers and manufacturers. The model-based approach to design, test, and integrate software systems is instrumental in achieving the necessary correctness guarantees for such complex systems. In this vein, several tools and methods have been developed and used over the years. In particular, the usage of MATLAB/Simulink [9] has become increasingly successful in a number of automotive companies [10].

A direct consequence of such software complexity is the possible presence of potentially dangerous edge cases, bugs due to subtle interactions, errors in software design and/or implementation. Although the automotive industry is constantly evolving, testing (including model-based testing [11]) is currently a prominent technique for software quality assurance [12]. However, an approach only based on testing is insufficient and partly infeasible to guarantee the correctness of autonomous vehicles [13]. Thus, there is a need for strict measures for quality assurance, and the use of formal methods in this regard promise to be beneficial [14,15].

Formal verification techniques can indeed be used to identify design errors in MATLAB/Simulink function block networks using Simulink Design Verifier (SDV) [16], and Polyspace [17] to perform static code analysis on C code generated from the function block networks. However, there are limitations in SDV; for instance in scalability and in the verification of temporal properties that cannot be expressed as assertions [18,19]. SDV also exclusively works with Simulink, which presents a challenge in reasoning about MATLAB code without Simulink function blocks. Additionally, SDV and Polyspace cannot be used to reason about function blocks where the internal implementation details are unavailable. Thus, in such cases there is a need to use complementary methods to guarantee the correctness of the complete system under design.

In [20], different formal verification methods were used to verify an existing decision making software (developed using MATLAB) in an autonomous driving vehicle. Formal models of the code were manually constructed to perform formal verification and several insights were presented. Admittedly, formal methods can be more beneficial if introduced during the early stages of the software development workflow rather than being used for after-development verification. However, there are several obstacles that impede the widespread adoption of formal methods [14,21]. Significant trade-offs (e.g., tools compatible with formal methods, development cost and time) have to be made that disrupt current industrial best practice. Therefore, any work towards industrial adoption of formal methods in the automotive domain without significant disruptions on current practice is definitely rewarding.

Formal verification techniques like model checking [22]—to prove the absence of errors in software designs—or, formal synthesis techniques like supervisor synthesis [23]—to generate a controller/supervisor that is correct by construction—require a model that describes the behavior of the system. However, constructing a formal model that captures the behavior of the software under design is a challenging task and is one of several impediments in the industrial adoption of formal methods. Manual construction of models is expensive, prone to human errors, and even intractable for large systems. Constructing such a model manually is also time consuming, which further complicates things as the specification and consequently the implementation changes frequently, especially so for rapidly evolving systems typical for the autonomous driving domain.

Automating model construction could speed up industrial adoption of formal methods by reducing the burden of manually constructing the models. Such automated methods will help find potential errors, as they can automatically generate and verify production code at regular intervals. This will further strengthen the suitability of formal methods for industrial use [21,24]. Automatically constructing formal models can also help understand and reason about ill-documented legacy systems and black-box systems, which is crucial for quality assurance of large-scale and complex automotive systems.

Active automata learning [25–32] is a field of research that addresses the problem of automatic model construction. These approaches constitute a class of machine learning algorithms that actively interact with the target system to deduce a finite-state automaton [33] describing its behavior. This article is an extended version of [34], in which we reported results from a case study to automatically learn a formal model of the Lateral State Manager (LSM), a sub-component of an autonomous driving software (under development) programmed in MATLAB. In addition, we described an interface between the learning tool MIDES [35] and MATLAB, which was used to learn a model of the LSM using two active learning algorithms L^* and Modular Plant Learner (MPL). The results demonstrated the feasibility of our approach, but also the practical challenges encountered. This article extends our previous work [34] with:

- Extensive description of the learning algorithms L^* [36] and MPL [30] with illustrative examples.
- An evaluation on the practical applicability of L^* to automatically learn a model of the LSM is presented based on experiments using LearnLib, an open-source automata learning framework [32], in addition to MIDES.
- Analysis of the learning outcome is performed by investigating optimizations that can potentially help improve the practical applicability of the learning algorithms.
- New insights on the approach of using automata learning to enable formal automotive software development is presented.

Note that this article does not aim to compare the performance of the different algorithms, but to show the applicability of active automata learning in a MATLAB development environment. The experiments also showed that a known bug existing in the actual MATLAB code was present in the learned model as well. This is very much desired, because the analysis of a model can only reveal a bug in the system if the bug also manifests itself in the model.

This article is structured as follows. Section 1.1 presents a brief overview of related work. Section 2 describes the system under learning followed by the necessary preliminaries in Section 3. Section 4 illustrates the learning algorithms with an example. A description of the learning framework and the results from the learning are presented in Sections 5 and 6, respectively. Section 7 presents the evaluation of the results. Section 8 includes the validation of the formal model learned and the threats to validity. Section 9 presents some insights from the experiments, discusses practical challenges and possible directions for integrating automata learning into automotive software development workflow. The article is concluded in Section 10 with a summary and some ideas for future work.

1.1. Related Work

Automatically learning finite-state models for formal verification has been done previously, for instance, from Java source code in [37], and from C code in [38,39]. These methods rely on extracting an automaton by analyzing the program source code. Hence, they are specific to the particular programming language and strictly rely on well defined coding patterns and program annotations. Additionally, the approaches of [37–39] cannot extract models where the source code is not available, such as when dealing with black-box systems or binaries. Active automata learning mitigates these restrictions and learns models of black-box systems through interaction.

There exist works on integrating the MATLAB development environment with tools compatible for formal verification. For example, in [40], MATLAB/Simulink designs are translated to the intermediate language Boogie that can later enable the use of SMT solvers [41] for verification. Other works include developing MATLAB toolboxes to integrate with a theorem prover [42] and a hybrid model checker [43]. Such methods depend on considerable manual (and skilled) work to understand the semantics of the MATLAB commands and built-in functions to develop the respective toolboxes. In contrast, by actively interacting with the actual MATLAB code the work in this article learns a formal model, which allows us to use general purpose formal methods tools to asses properties of the code. In addition, knowledge about the semantics of MATLAB code are not needed by the learning tool.

Active automata learning has been successfully applied to learn and verify communication protocols using Mealy machines [26,44], and to obtain formal models of biometric passports [45] and bank cards [46]. In [47], automata learning is used to learn embedded software programs of printers. Though such research indicates the use of active automata learning for real-life systems, challenges exist to broaden its impact for practical use [29,48,49]. There are very limited examples on the use of active automata learning in an automotive context [50,51] and it is yet to find its place in automotive software development. To the best of the authors' knowledge, active automata learning has not been used previously to learn formal models from MATLAB.

2. System under Learning: The LSM

The system under learning (SUL), the LSM, is a sub-component of the decision making and planning module in an autonomous driving system and is responsible for managing modes during an autonomous lane change. The lane change module is implemented in MATLAB-code [9] using several classes with different responsibilities. A simplified overview of the system and the interaction of the LSM with a high level strategic *Planner* and a low level *Path Planner* is shown in Figure 1. The lane change module is cyclically updated with the current vehicle state (e.g., position, velocity), surrounding traffic state, and other reference signals.

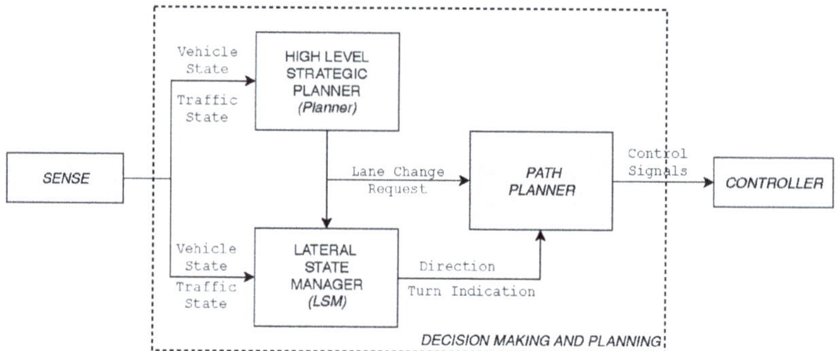

Figure 1. The lane change module system overview and interactions between the three components: *Planner*, *LSM*, and *Path Planner*.

The *Planner* in the lane change module is responsible for strategic decisions. Depending on the state of the vehicle, the *Planner* sends lane change requests to the *LSM*, indicating the desired lane to drive in. This request is sent in the form of a `laneChangeRequest` signal, which takes one of the three values: `noRequest`, `changeLeft`, or `changeRight` at any point in time. On receiving a request, the *LSM* keeps track of the lane change process by managing the different modes possible during the process, and issues commands to the *Path Planner*. If a lane change is requested, the *Path Planner* plans a path and sends required control signals to the low level controller to perform a safe and efficient lane change. Once a lane change is initiated, the LSM needs to remember where in the sequence it is, thus it is implemented as a finite state machine. A representation of the LSM, which consists of seven states, is shown in Figure 2. For confidentiality reasons, the state and event names are not detailed. An example of a state in the *LSM* state machine is *State_Finished*, abbreviated as S_F in Figure 2, which represents the completion of the lane change process.

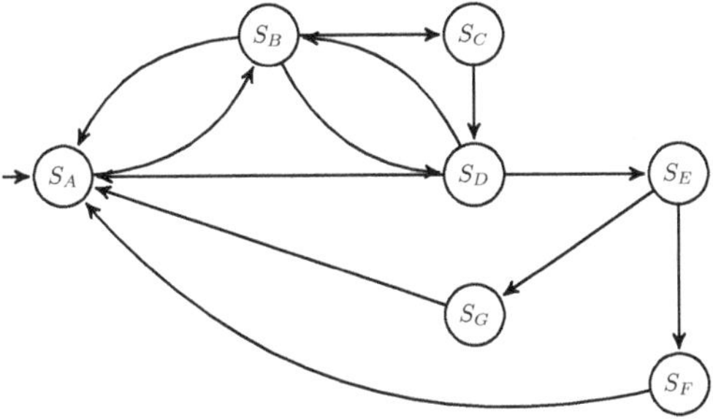

Figure 2. A finite state machine representing the LSM.

A call to the LSM is issued at every update cycle. During each call, the LSM undergoes three distinct execution stages. In the first stage, an associated function `updateState` is executed that updates all the inputs according to the function call arguments. Then, depending on the current state, code is executed to decide whether the system transits to a new state or not. This code also assigns outputs and internal variables. Finally, if a transition is performed, the last stage executes code corresponding to the new state entered and assigns new values to the variables.

3. Preliminaries

An *alphabet*, denoted by Σ, is a finite, nonempty set of events. A *string* is a finite sequence of events chosen from Σ. The empty string, denoted by ε, is the string with zero events. For two strings, s and t, their *concatenation* denoted by st, is the string formed such that s is followed by t. Let Σ^k be the set of all strings over Σ of length k. Then, Σ^2 is $\Sigma\Sigma$ and similarly $\Sigma^{(n+1)} = \Sigma^n\Sigma$. The set of all strings of finite length over an alphabet Σ, including $\Sigma^0 = \{\varepsilon\}$, is denoted by Σ^*.

A *language* $\mathcal{L} \subseteq \Sigma^*$ is a set of strings over Σ. A string s is a *prefix* of a string u, if there exists a string t such that $u = st$; t is then a *suffix* of u. For a string $s \in \mathcal{L} \subseteq \Sigma^*$, its *prefix-closure* $\bar{s}$ is the set of all prefixes of s, including s itself and ε. $\mathcal{L}$ is said to be *prefix-closed* if the prefix-closures of all its strings are also in $\mathcal{L}$, that is $\overline{\mathcal{L}} = \mathcal{L}$. Suffix-closure can be defined analogously.

Definition 1 (State). *Let* $V = \{v_1, v_2, ..., v_n\}$ *be an ordered set of* state variables, *where each variable* v_i *has a discrete finite domain defined as* v_i^D. *A* state *is then defined as the assignment of values to variables,* $\hat{V} \in V^D$ *where* $V^D = v_1^D \times v_2^D \times \cdots \times v_n^D$. $\hat{V}$ *is called a* valuation.

Definition 2 (DFA). *A* deterministic finite automaton *(DFA) is defined as a 5-tuple* $\langle Q, \Sigma, \delta, q_0, M \rangle$, *where:*

- Q *is the finite set of* states;
- Σ *is the* alphabet *of events;*
- $\delta : Q \times \Sigma \rightharpoonup Q$ *is the* partial transition function;
- $q_0 \in Q$ *is the* initial state;
- $M \subseteq Q$ *is the set of* marked states.

The set of all DFA is denoted $\mathcal{A}$. Every DFA $A \in \mathcal{A}$ determines a language *generated*, respectively, *marked*, by that DFA, defined with help of the extended transition function.

Definition 3 (Extended Transition Function). *Given a DFA* $\langle Q, \Sigma, \delta, q_0, M \rangle$, *the* extended transition function $\bar{\delta} : Q \times \Sigma^* \rightharpoonup Q$ *is defined as (with* $s \in \Sigma^*$, $a \in \Sigma$*):*

- $\bar{\delta}(q, \epsilon) = q$
- $\bar{\delta}(q, sa) = q'$ *if there exists* $q'' \in Q$ s.t. $\bar{\delta}(q, s) = q''$ *and* $\delta(q'', a) = q'$

Definition 4 (Generated and Marked Language). *Given a DFA $A = \langle Q, \Sigma, \delta, q_0, M \rangle$, the languages* generated *and* marked *by A, $\mathcal{L}(A)$ and $\mathcal{L}_m(A)$, respectively, are defined as:*

- $\mathcal{L}(A) = \{ s \in \Sigma^* \mid \bar{\delta}(q_0, s) \in Q \}$
- $\mathcal{L}_m(A) = \{ s \in \mathcal{L}(A) \mid \bar{\delta}(q_0, s) \in M \}$

Intuitively, the marked language is the set of all strings that lead to marked states. While the generated language denotes behavior that is possible but not necessarily accepted, the marked language denotes possible behavior that is accepted. A language is said to be *regular* if it is marked by some DFA. It is well-known [33] that for a given regular language, there exists a *minimal* automaton, in the sense of least number of states and transitions, that accepts that language.

4. The Learning Algorithms

This section introduces and illustrates the learning algorithms used in this article. An example consisting of two robots, R1 and R2, is used as the SUL. Each robot can perform two operations, *load* and *unload*, represented by the events l_1 and u_1, respectively, for R1, and l_2 and u_2 for R2. The marked languages of R1 and R2 are $\mathcal{L}_m(R1) = (l_1 u_1)^*$ and $\mathcal{L}_m(R2) = (l_2 u_2)^*$, respectively. The behaviors of the robots represented as automata are shown in Figure 3, to the left. Each robot starts in its respective initial state `i`, and moves to the working state `w` on the occurrence of a *load* event. Then the robot transits back to its initial state on the occurrence of an *unload* event. An automaton representing their joint behavior is given in Figure 3c.

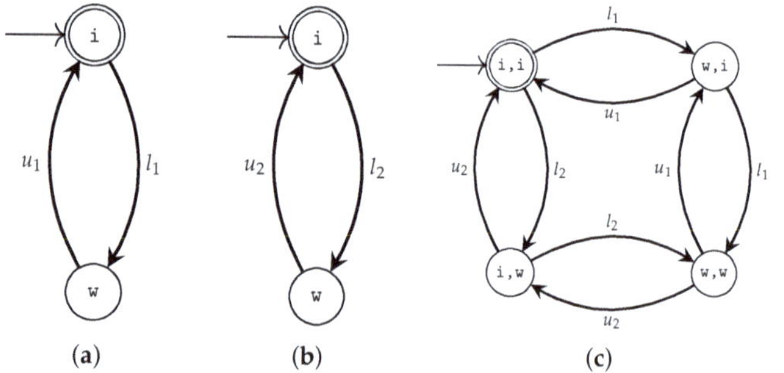

Figure 3. The example system. (**a**) R1; (**b**) R2; (**c**) joint behavior of R1 and R2.

4.1. The L^* Algorithm

The L^* algorithm [25] is a well-known active automata learning algorithm that has inspired a tremendous amount of work [26–29,32,36,46,52]. It learns a minimal automaton $\hat{\mathcal{M}}$ that generates the regular language $\mathcal{L}(\hat{\mathcal{M}}) \subseteq \Sigma^*$ representing the behavior of the SUL. L^* assumes access to an *oracle* that has complete knowledge of the system, and works by posing *queries* to the oracle. The modified L^* from [36,52] is used in this work. The learning algorithm interacts with the SUL to answer two types of queries:

Membership Queries: Given a string $s \in \Sigma^*$, a membership query for s returns 2 if the string can be executed by the SUL and takes the system (from the initial state) to a marked state. If the string can be executed, but does not reach a marked state, 1 is returned. Otherwise, 0 is returned. The membership query has the signature: $T : \mathcal{A} \times \Sigma^* \to \{0, 1, 2\}$, and for $A \in \mathcal{A}$ and $s \in \Sigma^*$:

$$T(A, s) = \begin{cases} 2, & s \in \mathcal{L}_m(A) \\ 1, & s \in \mathcal{L}(A) \setminus \mathcal{L}_m(A) \\ 0, & s \notin \mathcal{L}(A) \end{cases} \tag{1}$$

Equivalence Queries: Given a *hypothesis* automaton $\mathcal{H}$, an algorithm verifies if $\mathcal{H}$ represents the language $\mathcal{L}(\hat{\mathcal{M}})$. If not, a *counterexample* $c \in \Sigma^*$ must be provided, such that, either c is incorrectly generated (that is, $c \in \mathcal{L}(\mathcal{H})$ but $c \notin \mathcal{L}(\hat{\mathcal{M}})$), or incorrectly rejected (that is, $c \notin \mathcal{L}(\mathcal{H})$ but $c \in \mathcal{L}(\hat{\mathcal{M}})$) by $\mathcal{H}$. In this work, equivalence queries are performed using the W-method [53].

Let $\hat{\mathcal{M}}$ have n states. Given a hypothesis $\mathcal{H}$, with $m \leq n$ states, the W-method creates test strings to iteratively extend the hypothesis until it has $n \geq m$ states.

The learner constantly updates its knowledge about the SUL's language as an *observation table*. The observation table $O(S, E, T)$ is a 2-dimensional table, where S is a set of prefix-closed strings, and E is the set of suffix-closed strings. The table has rows indexed by elements of $S \cup (S\,\Sigma)$, and the columns indexed by elements of E. The value of a cell (s, e) (for $s \in S \cup (S\,\Sigma)$ and $e \in E$) is populated using membership queries. The algorithm ensures that the observation table is closed and consistent [36]. The observation table is used to obtain a deterministic finite-state automaton, the hypothesis. Then, the learner performs an equivalence query on the hypothesis automaton. If a counterexample is found, it is added to the observation table together with all its prefixes. The algorithm loops until no counterexample can be found.

Example 1. *To illustrate the working of the L* algorithm, consider the example with the two robots. L* initializes the observation table as seen in Figure 4. The empty event ε is related to the initial state, which is marked, and hence its membership query results in a value of* 2*. On the other hand, the table entries for the* l_1 *and* l_2 *events are* 1*, as these events are defined from the initial state but do not reach marked states. Additionally, membership queries for strings that begin with* u_1 *and* u_2 *result in a value of* 0*, as they are not defined from the initial states. Hence, rows corresponding to such strings are not included in subsequent observation tables. The rows corresponding to the two sets of elements belonging to S and S Σ are separated by a horizontal line in the table. For the sake of compactness, the ε prefix is omitted for non-empty strings.*

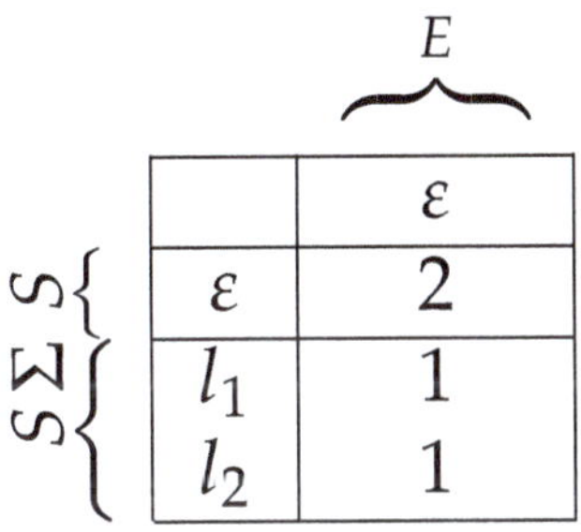

	ε
ε	2
l_1	1
l_2	1

Figure 4. L^*—The initial table.

The initial table is made closed, and consistent using membership queries, and the resulting table is shown in Figure 5a and its corresponding automaton in Figure 5b. States in the automaton correspond to the row values of the table.

Given the first hypothesis, L now makes an equivalence query resulting in the counterexample* $l_1 l_2 u_1$ *According to the hypothesis, this string is rejected. However, the membership query for this string will result in a value* 1*, as this string is possible in the system, as seen in Figure 3c. This counterexample and its prefixes are incorporated into the set S, and the learning continues until a new closed and consistent table, Figure 6a, is obtained. The corresponding hypothesis is seen in Figure 6b. Since no counterexample can be found at this stage, the algorithm terminates, returning the hypothesis as the learned model.*

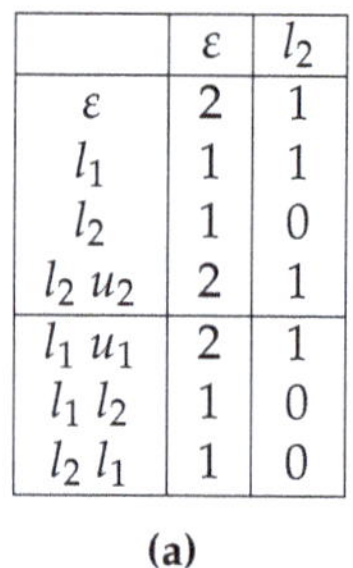

	ε	l_2
ε	2	1
l_1	1	1
l_2	1	0
$l_2\ u_2$	2	1
$l_1\ u_1$	2	1
$l_1\ l_2$	1	0
$l_2\ l_1$	1	0

(a)

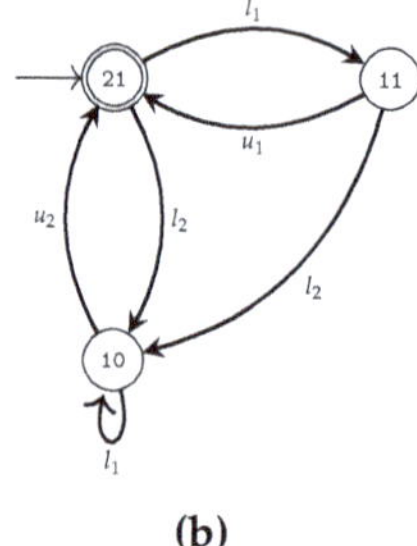

(b)

Figure 5. L^*—First iteration. (**a**) The initial table made closed and consistent; (**b**) first hypothesis.

	ε	l_2	u_2
ε	2	1	0
l_1	1	1	0
l_2	1	0	2
$l_2\ u_2$	2	1	0
$l_1\ l_2$	1	0	1
$l_1\ l_2\ u_1$	1	0	2
$l_1\ u_1$	2	1	0
$l_2\ l_1$	1	0	1
$l_2\ u_2\ l_1$	1	1	0
$l_2\ u_2\ l_2$	1	0	2
$l_1\ l_2\ u_2$	1	1	0
$l_1\ l_2\ u_1\ l_1$	1	0	1
$l_1\ l_2\ u_1\ u_2$	2	1	0

(a)

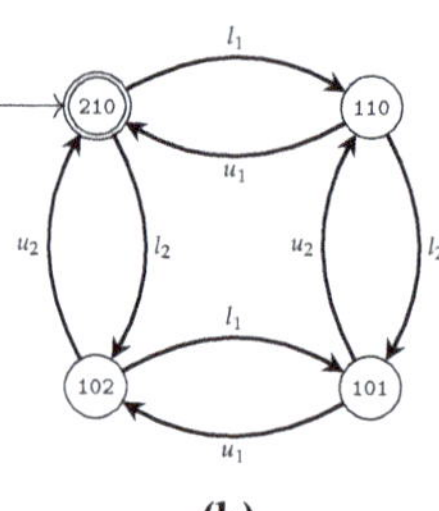

(b)

Figure 6. L^*—Second iteration. (**a**) The table updated with the counterexample and made closed and consistent; (**b**) second hypothesis.

4.2. The Modular Plant Learner

The Modular Plant Learner [30] is a state-based active learning algorithm developed to learn a *modular model*, that is, one composed of a set of interacting automata. These *modules* together define the behavior of the system. MPL does so by actively exploring the state-space of a program in a breadth-first search manner. It exploits structural knowledge of the SUL to search smartly. Hence, it requires access to the SUL's variables, and a Plant Structure Hypothesis (PSH) [30] defining the structure of the system. The PSH is a 3-tuple $H = \langle D, E, S \rangle$, where D is a set that provides a unique name for each module that is to be learned. The cardinality of D defines the number of modules to learn. E is an *event mapping* that defines which events belong to which module. S is a *state mapping* that defines the relationship between the modules and the variables in the SUL. The algorithm consists of the *Explorer*, which explores new states and a *ModuleBuilder* for each module to keep track of its module as it is learned.

The *Explorer* maintains a queue of states that need to be explored, terminating the algorithm when the queue is empty. The learning is initialized with the SUL's initial state in the queue, which becomes the search's starting state. For each state in the queue, the *Explorer* checks if an event from the alphabet Σ can be executed. If a transition is possible, the *Explorer* sends the current state (q), the event (σ), and the state reached (q') to all the *ModuleBuilders*.

Each of the *ModuleBuilders* evaluates if the received transition is relevant to its particular module. If it is, the transition is added to the module; otherwise it is discarded. The

ModuleBuilder tracks the learning of each module as an automaton. This is done by maintaining a set Q_m containing the module's states and a transition function $T_m : Q_m \times \Sigma_m \rightharpoonup Q_m$, for each module $m \in D$. Once the transition is processed, the *ModuleBuilder* waits for the *Explorer* to send the next transition. The algorithm terminates when all modules are waiting and the exploration queue is empty. Each *ModuleBuilder* can now construct and return an automaton based on Q_m and T_m.

Example 2. *Consider again the example with the two robots. Assume that the robots' states are stored in the variables $R1_{var}$ and $R2_{var}$, respectively, represented as a state vector $\langle R1_{var}, R2_{var}\rangle$. The initial state is then $\langle i, i\rangle$. A PSH for this example is defined as follows:*

- $D = \{R1, R2\}$
- $E(R1) = \{l_1, u_1\}$
- $E(R2) = \{l_2, u_2\}$
- $S(R1) = \{R1_{var}\}$
- $S(R2) = \{R2_{var}\}$

At the start, the Explorer *knows only about the initial state. Two* ModuleBuilders *are initialized, one for each robot. The* ModuleBuilders *use the known initial state and knowledge regarding the PSH to initialize themselves as seen in Figure 7. The state marked blue in the* Explorer *denotes the state that is to be explored next.*

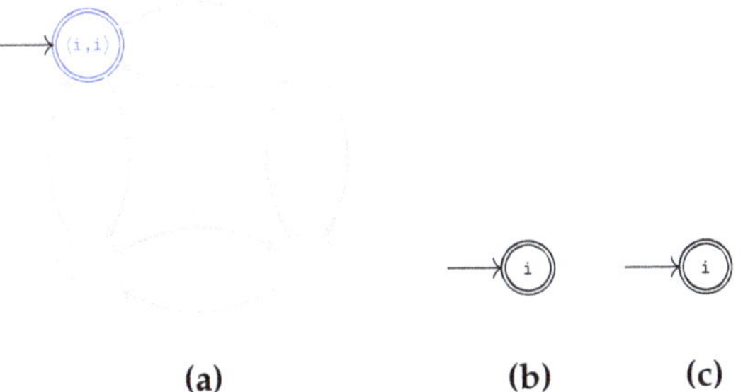

Figure 7. MPL—Initialization. (**a**) Explorer; (**b**) R1; (**c**) R2.

Once initialized, the Explorer *attempts to execute all the events in the alphabet from the initial state. Accepted states are reached only for the events l_1 and l_2; events u_1 and u_2 cannot be executed. Identified transitions are sent to the* ModuleBuilders, *where they are processed according to the PSH. Figure 8a shows the knowledge gained by the* Explorer. *The transitions in red show the current execution of the* Explorer, *and the blue states represent the states reached, but not yet explored. Figure 8b,c show the internal representation of the knowledge for each of the* ModuleBuilders. *The states reached in each of the modules are new. Hence, the corresponding global states are added to the exploration queue.*

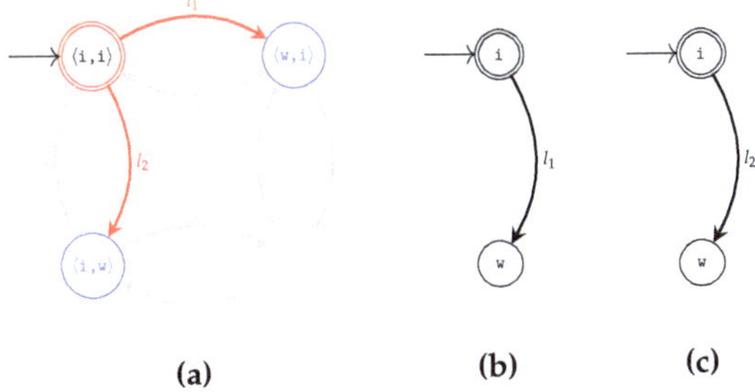

Figure 8. MPL—First Iteration. (**a**) Explorer; (**b**) R1; (**c**) R2.

For each of the states present in the queue, the Explorer *tries to execute all the events, and the obtained transitions, colored red in Figure 9a, are sent to the* ModuleBuilders. *The* ModuleBuilders *take only transitions relating to u_1 and u_2 to update their knowledge. As seen in Figure 9b,c the states reached by these (newly added) transitions are not new and have already been explored. Hence, no more states need to be explored, and the algorithm terminates. At termination the* Explorer *has explored only three states to learn a modular model describing the behavior of the two robots. In the worst-case scenario, though, the MPL must explore the entire state-space. This depends on the user-defined PSH and the possibility to decouple the system. Further details about the MPL are found in [30].*

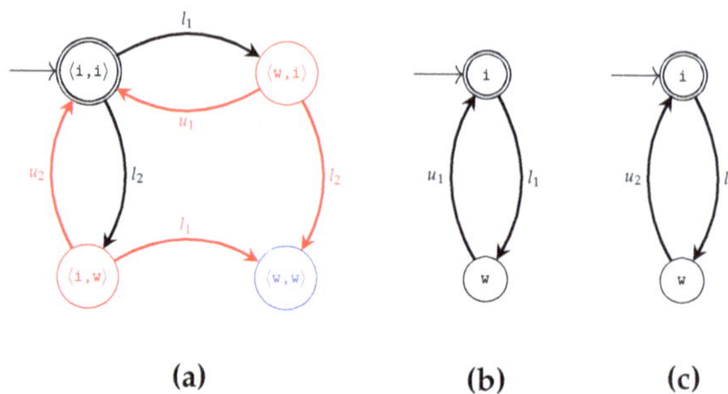

Figure 9. MPL—Final Iteration. (**a**) Explorer; (**b**) R1; (**c**) R2.

5. Method: The Learning Setup

To actively learn a DFA model of the SUL, an interface is necessary to execute (strings of) events, which represent the executable actions of the SUL. It should be possible to observe and set the state of the SUL. If an event is requested that is not executable by the SUL in its current state, the SUL should reply with an error message. Figure 10 presents an overview of the active automata learning setup used in this article. The learner refers to the learning tool MIDES [35] that implements the two learning algorithms described in Section 4. Furthermore, the learner can be replaced with other tools that follow a similar setup for automata learning, such as LearnLib [32]. The learning setup allows learning of automata models by (actively) interacting with the SUL. The following subsections describe the components and the learning outcome.

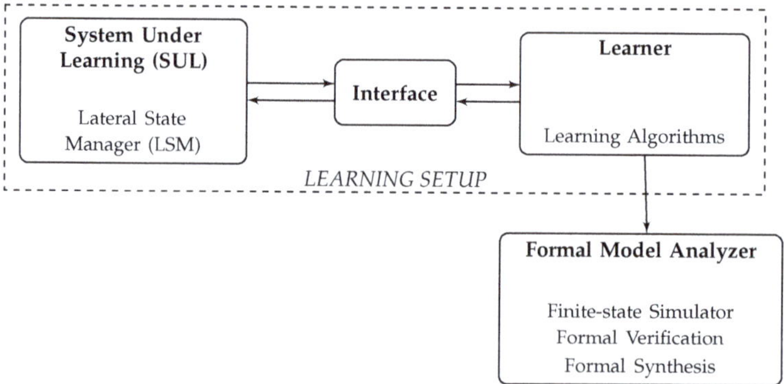

Figure 10. Overview of the learning setup.

5.1. Abstracting the Code

As described in Section 2, the LSM is a part of the lane change module, which is updated cyclically with the necessary signals. In order to decide whether the system transits to a new state or not, the LSM is dependent on external function calls. These interactions with external modules need to be abstracted away to learn a model of the LSM. Thus, the first step in the learning process is to abstract the MATLAB code such that all external dependencies are removed, which is described using Example 3.

Example 3. *Consider the small MATLAB code snippet shown in Listing 1. The function* `duringStateA` *decides whether the system transits to a new state or not depending on* `var1` *and* `var2`*. The values of these two variables are dependent on external function calls,* `function1` *and* `function2`*, respectively.*

Listing 1. A small illustrative example.

```
function duringStateA(self, laneChangeRequest)
    var1 = function1();
    var2 = function2(laneChangeRequest);
    if var1 && var2
        self.state = stateB;
    end
end
```

Listing 2 shows how the external function calls are replaced by the additional input argument `decisionVar`*. The two variables,* `decisionVar.var1` *and* `decisionVar.var2`*, have the domain* $\{True, False\}$*. While this abstraction is not universally valid and increases the number of input parameters of the function, it is possible in this particular context due to the way the different modules interact; the decision logic remains unchanged.*

Listing 2. Abstracted version of Listing 1.

```
function duringStateA(self,
                      laneChangeRequest,
                      decisionVar)
    if decisionVar.var1 && decisionVar.var2
        self.state = stateB;
    end
end
```

Similarly, all such external function calls are abstracted and the final abstracted function contains one additional input parameter, `decisionVar`, to the `updateState` function. The output of the `updateState` function is a set of internal variables, which includes the current state and the direction for the lane change among others. This set of variables is used by the learner to observe the behavior of the LSM during their interaction, as described in the following section.

5.2. Interaction with the SUL

The interaction between the learning tool MIDES and the LSM, implemented in MATLAB, is crucial, for which there is a need to:

1. Enable communication between MIDES and MATLAB,
2. Provide information to MIDES on how to execute the LSM and observe the output.

MIDES must be able to call MATLAB functions, evaluate MATLAB statements, and pass data to and get data from MATLAB. In this learning setup, the learner is compiled to Java bytecode, and the resulting executable code is run on a Java virtual machine. Therefore, the interface integrates Java with MATLAB using the MATLAB Engine API for Java [54], providing a suitable API for MIDES to interact with MATLAB.

With this interface established, the learner can now call the `updateState` function by providing an input assignment to the corresponding variables. However, to learn a model, the learner additionally requires, among other things, predicates over state valuations that define the marked states, the set of events, and event predicates that define when an event is enabled or disabled.

Since the interaction between the learner and the LSM is done via the `updateState` function, the input parameters define the alphabet of the model. Each unique valuation of the input parameters corresponds to one event in the alphabet. Since the abstracted LSM module is provided to the learner, each function that is abstracted into a decision variable potentially results in one additional input parameter. Following the abstraction described in Section 5.1, ten external function calls in the LSM resulted in ten Boolean valued `decisionVar`, in addition to one three-valued `laneChangeRequest`, as input parametersand a total of 3072 events. However, as state transitions in the LSM are defined only for a subset of these events, some of them would potentially not have any effect on the model behavior, and therefore their event predicates would be unsatisfiable.

The event predicates are defined over the state variables. The granularity of these predicates contribute to the performance of the learning algorithm. A very detailed predicate will potentially reduce the total number of strings to test in the SUL. A general rule of thumb for constructing these is to create one predicate for each abstracted variable. Taking the example in Listing 2, all events corresponding to `decisionVar.var1` and `decisionVar.var2` are enabled when the predicate, `self.state == stateA` evaluates to *True*. For an event to be enabled in a given state, all individual predicates corresponding to the different variables must evaluate to *True*. Events with unsatisfiable predicates can be discarded. Doing so for the LSM results in a total of 1536 events. Finally, to observe the behavior of the LSM, the learner requires a set of variables given by the output of the `updateState` function. Furthermore, initial valuation of the variables, which is then the initial state of the LSM, is known to the learner.

6. Results

This section discusses the learning outcome. The learning algorithms were run on an Intel i7 machine, with 8 GB ram, running Linux.

6.1. Learning with L^*

The L^* algorithm implemented in MIDES ran out of memory during the experiments and could not learn a full model. In our longest learning experiment, after 13 h of learning, it was observed that 6 iterations of the hypothesis involving about 500k membership queries resulted in a hypothesis model with 8 states and 231 transitions. On visual inspection, the automaton structure resembles parts of Figure 2. Each of the states in the partially learned model correspond to one or more states in the automaton of Figure 2. However, since L^* did not terminate successfully, further analysis is needed.

Two main obstacles were faced while learning using the L^*. Firstly, as the observation table grows in size, it takes longer to make the table closed and consistent. Furthermore, the memory used to store the table grows rapidly by a factor dependent on the size of the alphabet. Secondly, an exhaustive search for a counterexample using the W-method in the given setup is time consuming. The number of test strings grows rapidly due to the large alphabet size, which slow down the equivalence queries. A detailed analysis on learning using the L^* algorithm is discussed in Section 7.

6.2. Learning with MPL

Apart from the interface with the SUL the MPL requires information about the modules to learn from the SUL. The LSM is a monolithic system and cannot, in its current form, be divided into modules. Hence, the MPL, though specifically developed to learn a modular system consisting of several interacting automata, learns a monolithic model.

The resulting automaton consists of 37 states and 687 transitions. The learning took a total of 68 seconds. Furthermore, applying language minimization [33] to the learned model results in a model with 6 states and 114 transitions. The language minimized automation is shown in Figure 11, and its similarity to Figure 2 is obvious. Multiple transitions between two states are indicated by a single transition in Figure 11. The two states S_G and S_F of Figure 2 have the same future behavior and hence are *bisimilar* [33], so they both correspond to the single state q_6 of Figure 11.

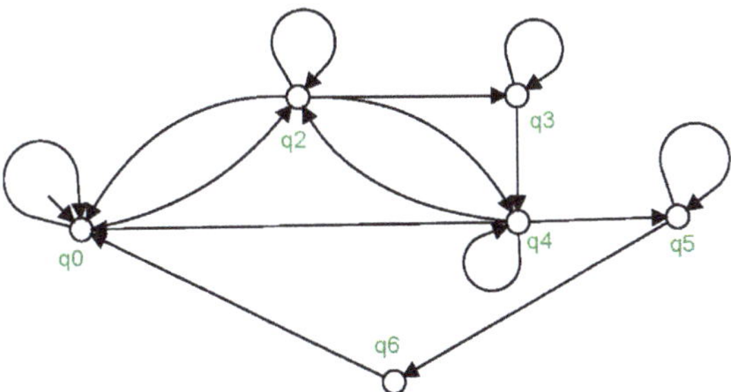

Figure 11. Learned model of the LSM.

The self-loops in the states of Figure 11 correspond to those events that are enabled in that particular state, but do not change the internal state of the LSM. For example, consider the code snippet in Listing 2. When `decisionVar.var1` is *True* and `decisionVar.var2` is *False*, the corresponding event is enabled in `stateA`, but when fired does not cause a change in the value of `self.state`, and thereby results in a self-loop. Similarly, all such enabled events that do not change the internal state become self-loops in the learned model. The state q_6 does not have a self-loop, as it is a transient state in the LSM. That is, irrespective of input parameters, when q_6 is reached, LSM transits to state q_0 for every enabled event.

7. Evaluation

As described in Section 6.1, the L^* algorithm implemented in MIDES did not learn a complete model. Prior to performing a detailed analysis, it is essential to eliminate any potential implementation specific causes for this negative outcome. Typical causes in this regard could be related to the use of inefficient data structures or non-optimized search strategies. LearnLib is an open-source library for active automata learning that has been shown to outperform other existing open-source libraries [32]. LearnLib features a variety of automata learning algorithms including the L^* algorithm. As LearnLib is implemented in Java, it is possible to use an interface similar to the one in Section 5.2 to actively learn a model of the LSM.

The L^* algorithm implemented in LearnLib was run on a standard computer with 2.7 GHz Intel i5 processor and 8 GB of ram storage to learn a model of the LSM. However, even in this case, the L^* algorithm from LearnLib failed to learn a complete model of the LSM before it ran out of memory. With these two experimental results, there is adequate evidence to rule out any implementation specific causes for the negative outcome and to warrant further analysis as described in the next section.

7.1. Learning Complexity

The complexity of L^* depends on the number of membership and equivalence queries as they require interaction with the system and storage of information in the observation table. In [25], the theoretical worst case size of the observation table is calculated to be $(k+1)(n+m(n-1))n$, where $k = |\Sigma|$ is the size of the alphabet, $n = |Q|$ is the number of states, and m is the maximum length of any counterexample presented by the oracle. This is also the upper bound for the number of membership queries and thus the complexity is $\mathcal{O}(m|\Sigma||Q|^2)$. Assuming that $m = |Q|$ in the worst case, the number of membership queries (and the size of the observation table) is $\mathcal{O}(|\Sigma||Q|^3)$. Note from Section 6.2 that the minimal automaton had 6 states. Since L^* learns a minimal model, it is safe to assume that it would learn a model with 6 states. Taking these values as an example would mean that learning the LSM would involve $1536 \times 6^3 \approx 3.3 \times 10^5$ membership queries. This corresponds with the observations in Section 6.1.

The efficiency of L^* for learning finite automata in practice has been investigated by empirical studies on some real-word examples and some randomly generated examples. For instance, [55] observes that the required number of membership queries grows quadratically in the number of transitions when learning prefix-closed languages. This is a challenge, as learning from autonomous driving (or automotive) software in general would typically involve models of reactive systems and therefore learning prefix-closed languages like the case with the LSM. Further empirical studies in [56] show that in general, more queries are required (i.e., harder) to learn a DFA with more marked states ($\approx|Q|$) and the number of membership queries per equivalence query grows linearly as a function of $|\Sigma|$ and $|Q|$. These results show the need for optimizations of the L^* algorithm for practical applicability, as is further shown by this study.

7.2. *Alphabet Reduction*

Based on the discussions so far, it is evident that optimizations are necessary to successfully learn a model of the LSM. The size of the alphabet directly affects the number of membership and equivalence queries. As also pointed out by [55], finding a counterexample to refute a hypothesis in an equivalence query becomes increasingly hard with large alphabets and prefix-closed languages. Therefore, any optimization that reduces the number of such queries is potentially beneficial. In this regard, it is valuable to investigate whether knowledge about the LSM could be exploited to reduce the size of the alphabet without loss of information during the learning process.

Through the abstraction method in Section 5, the alphabet size can be decreased by a reduction in the set of decision variables during the abstraction. For example, in Listing 2 the two variables `decisionVar.var1` and `decisionVar.var2` can potentially be combined into a single variable `decisionVar.var12`. Such a simplification is only possible if it is the case that `decisionVar.var1` and `decisionVar.var2` are exclusively used in `StateA` and do not affect other state changes in the LSM. This information could be obtained from knowledge about the design requirements of the LSM used to create the abstraction shown in Figure 2.

The above approach was implemented through a wrapper function before a call to the initially abstracted LSM from Section 5.2 is made. This ensures that the decision logic in the LSM remains unchanged. Thus, the number of Boolean decision variables was reduced from 10 to 6; bringing down the number of possible events to 192. However, even in this case, L^* failed to learn a complete model before running out of memory.

As further optimization was necessary, more knowledge about the LSM was used to reduce the alphabet size. The decision logic of the LSM was studied to find a correlation between the (now abstracted) 6 Boolean decision variables, showing that only 4 combinations were actually used, which could be encoded using only 2 Boolean variables. This resulted in 12 events in total, and L^* successfully learned a model similar to the one shown in Figure 11. The wrapper function described here to reduce the alphabet is similar to the approach of using event predicates described in Section 5.2 to remove unnecessary events from the alphabet. However, to reduce the alphabet in the proposed manner requires considerable knowledge about the LSM and is potentially prone to errors due to the manual abstraction involved. Though it is a limitation that the complete set of events is not queried in this way, it was adopted to explore the limits of L^*.

8. Model Validation

A formal model of the LSM was learned using the MPL and the optimized L^* as described in Section 6 and Section 7.2, respectively. To validate the learned model, similar to [45], it was compared to a model [57] manually constructed from the MATLAB code. This was done using the tool SUPREMICA [58], which includes an automata simulator. It is possible to view the current state, choose which event to execute, observe the resulting state changes, and step forwards and backwards through the simulation. Thus, a comparison of the simulations of the learned formal model and the simulation of the actual LSM code using MATLAB/Simulink is made.

Recall from Section 5.2 that the alphabet of the learned DFA model is constructed using the input parameters of the LSM code. Therefore, executing the LSM code with a set of input parameters can analogously be simulated by executing the corresponding string of events in the DFA model. Since this comparison is made between the results from simulating the actual LSM implementation and the automata simulation of the learned model, it also validates the abstraction choices described in Section 5.1. Furthermore, a known existing bug in the LSM development code manifested itself also in the learned model. This was validated by manually simulating the learned model with a sequence of input parameter changes known to provoke the bug.

Though no discrepancy was found between the code and the learned model, such manual inspection is not exhaustive and cannot guarantee completeness of the validation process. Alternatively, formal verification could be used to verify correctness. However, as only limited informal (natural language) specifications were available, this was not (easily) done. Still, the minimized model, together with the simulations in SUPREMICA strengthens the confidence in the results of the learning process.

Threats to Validity

This article investigated only one problem instance and so cannot give any concrete conclusions on the generalization or the scalability of the approach. Accidentally, a piece of MATLAB code could have been chosen that lent itself particularly well to automatic learning. Indeed, a piece of code was chosen that the authors were already familiar with. Furthermore, the validation of the learned model was admittedly rather superficial, visual inspection and comparison of simulation results between the learned model and the actual MATLAB code. Ideally, the learned model should have been used to assert functional properties of the MATLAB code. The closest in this respect was the known bug in the code, that could be shown to also be present in the learned model.

However, a general automata learning framework that was not tailored specifically to the SUL in this article was used; the only thing that was specifically implemented was the interface between the learning framework and MATLAB. Even so, that interface was intentionally kept general so that similar case studies of other pieces of code can be performed in the future to truly assess the validity of the presented approach.

9. Insights and Discussion

A formal model of the LSM was successfully learned, and validated in multiple ways. This section presents a discussion on the insights gained.

9.1. Towards Formal Software Development

The primary motivation for this work is to overcome the limitations in manual model construction so that techniques like formal verification and formal synthesis can be used to guarantee the correctness of software, without disrupting current industrial practices. The presented approach is independent of the semantics of the implementation languages One technical requirement to seamlessly integrate this approach with the daily engineering workflow is the possibility to establish an interface between the production code and the learning algorithms. Such a seamless integration makes it easier to use formal methods not only for safety-critical software, but also for other automotive software (e.g., infotainment). Though this article does not consider any kind of formal analysis on the learned model, formal analysis using different verification and synthesis tools can directly be done on the learned models with tools like SUPREMICA, similar to [20,57]. Of course, the learned models must be translated into an input format suitable for the particular tool.

Continuous Formal Development

With increasing complexity, software development in the automotive industry is adopting new model-based development approaches in the software development life cycle (SDLC) [59–62]. Quality assurance in such approaches relies on continuous integration methods where continuous testing is vital. However, safety critical software requires strict measures and testing, and unlike formal methods cannot guarantee the absence of errors. Continuous formal verification [63] is a viable solution in this regard. Though there is a need for a significant amount of research to adopt a continuous formal verification process for automotive SDLC, insights from this article can be used to scale active learning to obtain formal models for safety-critical software development.

9.2. *Practical Challenges*

This section discusses practical challenges encountered in the course of the study.

9.2.1. Interaction with the SUL

The interaction with real-life systems and the construction of application-specific learning setups remain as challenges for the automata learning community despite the application of automata learning in different scenarios over the years [29,48,49]. A major aspect of the active learning process is to establish a proper interface between the learner and the SUL. In this article, the interface is achieved through MATLAB-Java integration using the MATLAB Engine API for Java [54] as described in Section 5.2. A challenge is to establish an appropriate abstraction such that the learner can obtain necessary information about the alphabet to actively interact with the LSM. In this study, all external dependencies were abstracted such that the learner can easily interact with the SUL. However, data dependencies between different methods and user defined classes could present additional challenges to scale this approach, for example to learn a model of the *Planner* and the LSM together. The effort needed to design and implement application specific learning setups can be reduced by creating test-drivers [48] in the form of standalone libraries, and/or automatically constructing abstractions [27] for seamless integration between the SUL and the learner.

9.2.2. Efficiency of the Learning Algorithms

The L^* algorithm in both the tools (MIDES and LearnLib) failed to learn a complete model of the LSM. From Section 7, it is evident that the number of membership and equivalence queries affect the efficiency of the L^* algorithm. Optimization by exploiting knowledge about the decision logic of the LSM was necessary to successfully learn a model of the LSM. While this corresponds to similar observations in other applications [29,55], it highlights the limitations in the practical applicability of language-based learning to autonomous driving software.

Though the MPL successfully learned a model of the LSM, more empirical case studies are needed to explore the limits of state-based learning in this context. The MPL is specifically developed to learn a modular system with several interacting automata. The main benefit here is the reduction in search space achieved by exploiting the structure of the SUL. Unfortunately, due to the structure of the LSM, a monolithic model had to be learned. However, the modular approach could potentially be helpful in tackling the complexity that arises in learning larger systems.

9.2.3. States vs. Events

Both L^* and the MPL require a definition of the events that are relevant to the SUL. Interestingly, there is a trade-off between the size of the alphabet and the size of the state-space; a small alphabet leads to a large state-space, and vice versa. This trade-off is thus important, as the well known state-space explosion is a real practical problem.

The current learning setup resulted in 3072 events, one for each unique valuation of the input parameter. However, it is possible to use each of the input parameters as an event.

This would result in a considerably smaller alphabet of only 23 events. Using the 23 events to learn leads to a huge state-space, however, and both algorithms failed to learn a model. Multiple interlaced lattice structures are seen in the partial models that were obtained and these relate to the various combinations of input parameters. The efficiency of the learning algorithms can be improved by leveraging this trade-off when abstracting the code as seen in Section 7.2.

9.3. Software Reengineering and Reverse Engineering

Reverse engineering, which involves extracting high level specifications from the original system can help to understand (ill-documented) legacy systems and black-box systems, and to reason about theircorrectness. In addition, the development of intelligent autonomous driving features typically undergoes several design iterations before public deployment. In such a case, the formal approaches used to guarantee correctness need to adapt to the software reengineering lifecycle. Reengineering embedded automotive software is different from software reengineering in other domains due to unique challenges [64,65]. The active learning approach in this article can help identify unintended changes between different software implementations and also help to obtain high-level models from legacy systems, thereby aiding in the reengineering and the reverse engineering phase, respectively.

10. Conclusions

This article describes an application to interact with and learn formal models of MATLAB code. MATLAB/Simulink is currently a main engineering tool in the automotive industry, by automatically learning models of MATLAB code a significant step is taken towards the industrial adoption of formal methods. This is especially important for the development of safety-critical systems, like autonomous vehicles.

Using an active automata learning tool MIDES, which interacts with MATLAB, two different learning algorithms were applied to the code of a lane change module, the LSM, being developed for autonomous vehicles. One of these, an adaption of L^*, was unable to learn a model due to memory issues. The other, MPL, being a state-based method designed for learning a modular model, had more information about the target system, and learned a model in roughly one minute. To rule out possible implementation issues, another version of L^* from the open-source LearnLib toolbox was used to learn a model of the LSM. Even in this case, L^* failed to learn a model. Investigation of this negative outcome lead to alphabet reduction to improve the performance of L^*.

The learned models were validated in four ways:

- The language minimization of the model learned by MPL is very similar to the original model of the LSM.
- Manual comparison of the learned models to a manually developed model of the LSM indicated close similarity.
- Simulating the learned automata in SUPREMICA and comparing to the simulation of the actual code in MATLAB/Simulink showed no obvious discrepancies.
- A known bug in the development code was found also in the learned models.

Though the validation of the learned models were performed informally, taken together, they make a strong argument for the benefits of active automata learning in an industrial setting within the automotive domain. Model validation is a well known problem within the active automata learning community [29,31] and in the future, we would like to investigate different formal/semi-formal methods to validate the learned models.

Learning a monolithic model is a bottleneck, as it scales badly. Learning modular models potentially allows us to learn models of larger systems, which is important for industrial acceptance, so this is clearly future research. Currently, the main obstacle is how to define the modules and partition the variables among the modules; if not done properly, the benefits of modular learning are lost.

Existing learning frameworks, like Tomte [27], could potentially help in the learning process by providing more efficient ways to abstract and reduce the alphabet. Furthermore, learning richer structures (but with the same expressive power), like extended finite state machines [66], is an interesting topic for further research. In addition, to further corroborate our findings, we plan to study several other software components of an autonomous vehicle, using the generic interface discussed in this article.

All in all, the goal is to make active automata learning a tool to aid widespread adoption of formal methods in day to day development within the automotive industry, in much the same way as MATLAB currently is.

Author Contributions: Conceptualization, Y.S. and A.F.; methodology, Y.S. and A.F.; validation, Y.S., A.F. and M.F.; formal analysis, Y.S. and A.F; writing—original draft preparation, Y.S. and A.F.; writing—review and editing, W.A. and M.F.; supervision, W.A. and M.F.; project administration and article review, G.P. All authors read and agreed to the published version of the manuscript.

Funding: This research was supported by FFI-VINNOVA, under grant number 2017-05519, *Automatically Assessing Correctness of Autonomous Vehicles—Auto-CAV*, and by the Wallenberg AI, Autonomous Systems and Software Program (WASP) funded by the Knut and Alice Wallenberg Foundation.

Institutional Review Board Statement: Not applicable.

Informed Consent Statement: Not applicable.

Data Availability Statement: Not applicable.

Acknowledgments: The authors thank Oskar Begic Johansson, Felix Millqvist, and Jakob Vinkås for their experiments with LearnLib and the work on alphabet reduction presented in Section 7.2.

Conflicts of Interest: The authors declare no conflict of interest.

References

1. Litman, T. *Autonomous Vehicle Implementation Predictions*; Victoria Transport Policy Institute: Victoria, BC, Canada, 2020.
2. Koopman, P.; Wagner, M. Autonomous vehicle safety: An interdisciplinary challenge. *IEEE Intell. Transp. Syst. Mag.* **2017**, *9*, 90–96. [CrossRef]
3. Broy, M.; Kruger, I.H.; Pretschner, A.; Salzmann, C. Engineering automotive software. *Proc. IEEE* **2007**, *95*, 356–373. [CrossRef]
4. Broy, M. Challenges in Automotive Software Engineering. In *ICSE '06: Proceedings of the 28th International Conference on Software engineering, Shanghai, China, 20–28 May 2006*; Association for Computing Machinery: New York, NY, USA, 2006; pp. 33–42. [CrossRef]
5. Liebel, G.; Marko, N.; Tichy, M.; Leitner, A.; Hansson, J. Model-based engineering in the embedded systems domain: An industrial survey on the state-of-practice. *Softw. Syst. Model.* **2018**, *17*, 91–113. [CrossRef]
6. Charfi Smaoui, A.; Liu, F.; Mraidha, C. A Model Based System Engineering Methodology for an Autonomous Driving System Design. In Proceedings of the 25th ITS World Congress, Copenhagen, Denmark, 17–21 September 2018; HAL: Copenhagen, Denmark, 2018.
7. Struss, P.; Price, C. Model-based systems in the automotive industry. *AI Mag.* **2003**, *24*, 17.
8. Di Sandro, A.; Kokaly, S.; Salay, R.; Chechik, M. Querying Automotive System Models and Safety Artifacts: Tool Support and Case Study. *J. Automot. Softw. Eng.* **2020**, *1*, 34–50. [CrossRef]
9. The MathWorks Inc. MATLAB. Available online: https://se.mathworks.com/products/matlab.html (accessed on 17 February 2020).
10. Friedman, J. MATLAB/Simulink for automotive systems design. In Proceedings of the Design Automation & Test in Europe Conference, Munich, Germany, 6–10 March 2006 ; Volume 1, pp. 1–2.
11. Utting, M.; Legeard, B. *Practical Model-Based Testing*; Morgan Kaufmann: San Francisco, CA, USA, 2007.
12. Altinger, H.; Wotawa, F.; Schurius, M. Testing Methods Used in the Automotive Industry: Results from a Survey. In Proceedings of the 2014 Workshop on Joining AcadeMiA and Industry Contributions to Test Automation and Model-Based Testing, JAMAICA 2014, San Jose, CA, USA, 21 July 2014; Association for Computing Machinery: New York, NY, USA, 2014; pp. 1–6. [CrossRef]
13. Kalra, N.; Paddock, S.M. Driving to safety: How many miles of driving would it take to demonstrate autonomous vehicle reliability? *Transp. Res. Part A Policy Pract.* **2016**, *94*, 182–193. [CrossRef]
14. Luckcuck, M.; Farrell, M.; Dennis, L.A.; Dixon, C.; Fisher, M. Formal specification and verification of autonomous robotic systems: A survey. *ACM Comput. Surv.* **2019**, *52*, 1–41. [CrossRef]
15. Guiochet, J.; Machin, M.; Waeselynck, H. Safety-critical advanced robots: A survey. *Robot. Auton. Syst.* **2017**, *94*, 43–52. [CrossRef]
16. The MathWorks Inc. Simulink Design Verifier. Available online: https://se.mathworks.com/products/simulink-design-verifier.html (accessed on 17 February 2020).

17. The MathWorks Inc. Polyspace Products. Available online: https://se.mathworks.com/products/polyspace.html (accessed on 17 February 2020).
18. Leitner-Fischer, F.; Leue, S. Simulink Design Verifier vs. SPIN: A Comparative Case Study. 2008. Available online: http://kops.uni-konstanz.de/handle/123456789/21292 (accessed on 17 February 2020).
19. Schürenberg, M. Scalability Analysis of the Simulink Design Verifier on an Avionic System. Bachelor Thesis, Hamburg University of Technology, Hamburg, Germany, 2012.
20. Selvaraj, Y.; Ahrendt, W.; Fabian, M. Verification of Decision Making Software in an Autonomous Vehicle: An Industrial Case Study. In *Formal Methods for Industrial Critical Systems*; Springer: Cham, Switzerland, 2019; pp. 143–159.
21. Mashkoor, A.; Kossak, F.; Egyed, A. Evaluating the suitability of state-based formal methods for industrial deployment. *Softw. Pract. Exp.* **2018**, *48*, 2350–2379. [CrossRef]
22. Baier, C.; Katoen, J.P. *Principles of Model Checking*; MIT Press: Cambridge, MA, USA, 2008.
23. Ramadge, P.J.; Wonham, W.M. The control of discrete event systems. *Proc. IEEE* **1989**, *77*, 81–98. [CrossRef]
24. Liu, X.; Yang, H.; Zedan, H. Formal methods for the re-engineering of computing systems: A comparison. In Proceedings Twenty-First Annual International Computer Software and Applications Conference (COMPSAC'97), Washington, DC, USA, 13–15 August 1997; pp. 409–414.
25. Angluin, D. Learning regular sets from queries and counterexamples. *Inf. Comput.* **1987**, *75*, 87–106. [CrossRef]
26. Steffen, B.; Howar, F.; Merten, M. Introduction to active automata learning from a practical perspective. In *International School on Formal Methods for the Design of Computer, Communication and Software Systems*; Springer: Berlin/Heidelberg, Germany, 2011; pp. 256–296.
27. Aarts, F. Tomte: Bridging the Gap between Active Learning and Real-World Systems Title of the Work. Ph.D. Thesis, Radboud University, Nijmegen, The Netherlands, 2014.
28. Cassel, S.; Howar, F.; Jonsson, B.; Steffen, B. Active learning for extended finite state machines. *Form. Asp. Comput.* **2016**, *28*, 233–263. [CrossRef]
29. Howar, F.; Steffen, B. Active automata learning in practice. In *Machine Learning for Dynamic Software Analysis: Potentials and Limits*; Springer: Cham, Switzerland, 2018; pp. 123–148.
30. Farooqui, A.; Hagebring, F.; Fabian, M. Active Learning of Modular Plant Models. *IFAC-PapersOnLine* **2020**, *53*, 296–302. [CrossRef]
31. de la Higuera, C. *Grammatical Inference: Learning Automata and Grammars*; Cambridge University Press: New York, NY, USA, 2010.
32. Isberner, M.; Howar, F.; Steffen, B. The open-source LearnLib: A Framework for Active Automata Learning. In Proceedings of the International Conference on Computer Aided Verification; Springer: Cham, Switzerland, 2015; pp. 487–495.
33. Cassandras, C.G.; Lafortune, S. *Introduction to Discrete Event Systems*; Springer: New York, NY, USA, 2009.
34. Selvaraj, Y.; Farooqui, A.; Panahandeh, G.; Fabian, M. Automatically Learning Formal Models: An Industrial Case from Autonomous Driving Development. In Proceedings of the 23rd ACM/IEEE International Conference on Model Driven Engineering Languages and Systems: Companion Proceedings, MODELS '20, Virtual Event Canada, 16–23 October 2020; Association for Computing Machinery: New York, NY, USA, 2020. [CrossRef]
35. Farooqui, A.; Hagebring, F.; Fabian, M. MIDES: A Tool for Supervisor Synthesis via Active Learning. In Proceedings of the 2021 IEEE 17th International Conference on Automation Science and Engineering (CASE), Lyon, France, 23–27 August 2021; pp. 792–797. [CrossRef]
36. Farooqui, A.; Fabian, M. Synthesis of Supervisors for Unknown Plant Models Using Active Learning. In Proceedings of the 2019 IEEE 15th International Conference on Automation Science and Engineering (CASE),Vancouver, BC, Canada, 22–26 August 2019; pp. 502–508.
37. Corbett, J.C.; Dwyer, M.B.; Hatcliff, J.; Laubach, S.; Pasareanu, C.S.; Zheng, H. Bandera. Extracting finite-state models from Java source code. In Proceedings of the 2000 International Conference on Software Engineering, ICSE 2000 the New Millennium, Limerick, Ireland, 4–11 June 2000; pp. 439–448.
38. Holzmann, G.J. From code to models. In Proceedings of the Second International Conference on Application of Concurrency to System Design, Newcastle upon Tyne, UK, 25–29 June 2001; pp. 3–10.
39. Holzmann, G.J.; Smith, M.H. A practical method for verifying event-driven software. In Proceedings of the 1999 International Conference on Software Engineering (IEEE Cat. No. 99CB37002), Los Angeles, CA, USA, 16–22 May 1999; pp. 597–607.
40. Reicherdt, R.; Glesner, S. Formal verification of discrete-time MATLAB/Simulink models using Boogie. In Proceedings of the International Conference on Software Engineering and Formal Methods, Grenoble, France, 1–5 September 2014; Springer: Cham, Switzerland, 2014; pp. 190–204.
41. de Moura, L.; Bjørner, N. Z3: An Efficient SMT Solver. In *Tools and Algorithms for the Construction and Analysis of Systems*; Ramakrishnan, C.R., Rehof, J., Eds.; Springer: Berlin/Heidelberg, Germany, 2008; pp. 337–340.
42. Araiza-Illan, D.; Eder, K.; Richards, A. Formal verification of control systems' properties with theorem proving. In Proceedings of the 2014 UKACC International Conference on Control (CONTROL), Loughborough, UK, 9–11 July 2014; pp. 244–249.
43. Fang, H.; Guo, J.; Zhu, H.; Shi, J. Formal verification and simulation: Co-verification for subway control systems. In Proceedings of the 2012 Sixth International Symposium on Theoretical Aspects of Software Engineering, Beijing, China, 4–6 July 2012; pp. 145–152.

44. Jonsson, B. Learning of Automata Models Extended with Data. In *Formal Methods for Eternal Networked Software Systems, Proceedings of the 11th International School on Formal Methods for the Design of Computer, Communication and Software Systems, SFM 2011, Bertinoro, Italy, 13–18 June 2011. Advanced Lectures*; Springer: Berlin/Heidelberg, Germany, 2011; pp. 327–349. [CrossRef]
45. Aarts, F.; Schmaltz, J.; Vaandrager, F. Inference and Abstraction of the Biometric Passport. In *Leveraging Applications of Formal Methods, Verification, and Validation*; Margaria, T., Steffen, B., Eds.; Springer: Berlin/Heidelberg, Germany, 2010; pp. 673–686.
46. Aarts, F.; De Ruiter, J.; Poll, E. Formal models of bank cards for free. In Proceedings of the 2013 IEEE Sixth International Conference on Software Testing, Verification and Validation Workshops, Luxembourg, 18–22 March 2013; pp. 461–468.
47. Smeenk, W.; Moerman, J.; Vaandrager, F.; Jansen, D.N. Applying Automata Learning to Embedded Control Software. In *Formal Methods and Software Engineering*; Butler, M., Conchon, S., Zaïdi, F., Eds.; Springer: Cham, Switzerland, 2015; pp. 67–83.
48. Merten, M.; Isberner, M.; Howar, F.; Steffen, B.; Margaria, T. Automated learning setups in automata learning. In Proceedings of the International Symposium on Leveraging Applications of Formal Methods, Verification and Validation, Heraklion, Greece, 15–18 October 2012; Springer: Berlin/Heidelberg, Germany, 2012; pp. 591–607.
49. Merten, M. Active Automata Learning for Real Life Applications. Ph.D Thesis, TU Dortmund University, Dortmund, Germany, 2013.
50. Kunze, S.; Mostowski, W.; Mousavi, M.R.; Varshosaz, M. Generation of failure models through automata learning. In Proceedings of the 2016 Workshop on Automotive Systems/Software Architectures (WASA), Venice, Italy, 5–8 April 2016; pp. 22–25.
51. Meinke, K.; Sindhu, M.A. LBTest: A learning-based testing tool for reactive systems. In Proceedings of the 2013 IEEE Sixth International Conference on Software Testing, Verification and Validation, Luxembourg, 18–20 March 2013; pp. 447–454.
52. Zhang, H.; Feng, L.; Li, Z. A Learning-Based Synthesis Approach to the Supremal Nonblocking Supervisor of Discrete-Event Systems. *IEEE Trans. Autom. Control* **2018**, *63*, 3345–3360. [CrossRef]
53. Chow, T. Testing Software Design Modeled by Finite-State Machines. *IEEE Trans. Softw. Eng.* **1978**, *4*, 178–187. [CrossRef]
54. The MathWorks Inc. Java Engine API Summary. Available online: https://se.mathworks.com/help/matlab/matlab_external/get-started-with-matlab-engine-api-for-java.html (accessed on 17 February 2020).
55. Berg, T. Regular Inference for Reactive Systems. Ph.D. Thesis, Uppsala University, Uppsala, Sweden, 2006.
56. Czerny, M.X. Learning-Based Software Testing: Evaluation of Angluin's L* Algorithm and Adaptations in Practice. Bachelor Thesis, Karlsruhe Institute of Technology, Department of Informatics Institute for Theoretical Computer Science, Karlsruhe, Germany, 2014.
57. Zita, A.; Mohajerani, S.; Fabian, M. Application of formal verification to the lane change module of an autonomous vehicle. In Proceedings of the 2017 13th IEEE Conference on Automation Science and Engineering (CASE), Xi'an, China, 20–23 August 2017; pp. 932–937.
58. Malik, R.; Akesson, K.; Flordal, H.; Fabian, M. Supremica–An Efficient Tool for Large-Scale Discrete Event Systems. *IFAC-PapersOnLine* **2017**, *50*, 5794–5799. [CrossRef]
59. Rausch, A.; Brox, O.; Grewe, A.; Ibe, M.; Jauns-Seyfried, S.; Knieke, C.; Körner, M.; Küpper, S.; Mauritz, M.; Peters, H.; .; Strasser, A.; Vogel, M.; Weiss, N. Managed and Continuous Evolution of Dependable Automotive Software Systems. In Proceedings of the 10th Symposium on Automotive Powertrain Control Systems, 7 December 2014; Cramer: Braunschweig, Germany, 2014; pp. 15–51.
60. Patil, M.; Annamaneni, S. Model Based System Engineering (MBSE) for Accelerating Software Development Cycle. Technical Report, L&T Technology Services White Paper. 2015. Available online: https://www.ltts.com/sites/default/files/whitepapers/2017-07/wp-model-based-sys-engg.pdf (accessed on 27 December 2021).
61. Kubíček, K.; Čech, M.; Škach, J. Continuous enhancement in model-based software development and recent trends. In Proceedings of the 2019 24th IEEE International Conference on Emerging Technologies and Factory Automation (ETFA), 10–13 September Zaragoza, Spain, 2019; pp. 71–78.
62. Cie, K.M. *Agile in Automotive—State of Practice 2015*; Study; Kugler Maag Cie: Kornwestheim , Germany, 2015; p. 58.
63. Monteiro, F.R.; Gadelha, M.Y.R.; Cordeiro, L.C. Boost the Impact of Continuous Formal Verification in Industry. *arXiv* **2019**, arXiv:1904.06152.
64. Thums, A.; Quante, J. Reengineering embedded automotive software. In Proceedings of the 2012 28th IEEE International Conference on Software Maintenance (ICSM), Trento, Italy, 23–28 September 2012; pp. 493–502.
65. Schulte-Coerne, V.; Thums, A.; Quante, J. Challenges in reengineering automotive software. In Proceedings of the 2009 13th European Conference on Software Maintenance and Reengineering, Kaiserslautern, Germany, 24–27 March 2009; pp. 315–316.
66. Malik, R.; Fabian, M.; Åkesson, K. Modelling Large-Scale Discrete-Event Systems Using Modules, Aliases, and Extended Finite-State Automata. *IFAC Proc. Vol.* **2011**, *44*, 7000–7005. [CrossRef]

Article

Practical Nonlinear Model Predictive Controller Design for Trajectory Tracking of Unmanned Vehicles

Hui Pang [1], Minhao Liu [1], Chuan Hu [2,*] and Nan Liu [1]

[1] School of Mechanical and Precision Instrument Engineering, Xi'an University of Technology, Xi'an 710048, China; panghui@xaut.edu.cn (H.P.); 2200220079@stu.xaut.edu.cn (M.L.); 2200220062@stu.xaut.edu.cn (N.L.)

[2] Department of Mechanical Engineering, University of Alaska Fairbanks, Fairbanks, AK 99775, USA

* Correspondence: chuan.hu.2013@gmail.com

Abstract: The trajectory tracking issue of unmanned vehicles has attracted much attention recently, with the rapid development and implementation of sensing, communication, and computing technologies. This paper proposes a nonlinear model predictive controller (NMPC) for the trajectory tracking application of an unmanned vehicle (UV). First, a two-degree-of-freedom (2-DOF) kinematics model of this UV is used to derive the desirable controller with two control variables as forward velocity and yaw angle. Next, the one-step Euler method is employed to establish the nonlinear prediction model, then a nonlinear optimization objective function is formulated to minimize the tracking errors of forward velocity and yaw angle from a preset time-varying reference road. Finally, the effectiveness of the proposed NMPC scheme is assessed under two different driving scenarios via MATLAB simulations. The simulation results confirm that the proposed NMPC scheme reveals better control accuracy and computational efficiency than the standard MPC controller under two different prescribed roads. Moreover, an outdoor field test is conducted to verify the performance of the proposed NMPC scheme, and the results show that the proposed NMPC can be applied to the real vehicle and can improve the tracking accuracy and the driving stability of trajectory tracking.

Keywords: unmanned vehicle; nonlinear model prediction controller; trajectory tracking; outdoor field test

Citation: Pang, H.; Liu, M.; Hu, C.; Liu, N. Practical Nonlinear Model Predictive Controller Design for Trajectory Tracking of Unmanned Vehicles. *Electronics* **2022**, *11*, 1110. https://doi.org/10.3390/electronics11071110

Academic Editor: Mahmut Reyhanoglu

Received: 27 February 2022
Accepted: 28 March 2022
Published: 31 March 2022

Publisher's Note: MDPI stays neutral with regard to jurisdictional claims in published maps and institutional affiliations.

1. Introduction

Currently, the driverless technology of unmanned vehicles (UVs) has attracted more and more attention from industrial and academic circles because the efficient control schemes of UVs have great potential in guaranteeing vehicle safety performance and traffic efficiency [1,2]. Trajectory tracking control [3–5] is a key link connecting environment perception and motion control for the implementation of UVs, and efficient trajectory tracking can improve the safety performances and ride comforts of UVs.

In the current research on the trajectory tracking problems, many intelligent optimization algorithms have been widely proposed for the trajectory tracking of UVs, such as the ant colony algorithm [6], particle swarm optimization algorithm [7], genetic algorithm [8], graphic algorithm [9,10], etc. Although these algorithms can track the prescribed path, they mainly adopt the infinite iteration method that requires a large amount of computation to obtain the optimal solutions; therefore, the real-time performance and tracking accuracy of those methods may not be guaranteed. Herein, as a popular iterative optimized control approach, model predictive control (MPC) [11,12] can simultaneously incorporate system state constraints and control input constraints into the controller design process, i.e., at each sample time, the future system inputs and outputs can be obtained by updating the control plant and can be optimized via an appropriate optimization algorithm in the predictive horizon, wherein the system constraints are easily put in an explicit form.

As reported in the literature [13–20], the MPC method has been widely employed in the trajectory tracking controller development for different types of UVs. For instance, in the literature [18], a control plant integrating the kinematic motions and the dynamic behaviors of an automated guided vehicle was established, and an MPC-based trajectory tracking controller was then designed. In [19], an MPC method was presented to control the forward steering of autonomous vehicles by continuously linearizing nonlinear vehicle models. In addition, a direct data-driven MPC method has been proposed in [20] to relax the laborious model identification procedure. A linear-parameter-varying MPC approach was proposed in [21] for UVs, and the control scheme was validated through simulation and experiment investigations. Similarly, in order to estimate the driver intentions and the underlying behaviors, an MPC approach integrating with recurrent neural network and memory cell was proposed for an unmanned vehicle [22], and the simulation results verified the improvements in trajectory tracking performances for this vehicle.

In short, most of the early literature [18–20,22] mainly focuses on the simulation investigations on trajectory tracking problems, and the real-platform or field test study is lacking. Moreover, due to the MPC execution load and extra modeling errors, it is hard to linearize the original (nonlinear) system around the current working point and then design an effective model-based predictive controller.

Fortunately, a nonlinear model predictive control (NMPC) approach has been extensively adopted in the practical path planning and trajectory tracking problem due to its inherent profits to deal with the nonlinear input constraints aiming at speeding up the online computation. In [23], an NMPC approach was presented to address the safety issues regarding collision avoidance and lateral stability of unmanned ground vehicles in high-speed conditions. Further, a trajectory tracking NMPC strategy was proposed in [24] to address the explicit state and input constraints for autonomous surface craft, and the real-time implementation of the NMPC was validated through the experimental results. In [25], a distributed control scheme was provided to achieve the accurate tracking control of the autonomous underwater vehicle motion by using NMPC techniques. It can be concluded from these two studies that the NMPC algorithm is very appropriate for solving the nonlinear optimization problem with lower computational cost. Besides, it is inevitably encountered with the regular and irregular roads in the real world. With the NMPC method, the computational efficiency and control accuracy for a nonlinear control system can be guaranteed [26].

Therefore, inspired by the literature mentioned above, this paper proposes a practical NMPC controller design for the trajectory tracking application of a UV considering the nonlinear road trajectory. The main contributions of this work are summarized as:

(1) A novel NMPC controller is proposed to achieve the accurate trajectory tracking of the UV under different prescribed roads, wherein the one-step Euler method is used to establish the nonlinear prediction model. As model predictive control is an iterative process, the Euler method has the advantages of a wide range of numerical solutions, a simple form that is easy to calculate, thus the tracking errors of forward velocity and yaw angle are minimized through a nonlinear optimization method;
(2) MATLAB simulations are carried out to verify the control performances of the proposed NMPC controller under two different driving scenarios, and the results show that the strategy can deal with the nonlinear road trajectory well, and can improve the tracking accuracy and the driving stability;
(3) A simple test platform consisting of a scaled-down real racing car, sensors, microcontroller, and host computer is built up to verify the effectiveness of the designed NMPC controller in UV trajectory tracking applications.

The rest of this article is arranged as follows. The UV's kinematic model formulation is described in Section 2. Next, the design procedure of this expected NMPC controller is provided in Section 3. Then, in order to verify the control performance of the presented NMPC controller, both simulation and field test verifications are orderly conducted and discussed in Sections 4 and 5. Finally, the conclusions and perspectives of this paper are presented in Section 6.

2. The Unmanned Vehicle's Kinematics Model

Here, a kinematic model of UV with two degree-of-freedoms (2-DOF) is adopted to perform the controller synthesis [21]. To mimic a real unmanned vehicle's kinematic behaviors, it is assumed that this UV is driven by a servo motor installed in a rear wheel, and is steered by a servo motor installed in a front wheel. By ignoring the side-slip angle when making the front-wheel steering operation, as well as considering the longitudinal speed to be a constant value, one can construct the 2-DOF kinematic model as shown in Figure 1, in which point $A(X_f, Y_f)$ and $B(X_r, Y_r)$ stand for the center positions of the front axle and rear axle, and the other symbols used in this UV kinematic model are listed in Table 1.

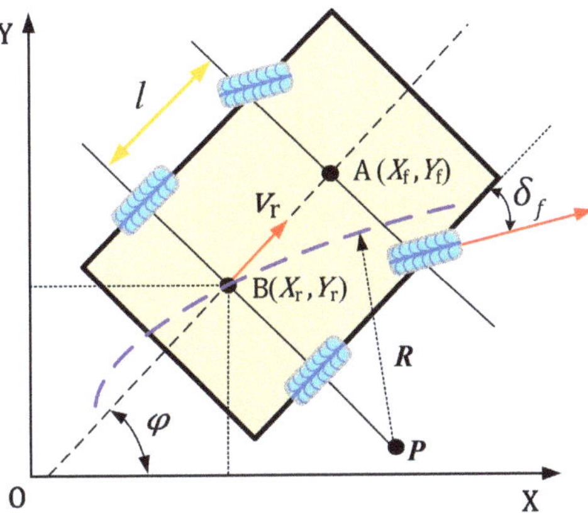

Figure 1. The 2-DOF kinematic model of UV.

Table 1. The used symbols and parameters of UV.

Symbol	Description	Symbol	Description
$A(X_f, Y_f)$	The center of the front axle	l	The distance from point A to B
$B(X_r, Y_r)$	The center of the rear axle	ω	The yaw rate of UV
δ_f	The deflection angle of the front axle	φ	The yaw angle of UV
v_r	The forward speed of point B		

In terms of the kinematic relationships of this UV, v_r can be expressed by

$$v_r = \dot{X}_r \cos\varphi + \dot{Y}_r \sin\varphi, \tag{1}$$

The kinematic constraints between the rear- and front- axles of this UV are easily obtained

$$\dot{X}_r \sin\varphi = \dot{Y}_r \cos\varphi, \tag{2}$$

$$\dot{X}_f \sin\left(\varphi + \delta_f\right) = \dot{Y}_f \cos\left(\varphi + \delta_f\right), \tag{3}$$

By integrating Equations (1) and (2), one obtains

$$\begin{cases} \dot{X}_r = v_r \cos\varphi \\ \dot{Y}_r = v_r \sin\varphi \end{cases}, \tag{4}$$

By further transformation, we have the geometric relationship between points A and B satisfying

$$\begin{cases} X_f = X_r + l\cos\varphi \\ Y_f = Y_r + l\cos\varphi \end{cases}, \tag{5}$$

For simplicity, let ω denote the derivative of φ, and by substituting Equation (4) into the derivative of Equation (5), we have

$$\begin{cases} \dot{X}_f = v_r\cos\varphi - l\dot{\varphi}\sin\varphi \\ \dot{Y}_r = v_r\sin\varphi + l\dot{\varphi}\cos\varphi \end{cases}. \tag{6}$$

By further integrating Equation (6) with Equation (3), we obtain

$$\dot{\varphi} = \omega = \frac{v_r}{l}\tan\delta_f, \tag{7}$$

The expressions for R and δ_f can be obtained as follows:

$$\begin{cases} R = \frac{v_r}{\omega} \\ \delta_f = \arctan\left(\frac{l}{R}\right) \end{cases}, \tag{8}$$

By combining Equations (4) and (7), we have

$$\begin{bmatrix} \dot{X}_r \\ \dot{Y}_r \\ \dot{\varphi} \end{bmatrix} = \begin{bmatrix} \cos\varphi \\ \sin\varphi \\ \tan\delta_f/l \end{bmatrix} v_r, \tag{9}$$

Thus, the state-space form of the kinematic Equation for this UV can be formulated as

$$\begin{bmatrix} \dot{X}_r \\ \dot{Y}_r \\ \dot{\varphi} \end{bmatrix} = \begin{bmatrix} \cos\varphi & 0 \\ \sin\varphi & 0 \\ 0 & 1 \end{bmatrix} \begin{bmatrix} v_r \\ \omega \end{bmatrix}. \tag{10}$$

Define $\boldsymbol{\xi}_{kout} = \begin{bmatrix} X_r & Y_r & \varphi \end{bmatrix}^T$ as the system output vector and $\boldsymbol{u}_{kin} = \begin{bmatrix} v_r & \omega \end{bmatrix}^T$ as the control input vector, then Equation (10) is further rewritten as

$$\dot{\xi}_{kout} = \begin{bmatrix} \cos\varphi & 0 \\ \sin\varphi & 0 \\ 0 & 1 \end{bmatrix} \boldsymbol{u}_{kin}. \tag{11}$$

It should be noted that the kinematics model can represent the relationship between the vehicle's state of motion and the control input, thus the model predictive controller can achieve the purpose of predetermined control. Now, the kinematics model of the UV is completed, and the proposed controller design will be illustrated in the following.

3. Design of Nonlinear Model Predictive Control

As a popular iterative optimization method, model predictive control has been extensively applied to the trajectory tracking and path-following control of UVs in recent years. For a common MPC, it consists predictive model, rolling optimization, and feedback correction. It should be noticed that a predictive model can predict the future system inputs and outputs according to the current system states, rolling optimization can generate the optimized control sequence, and feedback correction is usually to feed the system states at a certain current time back to the control system as the new input to perform the optimization calculation at the next iteration. Particularly, the NMPC can actually handle the linear and nonlinear trajectory tracking problems with the upper-lower limits, herein the one-step Euler method is used to derive the nonlinear model prediction controller with expecting to enhance the tracking accuracy and rate of the UV. The proposed NMPC framework is shown in Figure 2.

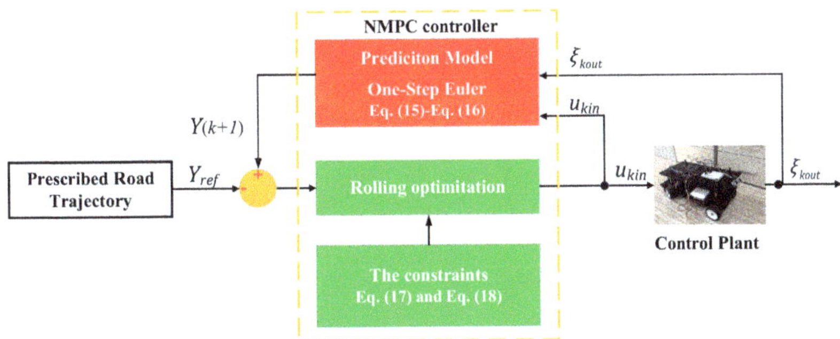

Figure 2. Block diagram of proposed NMPC controller.

3.1. Establishment of Nonlinear Prediction Model

To facilitate the expected nonlinear prediction model establishment for this UV, Equation (11) can be reconfigured as a nonlinear function expression as

$$\begin{cases} \dot{x} = f(x(t), u(t)) \\ y = g(x(t), u(t)) \end{cases}. \tag{12}$$

Define $x = [\varphi, X_r, Y_r]^T$ as the state vector, i.e., ξ_{kout}, $u(t)$ as the control input vector, i.e., u_{kin}, thus we can have

$$f(x(t), u(t)) = \begin{bmatrix} \omega \\ v_r \cos\varphi \\ v_r \sin\varphi \end{bmatrix}, \tag{13}$$

$$g(x(t), u(t)) = \begin{bmatrix} 0 & 1 & 0 \\ 0 & 0 & 1 \end{bmatrix} x(t). \tag{14}$$

In order to represent the NMPC controller design as a convex optimization problem, the discretization form of Equation (12) can be derived as

$$\begin{cases} x(k+1) = F(x(k), x(k)) \\ y(k) = G(x(k)) \end{cases}. \tag{15}$$

where $F(x(k), x(k))$ and $G(x(k))$ represent the discretized form of $f(x(t), u(t))$, and $g(x(t), u(t))$.

Furthermore, by introducing the sampling time T, the one-step Euler method is employed to describe the prediction model of Equation (15) as follows

$$\begin{cases} x(k+1) = x(k) + \begin{bmatrix} \omega(k) \\ v_r(k)\cos(\varphi(k)) \\ v_r(k)\sin(\varphi(k)) \end{bmatrix} T = x(k) + f(x(k), u(k))T \\ y(k) = g(x(k), u(k)) \end{cases}. \tag{16}$$

Remark 1. *If the sampling time T is too short, the calculation burden will increase and the tracking accuracy of the control system will be affected. In contrast, if T is too large, a large cumulative error will inevitably be encountered in the control process, which will impose negative impacts on the dynamical performances of the control system. According to the Shannon sampling theory and the control periods of the actuators, the sample time is selected as T = 0.02 s in this paper.*

For the control input $u(k)$, it is necessary to introduce its physical constraints, which is expressed as follows:

$$u_{min}(k) \leq u(k) \leq u_{max}(k), \tag{17}$$

Note that y_{ref} is the reference road trajectory, Δy_{min} and Δy_{max} represent for the minimal and maximal errors of the deviation between $y(k)$ and y_{ref}, both of them can be adjusted in the process of rolling optimization. The constraint relationship between them is expressed as follows:

$$\Delta y_{min} \leq y - y_{ref} \leq \Delta y_{max}. \tag{18}$$

To make a clear distinguishment, N_p and N_c are used to denote the control outputs in the prediction domain and the control domain, and the control outputs of Equation (16) in N_p are expressed by

$$\left.\begin{array}{r} y(k+1) = g(x(k+1), u(k+1)) \\ y(k+2) = g(x(k+2), u(k+2)) \\ \vdots \\ y(k+N_c) = g(x(k+N_c), u(k+N_c)) \\ \vdots \\ y(k+N_p) = g(x(k+N_p), u(k+N_p)) \end{array}\right\}. \tag{19}$$

where the condition of N_p and N_c should satisfy $N_p \geq N_c$. In this paper, we set the control horizon equal to the prediction horizon, $N_p = N_c$, in all simulations and experiments in order to predict future UV states as accurately as possible.

Additionally, the system outputs $y(k)$ and the control inputs $u(k)$ can be given by

$$Y(k+1) = \left[y(k+1), y(k+2), \cdots, y(k+N_p)\right]^T, \tag{20}$$

$$U(k) = [u(k+1), u(k+2), \cdots, u(k+N_c)]^T. \tag{21}$$

The main objective of our expected NMPC design in N_p is to minimize the tracking errors between the control output sequence and the reference trajectory sequence. To that end, the prescribed reference trajectory can be described as:

$$Y_{ref}(k+1) = \left[y_{ref}(k+1), y_{ref}(k+2), \cdots, y_{ref}(k+N_p)\right]^T. \tag{22}$$

Herein, our main goal is to minimize the tracking error in the prediction time domain N_p, which is expressed by

$$min\|Y(k+1) - Y_{ref}(k+1)\|. \tag{23}$$

3.2. Formulation of Objective Function

In order to find out the minimized solutions to the tracking error system described in (23), it is necessary to further transform (23) into a linear-quadratic-regulator (LQR) control problem. To this end, the weighting factors Q and R are introduced and the following LQR optimization problem can be formulated, which is expressed by

$$\begin{aligned} J_{\text{cost}}(k) = \textstyle\sum_{i=1}^{N_p} \left\{y(k+i) - y_{ref}(k+i)\right\}^T Q\left[y(k+i) - y_{ref}(k+i)\right] + \\ \textstyle\sum_{i=1}^{N_c} \{u(k+i)\}^T R u(k+i). \end{aligned} \tag{24}$$

where $J_{\text{cost}}(k)$ is the optimized objective function, i is the prediction step, Q is the weighting matrix that reveals the system's ability to follow up the reference road trajectory, R is the weighting matrix that manifests the stability performance of the trajectory tracking control problem, at time step k.

Therefore, the optimization formulation of the desirable NMPC controller can be expressed as

$$min\ J_{\text{cost}}(k). \tag{25}$$

Additionally, the physical constraints of the control input $u(k)$ described in (16) and (17) should be considered to find out the optimized solutions of (25). Finally, the nonlinear optimization problem of the NMPC controller design can be formulated as

$$\begin{aligned} x(k+i) = x(k+i-1) + f(x(k+i-1))T \\ +u(k+i-1), \end{aligned} \tag{26}$$

$$y(k+i-1) = g(x(k+i-1), u(k+i-1)), \tag{27}$$

$$u_{min}(k) \leq u(k+i) \leq u_{max}(k), \tag{28}$$

$$\Delta y_{min} \leq y(k+i) - y_{ref}(k+i) \leq \Delta y_{max}. \tag{29}$$

Practically, the optimal control sequence in N_c at the sample time k can be obtained as

$$u(k+i) = [u(k+1), u(k+2), \cdots u(k+N_c)]. \tag{30}$$

4. Simulation Verification

Here, comparative simulations have been performed under MATLAB software with the purpose of assessing the two MPC controllers: namely the traditional model predictive controller (TMPC) [27] and the proposed NMPC. The simulations are evaluated on this 2-DOF UV in the case of single-circle and twin-circle trajectory scenarios. Note that three indexes x, y, and φ are used to demonstrate the performances of the two MPC controllers.

4.1. Simulations under a Straight-Line Trajectory

First, we expect this UV to move on a preset straight-line(ST) trajectory from the starting point (0, 0) at a longitudinal velocity $v = 1$ m/s. The mathematical expression of the reference straight trajectory is $Y_{ref} = 3$, and the simulation time was set as 5 s. Figure 3 displays the tracking response curves of this UV from the TMPC and the proposed NMPC controllers.

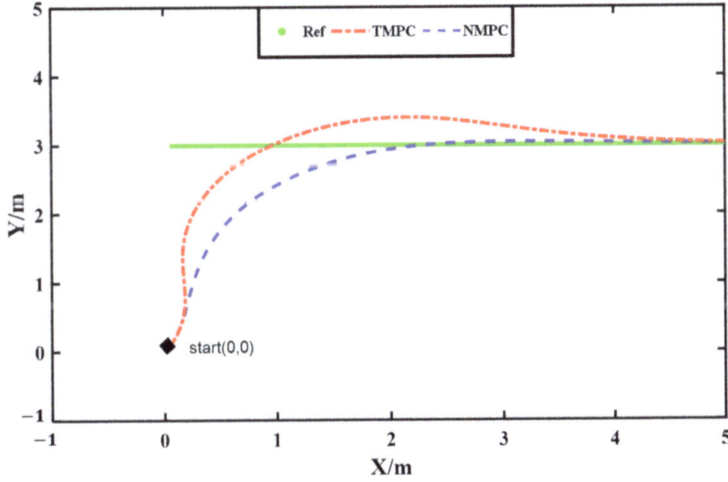

Figure 3. The simulation tracking results under the ST.

As shown in Figure 4, the three motion parameters x, y, and φ of this UV followed overall smooth trends for the two controllers. Meanwhile, the three motion parameters x, y, and φ converged to zero at about 2.5 s for our proposed NMPC and about 4.5 s for the TMPC.

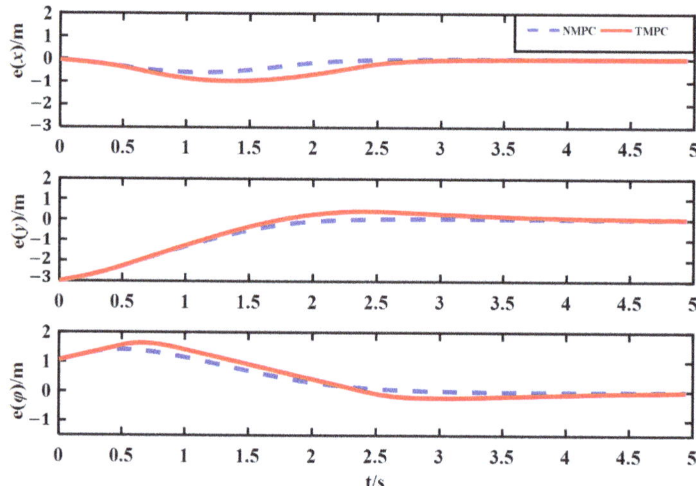

Figure 4. The tracking errors of x, y and φ under the ST.

Furthermore, the root mean square error (RMSE) value for the tracking errors of the three states is selected to quantitatively evaluate the control performance of the two MPC controllers, and based on the literature [28], the RMSE is defined by

$$\chi_{\text{RMSE}} = \frac{\|\chi_j - \chi_{ref,j}\|}{\sqrt{n}} = \sqrt{\frac{1}{n}\sum_{j=1}^{n}\left(\chi_j - \chi_{ref,j}\right)^2},\ j = 1\ldots n. \tag{31}$$

wherein χ is an n-dimensional state vector, and χ_j is the value of j-th element in χ, $\chi_{ref,j}$ is the value of j-th element in the reference road trajectory, n is the length of χ.

The results in Table 2 show that the RMSE values of x, y, and φ for the proposed NMPC were reduced by about 51.14%, 1.35%, and 18.16% compared to those of the TMPC.

Table 2. The RMSE comparisons of the tracking errors of x, y, and φ under the ST.

Type	x	y	φ
TMPC	0.559	1.11	0.8264
NMPC	0.2731 (↓51.14%)	1.095 (↓1.35%)	0.6763 (↓18.16%)

It can be seen from Figure 5 that under the ST condition, the change of the control input for the proposed NMPC is smoother than that of TMPC.

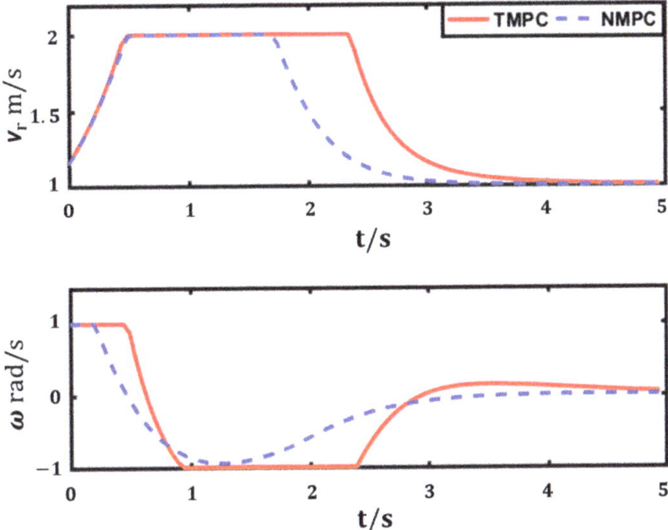

Figure 5. The control input of the TMPC and NMPC controllers under the ST.

4.2. Simulations under a Single-Circle Trajectory

A regular single-circle trajectory (SCT) road with a radius of 6 m is chosen as the reference trajectory road, and the corresponding simulations are conducted in this scenario. As shown in Figure 6, the simulation results generated by the TMPC and designed NMPC controller can well follow the prescribed reference road from the starting point (−10, 0).

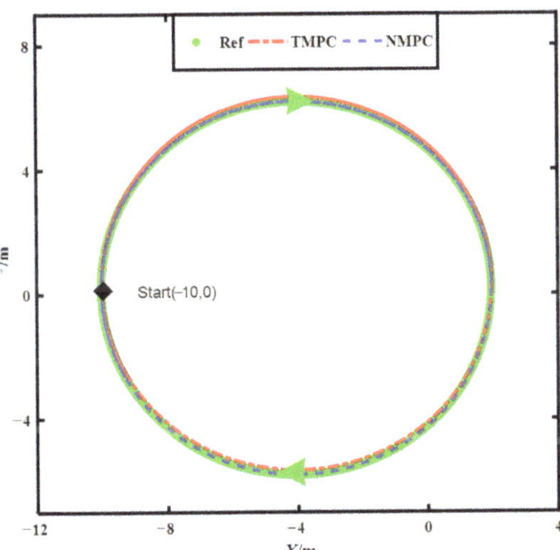

Figure 6. The simulation tracking results under the SCT.

Additionally, to examine the tracking accuracy of the two MPC controllers against the reference trajectory road, the tracking errors of x, y, and φ for the UV under the SCT scenario are provided in Figure 7. It is obvious that the tracking errors of these three states of x, y, and φ present a flatter trend with converging to zero. Moreover, in comparison with the TMPC controller, the tracking errors of x, y, and φ for the proposed NMPC controller have obviously fewer fluctuations on a whole and can reach a relative stability state in a shorter time. Particularly, the tracking errors of x, y, and φ can converge to zero at about 2 s for the NMPC controller, and at 6 s for the TMPC controller.

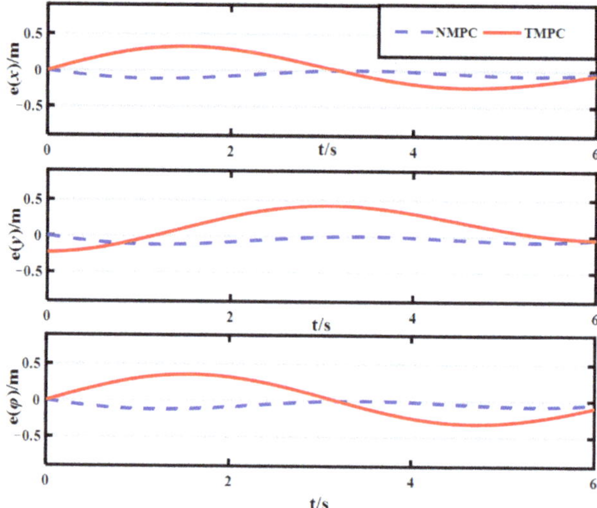

Figure 7. The tracking errors of x, y, and φ under the SCT.

Herein, Table 3 lists the RMSE comparisons of the tracking errors of x, y, and φ obtained by the TMPC and NMPC controllers under the SCT.

Table 3. The RMSE comparisons of the tracking errors of x, y, and φ under the SCT.

Type	x	y	φ
TMPC	0.2208	0.3734	0.2515
NMPC	0.1210 (↓45.20%)	0.2284 (↓40.44%)	0.1285 (↓48.95%)

It can be seen from Figure 8 that the oscillation of the control input v_r for the NMPC is much smaller than that of the TMPC under the SCT condition. Meanwhile, the control input ω of the NMPC can reach the target value faster.

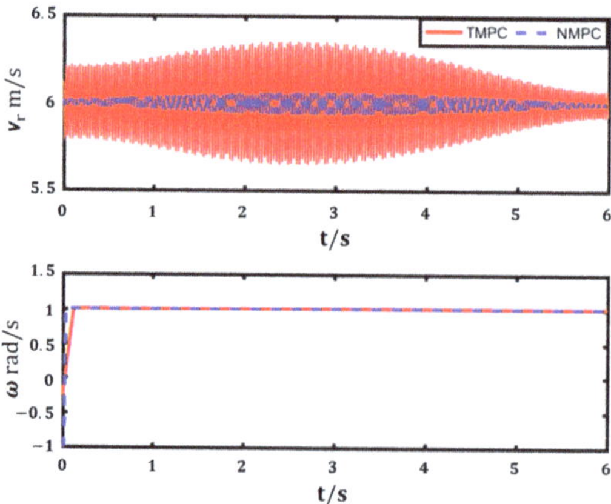

Figure 8. The control input of the TMPC and NMPC controllers under the SCT.

4.3. Simulations under a Twin-Circle Trajectory

To further evaluate the tracking performance of the NMPC controller, a twin-circle trajectory (TCT) road is used to conduct the simulations. The related tracking response curves are presented in Figure 9. It can be seen that both the TMPC and NMPC controllers can track the preset TCT starting from point (−10, −0.2) closely. The UV is first running counterclockwise and then clockwise, after the lane change, our designed NMPC controller can still track another circular trajectory road with relatively higher accuracy. Furthermore, Figure 10 shows the tracking errors of x, y, and φ for the two MPC controllers against the reference trajectory. It is obvious that the tracking errors of x, y, and φ are relatively flat in this TCT scenario. In particular, the three states of x, y, and φ can converge to zero at around 2 s for our presented NMPC controller, and 6 s for the TMPC controller.

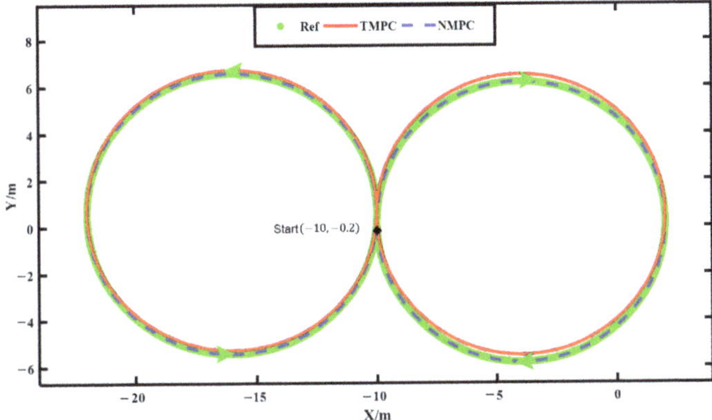

Figure 9. The simulation tracking results under the TCT.

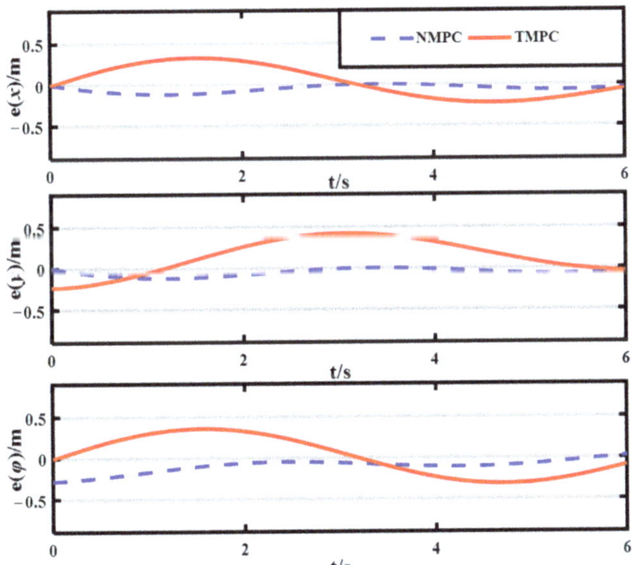

Figure 10. The tracking errors of x, y, and φ under the TCT.

Table 4 summarizes the RMSE comparisons of the three states x, y, and φ for the two MPC controllers under the TCT. Compared with the TMPC controller, the RMSE values of

x, y, and φ for the presented NMPC can be reduced by about 65.35%, 54.61%, and 71.10%, respectively.

Table 4. The RMSE comparisons of the tracking errors of x, y, and φ under the TCT.

Type	x	y	φ
TMPC	0.1622	0.2743	0.1848
NMPC	0.0562 (↓65.35%)	0.1245 (↓54.61%)	0.0543 (↓71.10%)

It can be seen from Figure 11 that the control input v_r for both MPC controllers have oscillations, but the oscillation amplitude of our proposed NMPC is significantly smaller than that of the TMPC. At the same time, the control input ω of the NMPC shows a faster response speed.

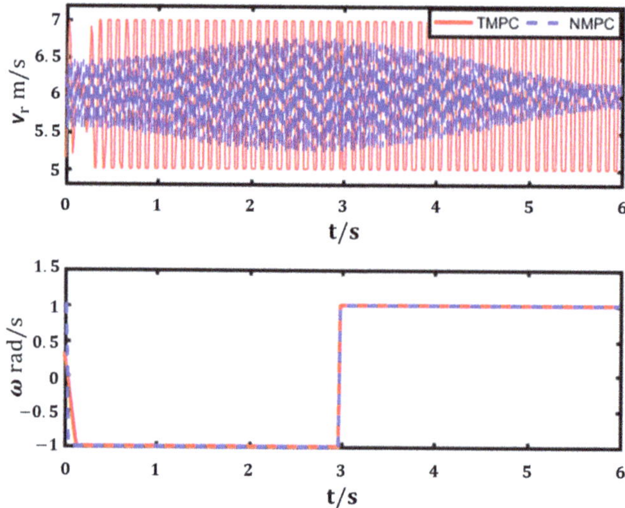

Figure 11. The control input of the TMPC and NMPC controllers under the TCT.

In general, our proposed NMPC controller exhibits better performance specifically for shorter tracking time and smaller fluctuation errors, in comparison with the TMPC controller. Additionally, according to the quantitative analysis of the tracking indexes as x, y, and φ, it is easily observed that our proposed NMPC controller has significant improvements in the tracking performance over the TMPC controller, which illustrates that the proposed NMPC has better control effects under three different driving scenarios.

5. Field Test Verification

In this section, an outdoor field test platform is constructed by ourselves in order to verify the effectiveness of those two MPC controllers. To facilitate the descriptions of this test platform, Figure 12 displays the schematic diagram of the experimental setup, and the outdoor field test photograph of this UV on the different prescribed road trajectories are provided in Figure 13. Different from the simulation verifications in Section 4, the field tests are conducted under two different driving scenarios.

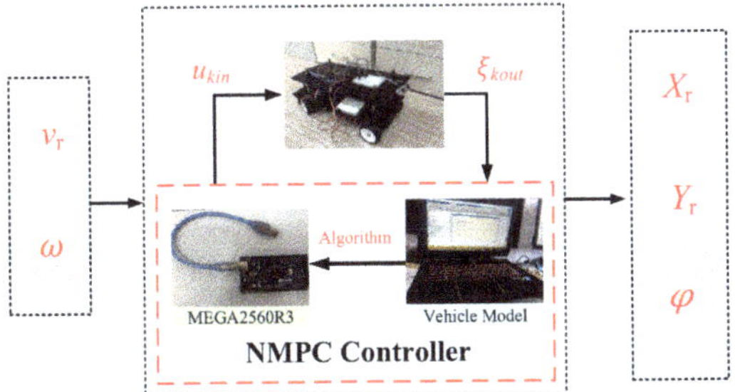

Figure 12. Schematic diagram of the experiment setup.

Figure 13. Outdoor field test setup.

Herein, a scaled unmanned vehicle—a BT-4 racing car, is used as the control plant, and an Arduino board (MEGA2560R3) is adopted to develop our proposed NMPC controller. Moreover, there are two DC (direct-current) motors in this BT-4 vehicle, one is used for steering, and another one is used to drive this UV at the rear wheel. Both DC motors are rooted in the Arduino (MEGA2560R3) board, using 54 digital I/O pins and 6-bit pulse width modulation (PWM) drivers. Because of the limitations of this self-established outdoor field test setup, only the measured points of (x, y) at the ground coordinates are collected to draw the tracking trajectories for this BT-4 vehicle, which are further utilized to compare and validate the tracking performances of the two different MPC controllers.

5.1. Test on Irregular Road

For the practical driving scenario, the irregular trajectory road (ITR) is often encountered. Thus, the outdoor field tests are conducted on a practical ITR, and the tracking response curves of this UV are shown in Figure 14. In which, the solid green line denotes the reference trajectory road, the blue-dotted and the red-dashed lines denote the tracking response curves generated by the TMPC and NMPC controllers, respectively. Moreover, Figure 15 reveals the tracking errors of x and y for the two MPC controllers regarding the reference trajectory road. It should be noticed that this UV runs starting from the point $x_0 = \begin{bmatrix} 285 & 280 \end{bmatrix}$ at v = 1 m/s under the ITR.

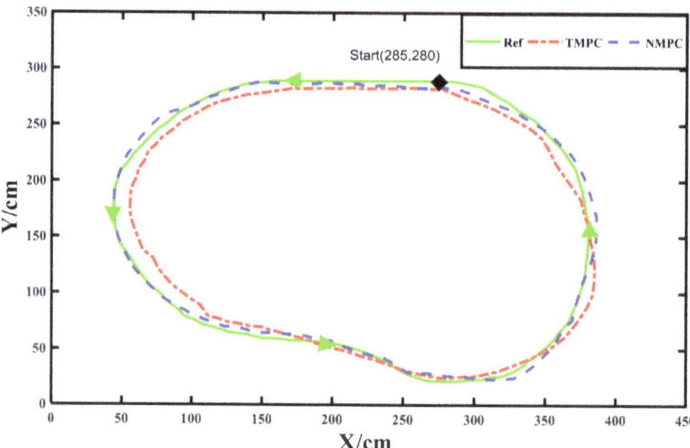

Figure 14. The tracking response curves of the two MPC controllers under the ITR.

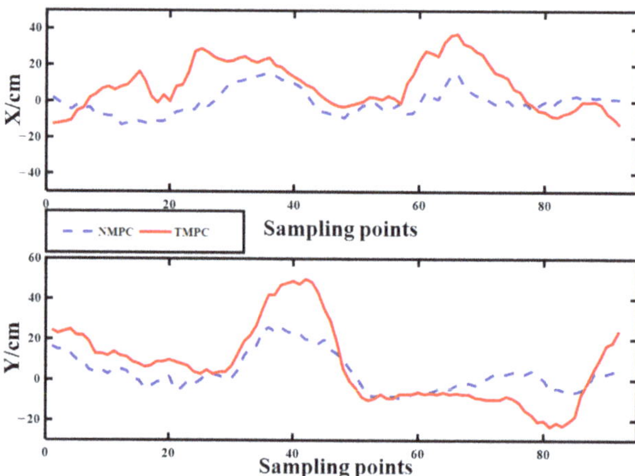

Figure 15. The tracking errors of x and y for the two MPC controllers under the ITR.

It is obvious from Figure 14 that the TMPC and proposed NMPC controller can track the ITR as possible as closely. Specifically, the blue-dashed line (denotes the tracking trajectory obtained by the NMPC controller) is much closer to the green-solid line (the reference trajectory) in comparison with the red-dashed line (denotes the tracking trajectory obtained by the TMPC controller). Furthermore, it is observed from Figure 15 that the tracking errors of x and y generated by the two MPC controllers could gradually converge to zero states, and the tracking error curves by the NMPC controller present a flatter appearance compared to the corresponding tracking curves by the TMPC controller. Particularly, the tracking error ranged from −10 cm to 15 cm for the NMPC controller and −15 cm to 35 cm for the TMPC controller in the X-direction. Additionally, in the Y-direction, the tracking errors of x and y are varied from 0 to 25 cm for the NMPC controller, and −20 cm to 50 cm for the TMPC controller, respectively.

Similar to the analysis of simulation verifications in Section 4, we herein use the RMSE comparisons of the tracking errors of x and y to evaluate the control performance of the designed controllers, which is presented in Table 5. In addition to this, the histogram comparisons of the tracking errors of x and y are also provided in Figure 16 to make a quantitative improvement of each tracking state for the TMPC and proposed NMPC controller under the ITR and the prescribed reference trajectory. Compared to the tracking performance of the TMPC controller, the UV using the NMPC controller can reduce its RMSE values of x and y by about 54.72% and 48.65%, respectively.

Table 5. The RMSE of the tracking errors of x, y under the single-circle road.

Type	x	y
TMPC	1.5800	2.0039
NMPC	0.7155 (↓54.72%)	1.0029 (↓48.65%)

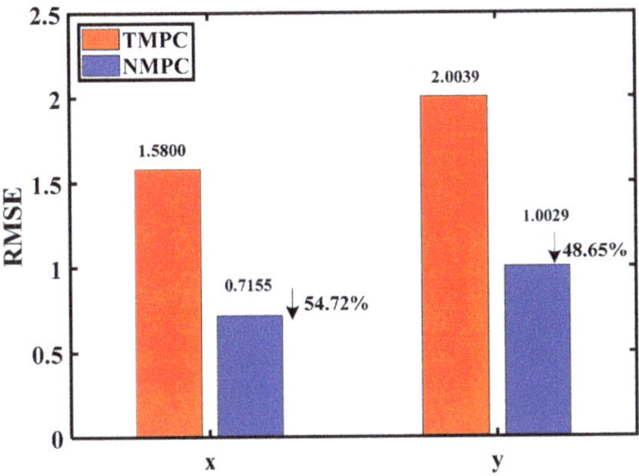

Figure 16. The histogram comparisons of two MPC controllers under the ITR.

5.2. Test on Double-Circle Road

Except for the irregular road trajectory, it is very necessary to validate the control performances of the proposed NMPC controller under a general double-circle trajectory road (DCTR). An outdoor field test is performed under a prescribed DCTR with a radius of 2.5 m, which starts from $x_0 = \begin{bmatrix} 250 & 50 \end{bmatrix}$ with $v = 1$ m/s. Figure 17 displays the tracking response curves obtained by the TMPC and NMPC controllers, and Figure 18 plots the tracking error curves of x and y for the two MPC controllers against the reference trajectory road.

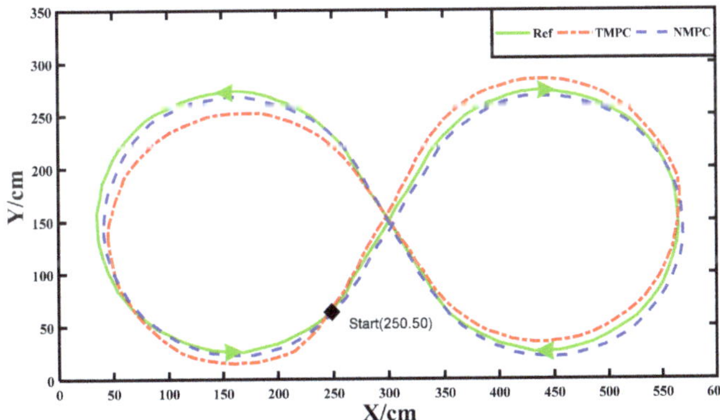

Figure 17. The tracking response curves of the two MPC controllers under the DCTR.

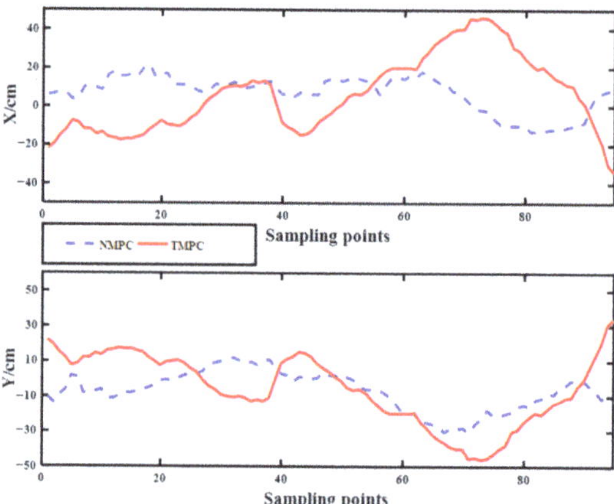

Figure 18. The tracking errors of x and y for the two MPC controllers under DCTR.

From Figure 17, it is easily seen that the two MPC controllers can nearly track the prescribed DCTR, and the tracking curves of the NMPC controller seem closer to the reference trajectory road. By a further observation from Figure 18, the tracking errors of x and y generated by the two MPC controllers are flat on a whole, yet the tracking errors by the NMPC controller can yield a smoother tendency and can enter into a relatively stable state at a faster rate compared to the TMPC controller. Besides, the tracking error of x is varied from −10 to 15 for the NMPC controller, and from −30 cm to 40 cm for the TMPC controller in the X-direction. Moreover, the tracking error of y is ranged from −20 cm to 10 cm for the NMPC controller and from −40 cm to 30 cm for the TMPC controller in the Y-direction.

Similarly, the RMSE comparisons of the tracking errors of x and y under this DCTR are quantitatively compared and provided in Table 6, and the histogram comparisons for the tracking errors of x and y are provided in Figure 16 to assess the control performance of the designed controller. It is clear from Table 6 and Figure 19 that the RMSE values of the tracking errors x and y for the NMPC controller are reduced by about 56.63% and 48.73%, respectively, in comparison with those of the TMPC controller, which further illustrates that the NMPC controller has better control effect under the DCTR scenario.

Table 6. The RMSE comparisons of the tracking errors of x and y under the DCTR.

Type	x	y
TMPC	2.7814	2.8215
NMPC	1.2061 (↓56.63%)	1.4465 (↓48.73%)

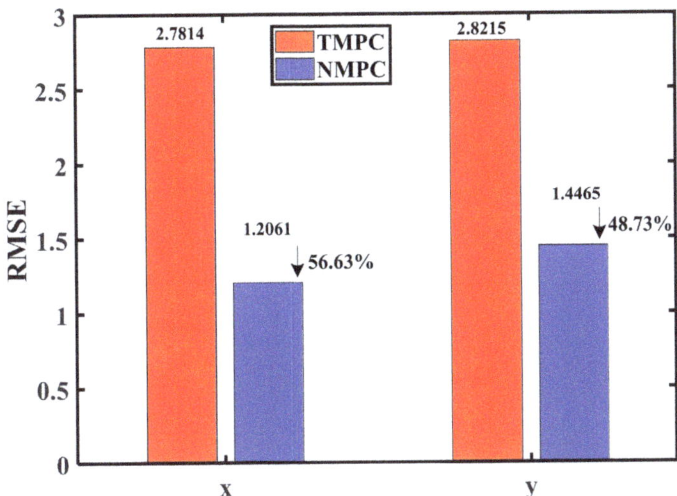

Figure 19. The histogram comparisons of two MPC controllers under the DCTR.

6. Conclusions

In this paper, a practical NMPC design method is proposed to achieve the accurate trajectory tracking application of a scaled UV. This desirable controller is developed based on a 2-DOF kinematics model of the UV, and in the framework of a standard MPC, the one-step Euler method is utilized to construct the nonlinear prediction model, then a nonlinear optimization objective function is formulated to minimize the tracking errors of forward velocity and yaw angle from a time-varying preset reference road. Finally, both the comparative simulations and the outdoor field tests are carried out to confirm the superior performances of the designed NMPC controller against the TMPC controller for this UV. The comparison of results demonstrates that the improvements of the tracking indexes x, y, and φ for the UV with the presented NMPC controller are at least 45%, 1.35%, and 18.16%, respectively. Moreover, the field outdoor test results show that the improvements of tracking indexes x and y for the UV with the proposed NMPC controller are about 54% and 48%, respectively, compared with those with the TMPC controller. On a whole, it can be concluded that the NMPC controller outperforms better control performances regarding the TMPC controller.

Future study will focus on the trajectory tracking controller design of this UV considering the kinematics and dynamics properties simultaneously, and the experimental validation of the proposed controller in a real-time outdoor field test platform.

Author Contributions: Conceptualization, H.P.; methodology, H.P.; validation, M.L. and N.L.; formal analysis, M.L. and N.L.; resources, H.P. and C.H.; funding acquisition, H.P.; supervision, H.P. and C.H.; writing—original draft preparation, H.P.; writing—review and editing, H.P. and C.H. All authors have read and agreed to the published version of the manuscript.

Funding: This research was funded by the National Natural Science Foundation of China under Grant 51675423.

Conflicts of Interest: The authors declare no conflict of interest.

References

1. Lin, Y.C.; Lin, C.L.; Huang, S.T.; Kuo, C.H. Implementation of an Autonomous Overtaking System Based on Time to Lane Crossing Estimation and Model Predictive Control. *Electronics* **2021**, *10*, 2293. [CrossRef]
2. Hu, C.; Qin, Y.; Cao, H. Lane keeping of autonomous vehicles based on differential steering with adaptive multivariable super-twisting control. *Mech. Syst. Signal Process.* **2019**, *125*, 330–346. [CrossRef]

3. Wang, Z.; Wang, J. Ultra-local model predictive control: A model-free approach and its application on automated vehicle trajectory tracking. *Control Eng. Pract.* **2020**, *101*, 104482. [CrossRef]
4. Wang, R.; Jing, H.; Hu, C. Robust H∞ path following control for autonomous ground vehicles with delay and data dropout. *IEEE Trans. Intell. Transp. Syst.* **2016**, *17*, 2042–2050. [CrossRef]
5. Wang, Y.; Gao, S.; Chu, H.; Wang, X. Planning of Electric Taxi Charging Stations Based on Travel Data Characteristics. *Electronics* **2021**, *10*, 1947. [CrossRef]
6. Li, J.; Xu, B.; Yang, Y. Three-phase qubits-based quantum ant colony optimization algorithm for path planning of automated guided vehicles. *Int. J. Robot. Autom.* **2019**, *34*, 156–163. [CrossRef]
7. Ma, Y.; Hu, M.; Yan, X. Multi-objective path planning for unmanned surface vehicle with currents effects. *ISA Trans.* **2018**, *75*, 137–156. [CrossRef]
8. Hu, L.; Gu, Z.Q.; Huang, J. Research and realization of optimum route planning in vehicle navigation systems based on a hybrid genetic algorithm. *Proc. Inst. Mech. Eng. Part D J. Automob. Eng.* **2008**, *222*, 757–763. [CrossRef]
9. Chalvet, V.; Haddab, Y.; Lutz, P. Trajectory planning for micromanipulation with a non-redundant digital microrobot: Shortest path algorithm optimization with a hypercube graph representation. *J. Mech. Robot.* **2016**, *8*, 021013. [CrossRef]
10. Ambike, S.; Schmiedeler, J.P.; Stanišić, M.M. Trajectory tracking via independent solutions to the geometric and temporal tracking subproblems. *J. Mech. Robot.* **2011**, *3*, 021008. [CrossRef]
11. Jeong, Y.; Yim, J. Model Predictive Control-Based Integrated Path Tracking and Velocity Control for Autonomous Vehicle with Four-Wheel Independent Steering and Driving. *Electronics* **2021**, *10*, 2812. [CrossRef]
12. Huang, Y.; Ding, H.; Zhang, Y. A motion planning and tracking framework for autonomous vehicles based on artificial potential field elaborated resistance network approach. *IEEE Trans. Ind. Electron.* **2019**, *67*, 1376–1386. [CrossRef]
13. Yang, J.A.; Kuo, C.H. Integrating Vehicle Positioning and Path Tracking Practices for an Autonomous Vehicle Prototype in Campus Environment. *Electronics* **2021**, *10*, 2703. [CrossRef]
14. Schwarting, W.; Alonso-Mora, J.; Paull, L. Safe nonlinear trajectory generation for parallel autonomy with a dynamic vehicle model. *IEEE Trans. Intell. Transp. Syst.* **2017**, *19*, 2994–3008. [CrossRef]
15. Guo, H.; Liu, F.; Xu, F. Nonlinear model predictive lateral stability control of active chassis for intelligent vehicles and its FPGA implementation. *IEEE Trans. Syst. Man Cybern. Syst.* **2017**, *49*, 2–13. [CrossRef]
16. Bujarbaruah, M.; Zhang, X.; Borrelli, F. Adaptive MPC with chance constraints for FIR systems. In Proceedings of the Annual American Control Conference, Milwaukee, WI, USA, 27–29 June 2018.
17. Kim, E.; Kim, J.; Sunwoo, M. Model predictive control strategy for smooth path tracking of autonomous vehicles with steering actuator dynamics. *Int. J. Automot. Technol.* **2014**, *15*, 1155–1164. [CrossRef]
18. Shen, C.; Guo, H.; Liu, F. MPC-based path tracking controller design for autonomous ground vehicles. In Proceedings of the 36th Chinese Control Conference, Dalian, China, 26–28 July 2017.
19. Law, C.K.; Dalal, D.; Shearrow, S. Robust model predictive control for autonomous vehicles/self driving Cars. *arXiv* **2018**, arXiv:1805.08551.
20. Piga, D.; Formentin, S.; Bemporad, A. Direct data-driven control of constrained systems. *IEEE Trans. Control Syst. Technol.* **2017**, *26*, 1422–1429. [CrossRef]
21. Pang, H.; Liu, N.; Hu, C.; Xu, Z. A practical trajectory tracking control of autonomous vehicles using linear time-varying MPC method. *Proc. Inst. Mech. Eng. Part D J. Automob. Eng.* **2022**, *236*, 709–723. [CrossRef]
22. Chen, Y.; Hu, C.; Wang, J. Human-centered trajectory tracking control for autonomous vehicles with driver cut-in behavior prediction. *IEEE Trans. Veh. Technol.* **2019**, *68*, 8461–8471. [CrossRef]
23. Hang, P.; Huang, S.; Chen, X. Path planning of collision avoidance for unmanned ground vehicles: A nonlinear model predictive control approach. *Proc. Inst. Mech. Eng. Part I J. Syst. Control Eng.* **2021**, *235*, 222–236. [CrossRef]
24. Guerreiro, B.J.; Silvestre, C.; Cunha, R. Trajectory tracking nonlinear model predictive control for autonomous surface craft. *IEEE Trans. Control Syst. Technol.* **2014**, *22*, 2160–2175. [CrossRef]
25. Shen, C.; Buckham, B.; Shi, Y. Modified C/GMRES algorithm for fast nonlinear model predictive tracking control of AUVs. *IEEE Trans. Control Syst. Technol.* **2016**, *25*, 1896–1904. [CrossRef]
26. Guo, N.; Lenzo, B.; Zhang, X. A real-time nonlinear model predictive controller for yaw motion optimization of distributed drive electric vehicles. *IEEE Trans. Veh. Technol.* **2020**, *69*, 4935–4946. [CrossRef]
27. Kuhne, F.; Lages, W.F.; da Silva, J.G., Jr. Model predictive control of a mobile robot using linearization. *Proc. Mechatron. Robot.* **2004**, *4*, 525–530.
28. Pang, H.; Zhang, X.; Yang, J. Adaptive backstepping-based control design for uncertain nonlinear active suspension system with input delay. *Int. J. Robust Nonlinear Control* **2019**, *29*, 5781–5800. [CrossRef]

MDPI

Article

Leader-Based Trajectory Following in Unstructured Environments—From Concept to Real-World Implementation

Georg Nestlinger [1,*], Johannes Rumetshofer [1,2] and Selim Solmaz [1]

1 Virtual Vehicle Research GmbH, Inffeldgasse 21a, 8010 Graz, Austria; johannes.rumetshofer@v2c2.at (J.R.); selim.solmaz@v2c2.at (S.S.)
2 Institute of Automation and Control, Graz University of Technology, 8010 Graz, Austria
* Correspondence: georg.nestlinger@v2c2.at

Abstract: In this paper, the problem of vehicle guidance by means of an external leader is described. The objective is to navigate a four-wheeled vehicle through unstructured environments, characterized by the lack of availability of typical guidance infrastructure like lane markings or HD maps. The trajectory-following approach is based on an estimate of the leader's path. For that, position measurements are stored over time with respect to an inertial frame. A new strategy is proposed to rate the significance of position measurements and ensure that a certain threshold of stored samples is not exceeded. Having an estimate of the leader path is essential to prevent the cutting-corner phenomenon and for exact path following in general. A spline-approximation technique is applied to obtain a smooth reference path for the underlying lateral and longitudinal motion controllers. For longitudinal tracking, a constant time-headway policy was implemented, to follow the leader with a constant time gap along the estimated path. The algorithm was first developed and tested in a simulation framework and then deployed in a demonstrator vehicle for validation under real operating conditions. The presented experimental results were achieved using only on-board sensors of the demonstrator vehicle, while high-accuracy differential GPS-based position measurements serve as the ground truth data for visualization.

Keywords: vehicle following; path following; path tracking; splines; spline approximation

Citation: Nestlinger, G.; Rumetshofer, J.; Solmaz, S. Leader-Based Trajectory Following in Unstructured Environments—From Concept to Real-World Implementation. *Electronics* **2022**, *11*, 1866. https://doi.org/10.3390/electronics11121866

Academic Editors: Giuseppe Prencipe and Umar Zakir Abdul Hamid

Received: 6 May 2022
Accepted: 9 June 2022
Published: 13 June 2022

1. Introduction

Automated-driving solutions and ADAS (Advanced Driver Assistance System) functions promise many advantages for driving comfort and safety, as a result of a reduced need for driver attention. While fully autonomous driving in the sense of SAE Level-5 [1] is far away in the distant future, many manufacturers are already bringing ADAS solutions such as emergency braking, adaptive cruise control, or lane keeping to their vehicles, due to the obvious safety and comfort benefits. However, there are other use cases for automated driving, which are not typically considered in everyday-driving scenarios. One such use case is the navigation of a vehicle by means of an external leader through an unstructured environment, such as a construction site, or in a convoying application, where trucks need to transfer loads several times. For an agriculture use case refer to [2]. In such environments, infrastructure elements like lane markings, vehicle-to-infrastructure communication or HD maps are not available to serve as a source for the automated vehicle's reference path. Lane markings also often exhibit the drawback of poor visibility due to wear and tear, occlusion by other vehicles, or lighting and weather conditions that result in an intermittent perception.

1.1. Problem Statement

The starting point is a convoy of two vehicles, where the preceding leader vehicle is manually driven, and its driven path defines the reference path for the automated follower vehicle. It is the leader's responsibility to avoid obstacles and take a collision-free path. Consequently, the follower's task is to track the leader's path as close as possible, to keep a safe distance to obstacles. The follower's longitudinal motion has to be controlled such that a safe distance is kept from the leader. For the special case of constant leader speed, a constant distance has to be maintained. For the practical implementation, two additional requirements need to be satisfied:

(R1) The vehicle-following system must not rely on vehicle-to-vehicle communication. Consequently, it was decided to only rely on on-board sensors for leader observations as well as follower-state estimation.

(R2) The vehicle-following system must not rely on GPS to avoid narrowing the driving function's operational-design domain to environments where undisturbed communication to GPS satellites can be guaranteed.

1.2. Literature Review

The problem statement of tracking a leader vehicle's trajectory dates back to at least 1998. In the framework of platooning, [3] discusses the drawbacks of direct-vehicle following, i.e., following the current leader position. Using this approach, the path between the follower and leader needs to be interpolated, causing the autonomous follower vehicle to deviate from the leader vehicle's path. Considering e.g., a straight-line interpolation [4], this effect scales with the distance from the leader vehicle. To overcome this issue, the authors propose an algorithm that makes use of the time history of stored leader positions. Results from field tests on a circular path, entered via a clothoid transition curve, showed a reduction of the maximum lateral deviation from approximately 1.2 m to 0.8 m, by applying the proposed algorithm. Unfortunately, a comparison of the leader/follower path as well as the exact path-tracking error model used for the implementation are not presented, making interpretation of and benchmarks against the results difficult.

A military convoying scenario in [5,6] motivates the goal of following a leader vehicle's trajectory with large inter-vehicle spacing without cutting corners. The authors propose following the leader vehicle by a constant time delay utilizing the stored trajectory of the leader obtained from on-board sensors consisting of a camera, a heading gyroscope, and wheel encoders. To deal with noisy measurements, cubic splines are proposed, motivated by their property of minimum curvature. The actual fitted curve is constructed as a weighted sum of identical splines and introduces parameters for the spline width, the separation between each spline, and the number of splines. Unfortunately, the strategy to obtain the leader trajectory in the first place is not covered in [5,6].

The issue of measurement dropouts that is often experienced with vision sensors is addressed in [7]. The authors apply a particle filter, utilizing observations from a vision sensor and the odometry data of the follower/leader pair, to estimate the leader's trajectory. A simple trajectory-reconstruction approach is presented in [8], basically implementing a first-in–first-out buffer where the oldest measurement is removed from the list of measurements before the current measurement is added. Unfortunately, both works [7,8] only present results obtained from simulations lacking experimental data.

With a focus on lateral-string stability in platooning applications, [9] also proposes a path-following approach over the direct-vehicle following approach, to avoid cutting corners. To estimate the path of the leader vehicle, position measurements are stored over time considering the following vehicle's moving reference frame. The actual reference path for the underlying lateral controller is constructed via polynomial fitting. A third order polynomial was chosen as this seemed "to be a good compromise between having enough degree of freedom to describe the actual path and filtering out the measurement noise" [9] (p. 61). In contrast to the works discussed above, the process to obtain the leader path is presented in quite a lot of detail. The stored leader measurements are maintained based on a first-in–first-out method similar to [8].

A sophisticated approach to generate a continuous-curvature trajectory using Sequential Quadratic Programming (SQP) is presented in [10]. The vehicle's maximum curvature, limits of the lateral and longitudinal acceleration, and the vehicle's maximum steering rate are considered in the optimization problem. The generated trajectory is represented as a sequence of clothoid arcs, with cartesian coordinates that are defined by Fresnel integrals. Unfortunately, Fresnel integrals cannot be solved analytically, which complicates the calculation of path-tracking errors. Although there are approximations for the Fresnel integrals [11], we decided to come up with another approach, also due to the fact that this topic was not addressed in [10].

An "integrated longitudinal and lateral control framework in vehicle following scenarios" is presented in [12]. In contrast to the work at hand, they use a radar sensor instead of a camera sensor to measure the inter-vehicle distance, velocity difference, and azimuth of the leader vehicle. Furthermore, the leader's velocity and yaw rate buffered at the host vehicle are obtained via vehicle-to-vehicle (V2V) communication. The leader path is then calculated backwards iteratively and tracked by a model-predictive controller (MPC).

There are further publications related to vehicle following. In contrast to the work at hand, they and the above discussed works differ at least in one of these points:

- They focus on the design of lateral and/or longitudinal controllers [5,12–15].
- They rely on V2V communication [12–16] or GPS [6].
- They lack experimental results [7,8,13–15].
- They focus on the design of a camera system for leader tracking [2], handling intermittent vision observations [7], or state-estimation techniques [16].

With this work, mainly two shortcomings observed in the literature are addressed with the following contributions:

1. A detailed explanation to obtain an estimate of the leader path from the on-board measurements, which are stored in a buffer, is given. To support the process of obtaining and maintaining the stored leader path, a new strategy to consider the importance of each measured sample for the estimated path is proposed. (Section 2.1)
2. To handle noisy position measurements, we propose smoothing the estimated path resulting from (1) via a computationally inexpensive spline-approximation approach. (Section 2.2)

In this respect, the most relevant existing works are [8,9,12] regarding contribution (1) and [9,10] regarding contribution (2).

1.3. *Structure of the Article*

The overall system architecture of the vehicle-following system is presented in Section 2. The main components are covered in detail in Section 2.1 (path estimation) and Section 2.2 (path smoothing). The control architecture is presented in Section 2.3. Section 3 deals with vehicle deployment of the proposed vehicle-following system, starting with the demonstrator vehicle in Section 3.1, state estimation in Section 3.2, leader-vehicle selection in Section 3.3, and experimental results in Section 3.4. The conclusion and outlook are given in Section 4.

2. System Architecture and Design

The overall system architecture of the developed vehicle-following system is shown in Figure 1. The *Object Sensor* attached to the follower vehicle is assumed to provide relative position and speed measurements $(\Delta s_x, \Delta s_y)$ and $(\Delta v_x, \Delta v_y)$ of the leader vehicle. The *Planner* component estimates the leader path and applies a spline-approximation algorithm to achieve a certain geometric continuity. Based on the state-of-the-art approach outlined in Section 1.2, the estimate of the leader path is obtained by storing the position measurements over time. Additionally, the relevance of each position measurement, with respect to the estimated leader path, is taken into account. Since the position measurements are in relation to the object sensor's frame, the motion of the follower needs to be taken into account. We decided to state the leader path with respect to an inertial frame, making it necessary to estimate the follower's position and orientation. As indicated by the *State Estimation* block and in compliance with the requirements from Section 1.1, position estimates $(\hat{x}, \hat{y})$ and orientation estimate $\hat{\psi}$ as well as estimates of lateral and longitudinal velocity $\hat{v}_x$ and $\hat{v}_y$ were obtained from the on-board vehicle measurements speed v, yaw rate $\dot{\psi}$, and steering angle δ.

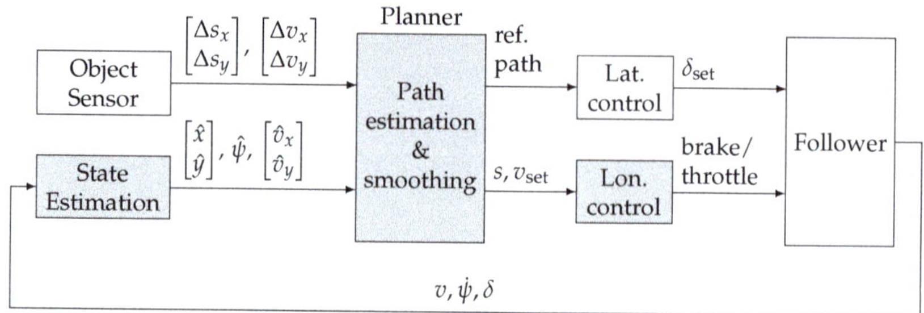

Figure 1. System architecture of the developed vehicle-following system. Shaded components are addressed in this paper, while the main contribution is regarding the Planner component.

The estimated leader path then servers as the reference path for the lateral and longitudinal controller. In addition to the control law, the lateral control component also implements the related error model, which calculates the control errors based on the reference path and the current vehicle position and orientation. The longitudinal controller maintains a constant time-gap to the leader by regulating the follower's speed v and the inter-vehicle distance s. Since the follower aims to track the estimated path as close as possible, it is reasonable to determine the distance s to the leader along the path. For that, a Frenet transformation is applied to the follower position with respect to the reference path.

The corresponding control signals, in terms of desired steering angle δ_{set} and pedal actuations, are then passed to the vehicle. For further details on the vehicle interfaces refer to Section 3.1.

2.1. Leader-Path Estimation

This section details the procedure to obtain the leader path from a series of position measurements. For that, the first step is to convert the position measurements from the sensor frame to an inertial frame, considering the motion of the follower vehicle. Given a follower position (x, y) and heading ψ, this is achieved by a rotation and translation according to

$$p = \begin{bmatrix} x \\ y \end{bmatrix} + \begin{bmatrix} \cos(\psi) & -\sin(\psi) \\ \sin(\psi) & \cos(\psi) \end{bmatrix} \begin{bmatrix} \Delta s_x \\ \Delta s_y \end{bmatrix}. \tag{1}$$

In the second step, these measurements are collected over time, resulting in a list

$$\boldsymbol{P}(t) = (\boldsymbol{p}_1, \boldsymbol{p}_2, \ldots, \boldsymbol{p}_N) \tag{2}$$

of N points $\boldsymbol{p}_i$, $i = 1, \ldots N$, at time t. The following is worth noting here:

- To guarantee the real-time capability of this strategy, the number N of points in the list $\boldsymbol{P}$ must not exceed an upper bound $\overline{N}$, i.e., $N \leq \overline{N}$, where $\overline{N}$ depends on the memory and computing resources of the target hardware.
- New position measurements might not contain relevant information in terms of the estimated path, due to low speed or standstill maneuvers and/or measurement noise. That is, a new point $\boldsymbol{p}_{N+1}$ is equal or very close to the most recent point $\boldsymbol{p}_N \in \boldsymbol{P}$.

The measure to address both constraints is motivated by a point-reduction algorithm [17] that determines the "importance" of a point within a polygonal chain, by the area of the triangle created by the point itself and its immediate neighbors. According to that, the point $\boldsymbol{p}_{N+1}$ is appended to the list $\boldsymbol{P}$ if the area

$$A = 0.5\left|\det\left(\begin{bmatrix} \boldsymbol{p}_{N-1} & \boldsymbol{p}_N & \boldsymbol{p}_{N+1} \\ 1 & 1 & 1 \end{bmatrix}\right)\right| \tag{3}$$

of the triangle formed by the points $\boldsymbol{p}_{N-1}, \dots, \boldsymbol{p}_{N+1}$ exceeds a certain threshold $\overline{A}$, otherwise $\boldsymbol{p}_N$ is replaced by $\boldsymbol{p}_{N+1}$. In the case of $N = \overline{N}$ and $A > \overline{A}$, $\boldsymbol{p}_{N+1}$ is appended to the list $\boldsymbol{P}$ after removing the point associated with the smallest area.

This procedure results in a list $\boldsymbol{P}(t)$ that represents the leader's path as a polygonal chain, with respect to an inertial frame at a certain time t. This polygonal path could already serve as the reference path for the underlying path-tracking controller. However, path-tracking controllers usually require geometric derivatives of the path such as heading and curvature. From that perspective, the polygonal path representation is disadvantageous as it is only G^0 continuous. For more details regarding the impact of specific path representations on the path-tracking performance, refer to [18].

2.2. Smoothing the Estimated Path

To achieve a specific continuity of the estimated path, an analytic representation satisfying the corresponding continuity requirements is required. For that, a parametric path

$$\gamma : \mathbb{R} \to \mathbb{R}^2, \quad \tau \mapsto \gamma(\tau) \tag{4}$$

with path parameter τ is advantageous, as it allows the definition of arbitrary paths. There are two basic possibilities to obtain $\gamma(\tau)$: interpolation or approximation of the underlying data points. The fact that an interpolant would pass through every point makes the first one an inappropriate choice, especially considering the presence of measurement noise. On the other hand, similar arguments can be made to justify the approximation approach.

Eventually, a spline-approximation algorithm [19] that relies on least-squares adjustment was implemented. Based on the estimated leader path given by (2), the algorithm calculates a parametric, two-dimensional spline

$$\boldsymbol{\Gamma}(\tau) = \begin{cases} \gamma_1(\tau) & \tau_0 \le \tau < \tau_1 \\ \vdots & \\ \gamma_n(\tau) & \tau_{n-1} \le \tau \le \tau_n \end{cases} \tag{5}$$

of n spline segments γ_i with $n+1$ strictly monotonic breaks $\tau_0 < \tau_1 < \cdots < \tau_n$ (also known as knots), where the degree k for the spline segments

$$\gamma_i(\tau) = \begin{bmatrix} \chi_{i,k}\,\tau^k + \cdots + \chi_{i,1}\,\tau + \chi_{i,0} \\ v_{i,k}\,\tau^k + \cdots + v_{i,1}\,\tau + v_{i,0} \end{bmatrix} \tag{6}$$

as well as the order l of geometric continuity G^l at the breaks τ_i can be specified. This requires solving a system

$$\boldsymbol{A}\boldsymbol{X} = \boldsymbol{B}, \quad \boldsymbol{A} \in \mathbb{R}^{\alpha \times \alpha}, \quad \boldsymbol{B} \in \mathbb{R}^{\alpha \times 2} \quad \text{and} \quad \alpha = (k+1)n + (l+1)(n-1) \tag{7}$$

of linear equations, which can be easily achieved under real-time requirements. Matrix $\boldsymbol{A}$ is composed of the independent variable of the underlying data, i.e., the path parameter τ, while $\boldsymbol{B}$ is composed of the corresponding dependent variable, i.e., $\boldsymbol{p}_i(\tau)$. The unknown polynomial coefficients $\chi_{i,j}$ and $v_{i,j}$, $j = 0, \dots, k$, are represented by $\boldsymbol{X}$. Notice that (7) could be extended to higher dimensional paths, e.g., 3D paths, at almost no computational cost, since the inversion of matrix $\boldsymbol{A}$ is independent of the number of columns of $\boldsymbol{B}$, i.e., the number of dimensions. Furthermore it is worth noting, that the size of $\boldsymbol{A}$ does not depend on the number of data samples per spline segment, but only on the number n of spline segments, the polynomial degree k and the order l of geometric continuity. In addition to [19], the spline-approximation algorithm was extended by the possibility to also specify boundary conditions for the differential continuity of the path's (5) terminal points, $\boldsymbol{\Gamma}(\tau_0)$ and $\boldsymbol{\Gamma}(\tau_n)$.

Choosing the breaks τ_i in (5) is a research topic on its own [20,21]. For the presented work, the calculation of a spline segment γ_i is triggered if the list (2) contains a certain number ν of points, i.e., $N = \nu \leq \overline{N}$. The points $\boldsymbol{p}_1, \ldots, \boldsymbol{p}_\nu$ are then removed from the list $\boldsymbol{P}$ and the procedure is repeated if the condition $N = \nu$ is satisfied again, and so on. This strategy has two advantages at the cost of one disadvantage over calculating the whole spline $\boldsymbol{\Gamma}$ for all N points each iteration:

- Unfortunately, the resulting spline from the spline-approximation algorithm, according to [19], lacks the local support property. That is, a variation of a single data point not only affects the related spline segment but all spline segments, if just a single point of the underlying approximation data changes. Considering that the spline $\boldsymbol{\Gamma}$ serves as the reference path for the path-tracking controller, as shown in Figure 1, this could result in jumps of the control reference and eventually of the control error every time the spline is updated. Although there are controllers that implement bumpless transfer functionality that could handle steps in the control reference to some extent [22], this is usually not the case; see, for example, the well known Stanley [23] and Pure-Pursuit [24] path-tracking controllers. Therefore, the iterative approach mentioned above was implemented, since it enables the extension of the spline path $\boldsymbol{\Gamma}$ by a spline segment γ_{i+1}, while letting the segments $\gamma_1, \ldots, \gamma_i$ be unaffected.
- According to (7), the computational effort mainly depends on the number n of spline segments, considering k and l as fixed. Therefore, the computational effort can be reduced compared to calculating the spline $\boldsymbol{\Gamma}$ for all points $\boldsymbol{p}_1, \ldots, \boldsymbol{p}_N$ during each iteration, depending on the actual values of ν and N.
- These improvements are at the cost of only achieving continuous zeroth derivatives at the spline's breaks. Simulation results showed that requesting higher order continuity at the breaks results in a spline approximation becoming unstable over time. During field tests, this disadvantage turned out to be neglectable.

This smoothing strategy is demonstrated in Figure 2 based on real-world measurement data, where the leader/follower convoy was driving from the top-right corner to the bottom-left corner of the graph.

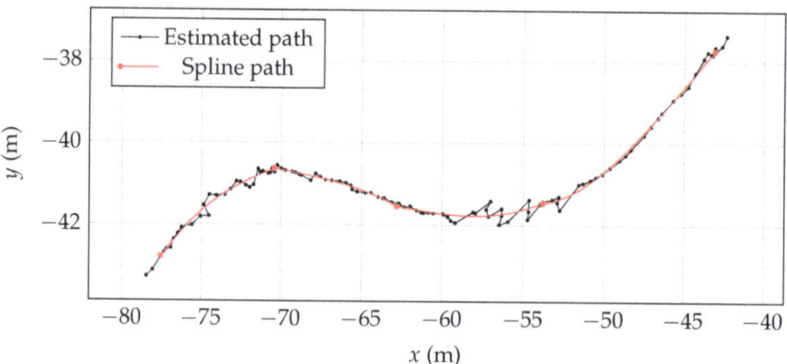

Figure 2. Estimated path from sensor measurements and resulting spline path after applying the spline-approximation algorithm.

The black dotted line shows the estimated path, according to Section 2.1, where each dot represents a position measurement obtained from a camera sensor. The red line shows the resulting spline consisting of four spline segments, where the segment's terminal points are highlighted by dots. Although the spline is only G^0 continuous at the breaks, the overall smoothness is significantly improved in contrast to the estimated path.

2.3. Vehicle Control

From a control-design point of view and according to Figure 1, the lateral and longitudinal motion of the follower vehicle was considered to be decoupled. The corresponding controllers are presented in the following.

2.3.1. Lateral Control

The theoretical background of the lateral control component used in the context of the presented vehicle-following system was published in [22,25]. Since then, it was successfully applied to several practical implementations such as lane keeping [26]. Therefore, only the essence is recapitulated here.

Based on the linear single-track model [27] and a linearized and time-discretized path-tracking error model [25,26], a state-feedback control law

$$\delta_{\text{set}} = -\begin{bmatrix} \hat{v}_y & \dot{\psi} & e_{\text{lat}} & e_{\psi} & \chi \end{bmatrix} \boldsymbol{k}_{\text{lat}} \tag{8}$$

is obtained via LQR design. The path-tracking error model, according to the classification introduced in [18], is as follows: The reference is located in the vehicle's center of gravity, the look-ahead distance is along the vehicle heading, while the lateral error e_{lat} and heading error e_{ψ}, with respect to the reference path, are perpendicular to the vehicle heading. To account for a varying vehicle speed and a speed-dependant look-ahead distance, the feedback law (8) was gain scheduled and implemented using a look-up table. The dynamics of the demonstrator vehicle's steering actuator were considered by a first-order transfer function [26], introducing one additional state χ.

The choice of the actual path-tracking controller was predetermined by the related error model, which should fit the problem statement. For the present application, the task is to follow the leader path as close as possible, to avoid cutting curves. As this requirement can only be fulfilled for a single point of a four-wheeled vehicle, a reasonable choice is the center of gravity along the vehicle's longitudinal axis, in conjunction with zero look-ahead distance. Given this configuration, a vanishing control error would imply that the center of gravity perfectly tracks the reference path. Unfortunately, the look-ahead distance cannot be decreased arbitrarily as this also reduces the closed-loop phase margin [28] and can eventually lead to instability. The actual values used during deployment are given in Section 3.

It is also worth noting, that with the presented design, it would have been easily possible to replace the current controller with any other controller based on the same error model. A classification of path-tracking controllers from the literature, regarding their underlying error model, is given in Table 2 of [18].

2.3.2. Longitudinal Control

The task of the longitudinal controller is to maintain a constant time gap t_{h} from the leader vehicle, according to the constant time-headway policy [29]

$$s_{\text{set}} = \max\{v_{\text{set}} t_{\text{h}}, s_{\min}\} \tag{9}$$

given a desired speed v_{set} and a minimum safety distance $s_{\min}$. The desired speed v_{set} refers to the leader vehicle's speed and was estimated according to

$$v_{\text{set}} = \left\| \begin{bmatrix} \hat{v}_x + \Delta v_x - \dot{\psi} r \sin(\theta) \\ \hat{v}_y + \Delta v_y + \dot{\psi} r \cos(\theta) \end{bmatrix} \right\| \tag{10}$$

from on-board measurements, where

$$r = \sqrt{\Delta s_x^2 + \Delta s_y^2} \tag{11a}$$

$$\theta = \text{atan2}(\Delta s_y, \Delta s_x) \tag{11b}$$

are the distance r and bearing θ to the leader.

Considering that the distance s from the follower to the leader vehicle is negative, proportional with respect to the follower speed v, the follower's longitudinal dynamics can be modeled as

$$\dot{x} = \begin{bmatrix} 0 & -1 \\ 0 & 0 \end{bmatrix} x + \begin{bmatrix} 0 \\ 1 \end{bmatrix} a, \tag{12}$$

with acceleration a and state $x = [s \quad v]^t$. The distance s to the leader is with respect to the reference path and determined via a Frenet transformation of the ego vehicle's position. The error model then reads as

$$e = x_{\text{set}} - x \tag{13}$$

with distance error $e_s = s_{\text{set}} - s$ and velocity error $e_v = v_{\text{set}} - v$. For the model (12), a controller

$$a_{\text{set}} = k_{\text{lon}}^t e \tag{14}$$

can be designed via, e.g., pole placement.

The follower's longitudinal-control strategy was eventually implemented as a cascaded control loop, according to Figure 3, where the desired acceleration (14) is tracked by the inner loop, implementing a PI controller.

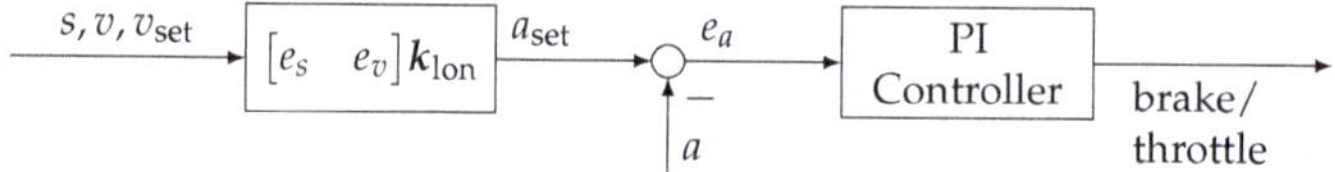

Figure 3. Longitudinal dynamics control structure consisting of an outer control loop that regulates spacing error e_s and speed error e_v as well as an inner loop regulating acceleration error e_a.

This architecture was mainly motivated by the brake and throttle interfaces available in the test vehicle, in terms of pedal positions.

3. Test-Vehicle Integration and Field Trials

Until here, the vehicle-following system was mainly presented from a simulation-based perspective. This section discusses practical aspects related to the deployment of the proposed following system and also presents test results obtained under real operating conditions utilizing a series-production demonstrator car.

3.1. Demonstrator Vehicle

The demonstrator vehicle is a Ford Mondeo 2.0 Hybrid equipped with additional sensors and hardware; an overview is shown in Figure 4.

A so-called *ADAS Kit* from Dataspeed Inc. gives access to the vehicle's CAN-bus. This allows for control of the steering wheel angle as well as brake and throttle via pedal-position commands. It also provides on-board sensor measurements like speed, acceleration, yaw rate, and series-production GPS readings.

The developed algorithms were deployed to a dSPACE MicroAutoBox II, equipped with a DS1401 processor board and 1511/1512 I/O boards. This setup served as the real-time control hardware executing the planning, state estimation, and vehicle-control algorithms. Additionally, RTMaps running on an industrial-grade PC was used as the interface to the Novatel differential GPS. This data was then transmitted to the MicroAutoBox via ethernet. Since the Mobileye 630 camera sensor provided the most robust object detection and classification, it was used during the field tests.

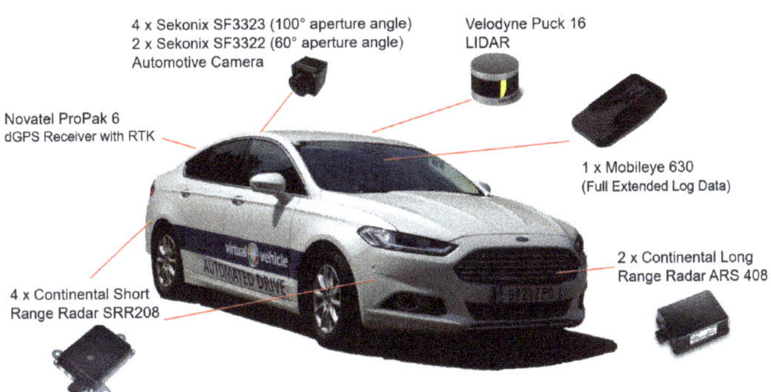

Figure 4. Ford Mondeo demonstrator vehicle with additionally installed sensors. For the implementation presented in this paper, only the Mobileye 630 was used.

3.2. State Estimation

As pointed out in Section 2, estimates of the followers position $(\hat{x}, \hat{y})$ and orientation $\hat{\psi}$ with respect to an inertial frame are required. This was accomplished by a dead-reckoning approach using a kinematic vehicle model [30] (p. 26)

$$\hat{x}(t) = \int_0^t v(\xi)\cos(\hat{\beta}(\xi) + \hat{\psi}(\xi))\,\mathrm{d}\xi \tag{15a}$$

$$\hat{y}(t) = \int_0^t v(\xi)\sin(\hat{\beta}(\xi) + \hat{\psi}(\xi))\,\mathrm{d}\xi \tag{15b}$$

$$\hat{\psi}(t) = \int_0^t \dot{\psi}(\xi)\,\mathrm{d}\xi \tag{15c}$$

$$\hat{\beta}(t) = \mathrm{atan2}(l_\mathrm{r}\tan(\delta(t)),\, l_\mathrm{f} + l_\mathrm{r}). \tag{15d}$$

In compliance with the requirements from Section 1.1, the implementation of (15) only requires measurements from on-board sensors, namely the vehicle speed v, the front-wheel steering angle δ, and the yaw rate $\dot{\psi}$. Additionally, the front and rear axle offset from the center of gravity, denoted by l_f and l_r, are required.

Over time, this approach causes the estimates to deviate from the true position (x, y) and true orientation ψ. Assuming that the follower is tracking the leader vehicle with a steady-state time gap t_h, the error between the actual and estimated values must be sufficiently small. In other words, the estimation error that accumulated during the time between obtaining a specific leader position until reaching this position is required to be small, to ensure satisfactory path tracking. This also implies that an upper bound for the time gap t_h exists, depending on the estimation accuracy of the follower's position and orientation.

3.3. Leader Selection

To simplify the development of the vehicle-following system, the task of identifying the leader vehicle from a list of multiple objects was neglected in simulations. Instead, just a single object was used as the leader vehicle. For the practical implementation during the field tests, object sensors mounted to the test vehicle were used for object detection. These sensors, like a camera, radar, or lidar, typically provide object lists containing not only the leader vehicle but also stationary objects or ghost objects. Therefore, a measure to identify the actual leader vehicle utilizing the spline path (5) was implemented.

From the path estimation and smoothing, an analytic formula of the leader path is available, given by the spline (5). Taking advantage of its analytic form, the path can easily be extrapolated by means of path parameter τ. For the actual implementation, the path parameter τ, referring to the arc length and the breaks $\tau_0 < \tau_1 < \cdots < \tau_n$, was obtained from the length of the polygonal path formed by the position measurements (2).

For the field tests, the actual path extrapolation distance was obtained by overestimating the traveled distance Δs of the leader vehicle, according to $\Delta s = v_{\text{set}} T_{\text{s}}$, where v_{set} is the estimated leader speed (10) and T_{s} is the sample time of the vehicle-following system. Given the extrapolated leader path and a list of objects, the leader vehicle was chosen as the closest object to the follower vehicle that intersects the extrapolated path. At initialization of the vehicle-following system, the estimated path is not available and the leader vehicle is selected according to the object type.

3.4. Experimental Results

The vehicle-following system was verified on a graveled land area at the campus of the Technical University of Graz, Austria. The tests were performed using two of Virtual Vehicle's Automated Drive demonstrators; the automated follower vehicle is shown in Figure 4 and a bird's-eye view of the test setup is shown in Figure 5.

Figure 5. Bird's-eye view of the leader (left) and follower (right) vehicle setup, taken during the field tests. Video footage is available online at https://youtu.be/0EnHqTouluc (Accessed on 1 May 2022).

Both the leader vehicle and the follower vehicle were equipped with DGPS to provide ground truth data. The vehicle-following system was executed on real-time hardware, as mentioned in Section 3.1, executing at a sample time of 20 ms. The complete list of parameters is stated in Table 1.

Table 1. Parameter values used throughout the field trials.

Name	Symbol	Value
Sample time	T_{s}	20 ms
Max. number points	$\overline{N}$	100
Area threshold	$\overline{A}$	1×10^{-4} m^2
Min. clearance	s_{min}	5 m
Time gap	t_{h}	2 s
Look-ahead time	–	300 ms
Samples per spline segment	ν	12
Polynomial degree	k	3
Geometric continuity	l	2

The test procedure was as follows: both vehicles were at a standstill, with the leader vehicle in front of the follower vehicle and within the object sensor's field of view. After confirming leader detection, both vehicles started driving manually, meaning the follower vehicle needed to start idling before enabling the drive-by-wire mode.

Results of one exemplary trial are shown in Figure 6. Figure 6a shows a comparison of the leader and follower paths, while Figure 6b shows qualitative lateral and longitudinal control signals.

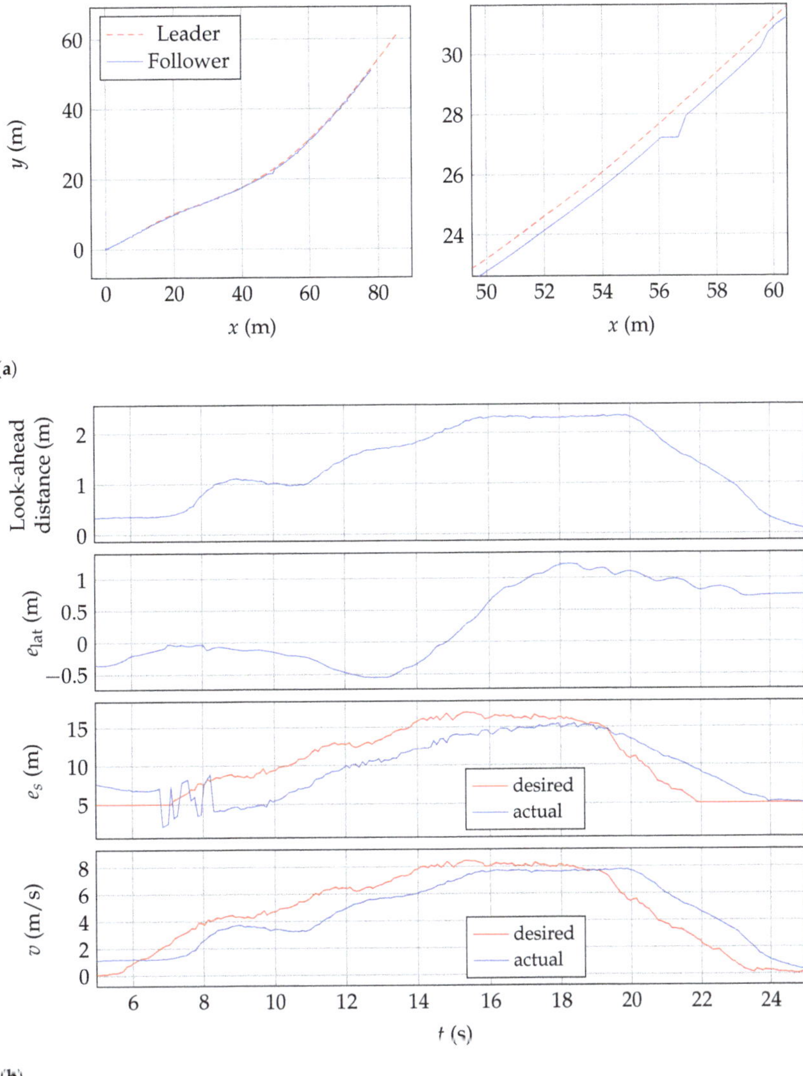

Figure 6. Results from an exemplary field trial run. (**a**) Leader and follower path obtained from differential GPS. The close-up (right figure) also shows an outlier around coordinate (57, 27). Due to its minor character and time constraints, we did not investigate the root cause. (**b**) Speed-dependent look-ahead distance, related lateral-tracking error, and longitudinal control signals in terms of distance and speed.

It should be pointed out, that the closed-loop lateral error e_{lat} in Figure 6b is with respect to the reference path obtained from the spline-approximation algorithm. Since this path is an approximation of the estimated leader path, which is itself an estimate of the true leader path, the lateral error e_{lat} does not reflect the follower's offset from the true leader path. However, it reflects the performance of the path-tracking controller. According to the follower's speed, the look-ahead distance varies during the trial. Although the related lateral-path tracking error e_{lat} varies between roughly -0.5 m to 1 m, the lateral offset with respect to the follower's center of gravity stayed well between -0.4 m to 0.4 m (Figure 6a). Around 7 s to 8 s, the actual inter-vehicle distance s shows noisy characteristics. At this point in time, the follower reached the leader's initial position, which was starting from a standstill. Due to that, several similar position measurements were added to the list (2), although the leader was not moving. This could have been avoided by tuning the area threshold parameter $\overline{A}$ with respect to the noise characteristics of the object sensor as well as the follower state estimation. Unfortunately, time constraints did not allow further investigation of this topic, as these effects are not easily reproducible during field trials. For future tests, recreating these effects in simulation to tune the related parameters beforehand is planned.

4. Conclusions and Outlook

In this work we have presented the concept, architecture, and real-world implementation of a state-of-the-art vehicle following a system relying on on-board sensors only. For detection of the leader vehicle, an optical sensor mounted to the follower vehicle was used. To obtain an estimate of the leader path, a new algorithm was proposed that considers the importance of new leader measurements with respect to the currently estimated path.

The smoothness of this estimated path was improved by a spline-approximation algorithm, which closes the gap between simple polynomial-fitting approaches [9] and computational-demanding ones, like presented in [10]. For the proof of concept of the proposed algorithms, field trials were performed on a graveled area achieving path-tracking errors between $-0.4\,\mathrm{m}$ to $0.4\,\mathrm{m}$. We assume that these values can be improved by either tuning the existing path-tracking controller or implementing a more sophisticated one, which would be easily possible with the presented architecture.

For the future, it is planned to perform more extensive field trials utilizing various sensors for object detection. Another branch of investigation is regarding the path-tracking controller. The generic interface between the planning and the path-tracking component allows to benchmark various controllers from the literature and investigate the influence of their specific error models.

Author Contributions: Conceptualization, G.N. and J.R.; software, G.N.; validation, G.N.; writing—original draft preparation, G.N.; writing—review and editing, G.N., J.R. and S.S.; visualization, G.N.; supervision, G.N. All authors have read and agreed to the published version of the manuscript.

Funding: The publication was written at Virtual Vehicle Research GmbH in Graz, Austria. The authors would like to acknowledge the financial support within the COMET K2 Competence Centers for Excellent Technologies from the Austrian Federal Ministry for Climate Action (BMK), the Austrian Federal Ministry for Digital and Economic Affairs (BMDW), the Province of Styria (Dept. 12), and the Styrian Business Promotion Agency (SFG). The Austrian Research Promotion Agency (FFG) has been authorized for the program management.

Acknowledgments: The authors would like to express their thanks to the supporting industrial partner Infineon Austria.

Conflicts of Interest: The authors declare no conflict of interest.

Abbreviations

The following abbreviations are used in this manuscript:

CAN	Controller Area Network
DGPS	Differential Global Positioning System
GPS	Global Positioning System
HD	High Definition
LQR	Linear Quadratic Regulator
MPC	Model Predictive Control
PI	Proportional-Integral
SQP	Sequential Quadratic Programming
V2V	Vehicle-to-Vehicle

References

1. On-Road Automated Driving (ORAD) Committee. *Taxonomy and Definitions for Terms Related to Driving Automation Systems for On-Road Motor Vehicles*; Technical Report; SAE International: Warrendale, PA, USA, 2021. [CrossRef]
2. Zhang, L.; Ahamed, T.; Zhang, Y.; Gao, P.; Takigawa, T. Vision-Based Leader Vehicle Trajectory Tracking for Multiple Agricultural Vehicles. *Sensors* **2016**, *16*, 578. [CrossRef] [PubMed]
3. Gehrig, S.K.; Stein, F.J. A trajectory-based approach for the lateral control of car following systems. SMC'98 Conference Proceedings. In Proceedings of the 1998 IEEE International Conference on Systems, Man, and Cybernetics (Cat. No.98CH36218), San Diego, CA, USA, 14 October 1998; IEEE: Piscataway, NJ, USA, 1998; Volume 4, pp. 3596–3601. [CrossRef]
4. Daviet, P.; Parent, M. Longitudinal and lateral servoing of vehicles in a platoon. In Proceedings of the Conference on Intelligent Vehicles, Tokyo, Japan, 19–20 September 1996; IEEE: Piscataway, NJ, USA, 1996; pp. 41–46. [CrossRef]
5. Goi, H.K.; Giesbrecht, J.L.; Barfoot, T.D.; Francis, B.A. Vision-Based Autonomous Convoying with Constant Time Delay. *J. Field Robot.* **2010**, *27*, 430–449. [CrossRef]
6. Goi, H.K.; Barfoot, T.D.; Francis, B.A.; Giesbrecht, J.L. Vision-Based Vehicle Trajectory Following with Constant Time Delay. In *Field and Service Robotics*; Howard, A., Iagnemma, K., Kelly, A., Eds.; Springer: Berlin/Heidelberg, Germany, 2010; pp. 137–147. [CrossRef]
7. Shan, M.; Zou, Y.; Guan, M.; Wen, C.; Lim, K.Y.; Ng, C.L.; Tan, P. Probabilistic trajectory estimation based leader following for multi-robot systems. In Proceedings of the 2016 14th International Conference on Control, Automation, Robotics and Vision (ICARCV), Phuket, Thailand, 13–15 November 2016; IEEE: Piscataway, NJ, USA, 2016; pp. 1–6. [CrossRef]
8. Zou, Y.; Shan, M.; Guan, M.; Wen, C.; Lim, K.Y. A trajectory reconstruction approach for leader-following of multi-robot system. In Proceedings of the 2017 12th IEEE Conference on Industrial Electronics and Applications (ICIEA), Siem Reap, Cambodia, 18–20 June 2017; IEEE: Piscataway, NJ, USA, 2017; pp. 1534–1539. [CrossRef]
9. Jansen, W.L. Lateral Path-Following Control of Automated Vehicle Platoons. Master's Thesis, Delft University of Technology, Delft, The Netherlands, 2016.
10. Fassbender, D.; Heinrich, B.C.; Luettel, T.; Wuensche, H.J. An optimization approach to trajectory generation for autonomous vehicle following. In Proceedings of the 2017 IEEE/RSJ International Conference on Intelligent Robots and Systems (IROS), Vancouver, BC, Canada, 24–28 September 2017. [CrossRef]
11. Heald, M.A. Rational approximations for the Fresnel integrals. *Math. Comput.* **1985**, *44*, 459–461. [CrossRef]
12. Wei, S.; Zou, Y.; Zhang, X.; Zhang, T.; Li, X. An Integrated Longitudinal and Lateral Vehicle Following Control System With Radar and Vehicle-to-Vehicle Communication. *IEEE Trans. Veh. Technol.* **2019**, *68*, 1116–1127. [CrossRef]
13. Wang, Y.; Bian, N.; Zhang, L.; Chen, H. Coordinated Lateral and Longitudinal Vehicle-Following Control of Connected and Automated Vehicles Considering Nonlinear Dynamics. *IEEE Control. Syst. Lett.* **2020**, *4*, 1054–1059. [CrossRef]
14. Floren, M.; Khajepour, A.; Hashemi, E. An Integrated Control Approach for the Combined Longitudinal and Lateral Vehicle Following Problem. In Proceedings of the 2021 American Control Conference (ACC), New Orleans, LA, USA, 25–28 May 2021; IEEE: Piscataway, NJ, USA, 2021; pp. 436–441. [CrossRef]
15. Liu, J.; Wang, Z.; Zhang, L. Event-Triggered Vehicle-Following Control for Connected and Automated Vehicles under Nonideal Vehicle-to-Vehicle Communications. In Proceedings of the 2021 IEEE Intelligent Vehicles Symposium (IV), Nagoya, Japan, 11–17 July 2021; IEEE: Piscataway, NJ, USA, 2021; pp. 342–347. [CrossRef]
16. Schinkel, W.; van der Sande, T.; Nijmeijer, H. State Estimation for Cooperative Lateral Vehicle Following Using Vehicle-to-Vehicle Communication. *Electronics* **2021**, *10*, 651. [CrossRef]
17. Visvalingam, M.; Whyatt, J.D. Line generalisation by repeated elimination of points. *Cartogr. J.* **1993**, *30*, 46–51. [CrossRef]
18. Rumetshofer, J.; Stolz, M.; Watzenig, D. A Generic Interface Enabling Combinations of State-of-the-Art Path Planning and Tracking Algorithms. *Electronics* **2021**, *10*, 788. [CrossRef]
19. Ezhov, N.; Neitzel, F.; Petrovic, S. Spline approximation, Part 1: Basic methodology. *J. Appl. Geod.* **2018**, *12*, 139–155. [CrossRef]
20. Eilers, P.H.C.; Marx, B.D. Flexible smoothing with B-splines and penalties. *Stat. Sci.* **1996**, *11*, 89–121. [CrossRef]
21. Goepp, V.; Bouaziz, O.; Nuel, G. Spline Regression with Automatic Knot Selection. *arXiv* **2018**, arXiv:1808.01770.
22. Nestlinger, G.; Stolz, M. Bumpless Transfer for Convenient Lateral Car Control Handover. *IFAC-PapersOnLine* **2016**, *49*, 132–138. Special Issue: In Proceedings of the 9th IFAC Symposium on Intelligent Autonomous Vehicles IAV 2016, Leipzig, Germany, 29 June–1 July 2016. [CrossRef]
23. Hoffmann, G.M.; Tomlin, C.J.; Montemerlo, M.; Thrun, S. Autonomous Automobile Trajectory Tracking for Off-Road Driving: Controller Design, Experimental Validation and Racing. In Proceedings of the 2007 American Control Conference, New York, NY, USA, 9–13 July 2007; IEEE: Piscataway, NJ, USA, 2007; pp. 2296–2301. [CrossRef]
24. Coulter, R.C. *Implementation of the Pure Pursuit Path Tracking Algorithm*; Technical Report; Camegie Mellon University: Pittsburgh, PA, USA, 1992.
25. Nestlinger, G. Modellbildung und Simulation Eines Spurhalte-Assistenzsystems. Master's Thesis, Graz University of Technology, Graz, Austria, 2013.
26. Solmaz, S.; Nestlinger, G.; Stettinger, G. Compensation of Sensor and Actuator Imperfections for Lane-Keeping Control Using a Kalman Filter Predictor. *SAE Int. J. Connect. Autom. Veh.* **2021**, *4*, 8. [CrossRef]
27. Riekert, P.; Schunck, T. Zur Fahrmechanik des gummibereiften Kraftfahrzeugs. *Ingenieur-Archiv* **1940**, *11*, 210–224. [CrossRef]

28. Hsu, J.C.; Tomizuka, M. Analyses of Vision-based Lateral Control for Automated Highway System. *Veh. Syst. Dyn.* **1998**, *30*, 345–373. [CrossRef]
29. Swaroop, D.; Rajagopal, K. A review of constant time headway policy for automatic vehicle following. In Proceedings of the 2001 IEEE Intelligent Transportation Systems, Oakland, CA, USA, 25–29 August 2001; IEEE: Piscataway, NJ, USA, 2001; pp. 65–69. [CrossRef]
30. Rajamani, R. Lateral Vehicle Dynamics. In *Vehicle Dynamics and Control*; Mechanical Engineering Series; Springer: New York, NY, USA, 2012; Chapter 2, pp. 15–46. [CrossRef]